Quantitative Chemical Analysis: An Introduction

Quantitative Chemical Analysis: An Introduction

Contributors

Yan-Hong Shi, Zhi-Yong Xie et al.

AURIS
Reference

www.aurisreference.com

Quantitative Chemical Analysis: An Introduction

Contributors: Yan-Hong Shi, Zhi-Yong Xie et al.

Published by Auris Reference Limited
www.aurisreference.com

United Kingdom

Quantitative Chemical Analysis: An Introduction

ISBN: 978-1-78154-896-7

British Library Cataloguing in Publication Data
A CIP record for this book is available from the British Library

Printed in the United Kingdom

Exclusively distributed by CBS Publishers & Distributors Pvt. Ltd.

Sales & Distribution Rights only for India, Pakistan, Bangladesh, Sri Lanka, Nepal and Bhutan.This book is not to be sold outside these territories.

Contents

List of Abbreviations

CCD	Charge Coupled Device
CID	Collision Induced Dissociation
CID	Charge Injection Devices
DAD	Diode Array Detectors
DSDME	Direct Suspended Droplet Microextraction
EIC	Extracted Ion Chromatograms
ESI	Electrospray Ionization
FSH	Follicle Stimulating Hormone
HCA	Hierarchical Clustering Analysis
HEX	High Energy X ray system
HPLC	High Performance Liquid Chromatography
ICH	International Conference on Harmonization
IRD	International Radiation Detectors
IUPAC	International Union of Pure and Applied Chemistry
LH	Lutropin Hormone
LLNL	Lawrence Livermore National Laboratory
LOD	Limit of Detection
LOQ	Limits of Quantification
LR	Lidar Ratio
MRM	Multiple Reaction Monitoring
MS	Mass Spectrometry
NIF	National Ignition Facility
NIST	National Institute for Standards and Technologies
NPD	Neurophysiologic Detectors
NVLAP	National Voluntary Laboratory Accreditation Program
PBL	Planetary Boundary Layer
PCA	Principal Component Analysis
PDA	Photodiode Array
PSL	Photostimulated Luminescence
RSD	Relative Standard Deviation
SA	Similarity Analysis
SFDA	State Food and Drug Administration
SIM	Selected Ion Monitoring
SIMCA	Soft Independent Modeling of Class Analogy
SMPS	Scanning Mobility Particle Sizer
SPE	Solid Phase Extraction
SRM	Selected Reaction Monitoring
SXI	Static X-Ray Imager
TCM	Traditional Chinese medicine
TCMP	Traditional Chinese medicine Preparations
TLC	Thin Layer Chromatography

TSH	Thyroid Stimulating Hormone
UHPLC	Ultra High Pressure Liquid Chromatography
VOC	Volatile Organic Compounds
WSOC	Water Soluble Organic Carbon

List of Contributors

Yan-Hong Shi
School of Pharmacy, Shanghai University of Traditional Chinese Medicine, Shanghai 201203, China

Zhi-Yong Xie
School of Pharmacy, Shanghai University of Traditional Chinese Medicine, Shanghai 201203, China

Rui Wang
School of Pharmacy, Shanghai University of Traditional Chinese Medicine, Shanghai 201203, China

Shan-Jun Huang
School of Pharmacy, Shanghai University of Traditional Chinese Medicine, Shanghai 201203, China

Yi-Ming Li
School of Pharmacy, Shanghai University of Traditional Chinese Medicine, Shanghai 201203, China

Zheng-Tao Wang
Institute of Chinese Materia Medica, Shanghai University of Traditional Chinese Medicine, Shanghai 201203, China

Yan-Bin Wu
Academy of Integrative Medicine, Fujian University of Traditional Chinese Medicine, Fuzhou 350122, China

Li-Jun Zheng
Academy of Integrative Medicine, Fujian University of Traditional Chinese Medicine, Fuzhou 350122, China

Jun Yi
Department of Science, Fujian Institute of Education, Fuzhou 350001, China

Jian-Guo Wu
Academy of Integrative Medicine, Fujian University of Traditional Chinese Medicine, Fuzhou 350122, China

Ti-Qiang Chen
Fujian Academy of Agricultural Sciences, Fuzhou 350013, China

Jin-Zhong Wu
Academy of Integrative Medicine, Fujian University of Traditional Chinese Medicine, Fuzhou 350122, China

Teng-Hua Wang
Lab of Chinese Materia Medica Preparation, the Second College of Clinic Medicine, Guangzhou University of Chinese Medicine; Guangdong Province Institute of TCM, Guangzhou 510006, China
School of Chinese Materia Medica, Guangzhou University of Chinese Medicine, Guangzhou 510006, China

Jing Zhang
Lab of Chinese Materia Medica Preparation, the Second College of Clinic Medicine, Guangzhou University of Chinese Medicine; Guangdong Province Institute of TCM, Guangzhou 510006, China

Xiao-Hui Qiu
Lab of Chinese Materia Medica Preparation, the Second College of Clinic Medicine, Guangzhou University of Chinese Medicine; Guangdong Province Institute of TCM, Guangzhou 510006, China

Jun-Qi Bai
Lab of Chinese Materia Medica Preparation, the Second College of Clinic Medicine, Guangzhou University of Chinese Medicine; Guangdong Province Institute of TCM, Guangzhou 510006, China

You-Heng Gao
School of Chinese Materia Medica, Guangzhou University of Chinese Medicine, Guangzhou 510006, China

Wen Xu
Lab of Chinese Materia Medica Preparation, the Second College of Clinic Medicine, Guangzhou University of Chinese Medicine; Guangdong Province Institute of TCM, Guangzhou 510006, China
School of Chinese Materia Medica, Guangzhou University of Chinese Medicine, Guangzhou 510006, China

Shao-Dan Chen
The Second College of Clinic Medicine, Guangzhou University of Chinese Medicine, Guangzhou 510000, China

Guangdong Provincial Hospital of Chinese Medicine, Guangzhou 510000, China

Chuan-Jian Lu
The Second College of Clinic Medicine, Guangzhou University of Chinese Medicine, Guangzhou 510000, China
Guangdong Provincial Hospital of Chinese Medicine, Guangzhou 510000, China

Rui-Zhi Zhao
The Second College of Clinic Medicine, Guangzhou University of Chinese Medicine, Guangzhou 510000, China
Guangdong Provincial Hospital of Chinese Medicine, Guangzhou 510000, China

Zarrin Es'haghi
Department of Chemistry, Payame Noor University, 19395-4697 Tehran, I.R. of IRAN

Mohammed Mahabubur Rahman
United Graduate School of Agricultural Science, Ehime University, Matsuyama, Ehime, Japan
Education and Research Center for Subtropical Field Science, Faculty of Agriculture, Kochi University, Kochi, Japan

Rahman Md. Motiur
Silvacom Ltd., Edmonton, Canada

Lay-Harn Gam
School of Pharmaceutical Sciences, Universiti Sains Malaysia, Malaysia

Michael J. Haugh
National Security Technologies, LLC, USA

Marilyn Schneider
Lawrence Livermore National Laboratory, USA

Gerhard Held
Instituto de Pesquisas Meteorológicas, Universidade Estadual Paulista, Bauru, S.P., Brazil

Ana Maria Gomes
Instituto de Pesquisas Meteorológicas, Universidade Estadual Paulista, Bauru, S.P., Brazil

Andrew G. Allen
Instituto de Química, Universidade Estadual Paulista, Araraquara, S.P., Brazil

Arnaldo A. Cardoso
Instituto de Química, Universidade Estadual Paulista, Araraquara, S.P., Brazil

Fabio J.S. Lopes
Centro de Lasers e Aplicações, Instituto de Pesquisas Energéticas e Nucleares, Universidade de São Paulo, São Paulo, S.P., Brazil

Eduardo Landulfo
Centro de Lasers e Aplicações, Instituto de Pesquisas Energéticas e Nucleares, Universidade de São Paulo, São Paulo, S.P., Brazil

Preface

Quantitative chemical analysis, branch of chemistry that deals with the determination of the amount or percentage of one or more constituents of a sample. A variety of methods is employed for quantitative analyses, which for convenience may be broadly classified as chemical or physical, depending upon which properties are utilized. Quantitative Chemical Analysis: An Introduction provides a sound physical understanding of the principles of analytical chemistry, showing how these principles are applied in chemistry and related disciplines—especially in life sciences and environmental science. First chapter aims to reveal the correlation and consistency of quality control in crude herbs, prepared slices, and Radix Isatidis preparations. Second chapter presents a study the indicated that the combination of quantitative and chromatographic fingerprint analysis can be readily utilized as a quality control method for Receptaculum Nelumbinis and its related traditional Chinese medicinal preparations. Third chapter presents the application of ultra-high-performance liquid chromatography coupled with LTQ-orbitrap mass spectrometry for the qualitative and quantitative analysis of polygonum multiflorum thumb. and its processed products. Fourth chapter focuses on qualitative and quantitative analysis of rhizoma smilacis glabrae by ultra-high performance liquid chromatography coupled with LTQ orbitrapxlhybrid mass spectrometry. Photodiode array detection in clinical applications; quantitative analyte assay advantages, limitations and disadvantages are discussed in fifth chapter. Sixth chapter discusses the quantitative defense traits of forest litter, their effects on litter decomposition and ecosystem functioning. Tandem mass spectrometry for simultaneous qualitative and quantitative analysis of protein has been presented in seventh chapter. Eighth chapter describes the characterization of several X-ray sources and their use in calibrating different types of X-ray cameras at National Security Technologies, LLC (NSTec). Ninth chapter provides a review of all relevant historical data concerning the nature, concentrations and impacts of atmospheric aerosols in southeast Brazil.

Chapter 1

QUANTITATIVE AND CHEMICAL FINGERPRINT ANALYSIS FOR THE QUALITY EVALUATION OF ISATIS INDIGOTICA BASED ON ULTRA-PERFORMANCE LIQUID CHROMATOGRAPHY WITH PHOTODIODE ARRAY DETECTOR COMBINED WITH CHEMOMETRIC METHODS

Yan-Hong Shi [1], Zhi-Yong Xie [1], Rui Wang [1], Shan-Jun Huang [1], Yi-Ming Li [1] and Zheng-Tao Wang [2]

[1]School of Pharmacy, Shanghai University of Traditional Chinese Medicine, Shanghai 201203, China

[2]Institute of Chinese Materia Medica, Shanghai University of Traditional Chinese Medicine, Shanghai 201203, China

ABSTRACT

A simple and reliable method of ultra-performance liquid chromatography with photodiode array detector (UPLC-PDA) was developed to control the quality of *Radix Isatidis* (dried root of *Isatis indigotica*) for chemical fingerprint analysis and quantitative analysis of eight bioactive constituents, including *R,S*-goitrin, progoitrin, epiprogoitrin, gluconapin, adenosine, uridine, guanosine, and hypoxanthine. In quantitative analysis, the eight components showed good regression ($R > 0.9997$) within test ranges, and the recovery method ranged from 99.5% to 103.0%. The UPLC fingerprints of the *Radix Isatidis* samples were compared by performing chemometric procedures, including similarity analysis, hierarchical clustering analysis, and principal component analysis. The chemometric procedures classified *Radix Isatidis* and its finished products such that all samples could be successfully grouped according to crude herbs, prepared slices, and adulterant *Baphicacanthis cusiae* Rhizoma et Radix. The combination of quantitative and chromatographic fingerprint analysis can be used for the quality assessment of *Radix Isatidis* and its finished products.

INTRODUCTION

Radix Isatidis is the dried root of *Isatis indigotica* Fort. (Fam. Cruciferae), which is known as *Banlangen* (BLG) or *Bei-Banlangen*. *Baphicacanthis cusiae* Rhizoma et Radix is the dried rhizome and root of *Baphicacanthus cusia* (Nees) Bremek. (Fam. Acanthaceae), which is known as *Nan-Banlangen* (NBLG). In several southern regions in China, NBLG has been improperly used as *Radix Isatidis* even though both have been officially listed in Chinese Pharmacopoeia (State Pharmacopoeia 2000) as two different crude herbs [1–4].

Radix Isatidis and its finished products have important functions in preventing and treating influenza, tonsillitis, and malignant infectious diseases [5–7], especially severe acute respiratory syndrome (SARS) and H1N1-influenza [8–11] because of its anti-viral, anti-bacterial, anti-inflammatory, anti-tumor, and immune regulatory functions. Moreover, numerous gratifying successes of the *Radix Isatidis* anti-viral effect have been reported [12–15]. *Radix Isatidis* has become an important component in various traditional Chinese medicine preparations (TCMPs), of which Banlangen Granules (BLGG) is widely used for toxic-heat removal and as an anti-viral drug in clinical practice.

The chemical constituents in *Radix Isatidis* are as follows: nucleosides, glucosinolates, amino acids, polysaccharides, alkaloids, organic acid, trace elements, and so on [16–18]. Compared with other *Radix Isatidis* extraction methods, the water extraction method has the most significant anti-viral, anti-bacterial, and anti-endotoxic effects, and the chemical constituents of glucosinolates (*R,S*-gotrin, progoitrin, epiproguotrin, and gluconapin) and nucleosides (hypoxanthine, adenosine, uridine, and guanosine) are the major bioactive components [19–21]. Therefore, using the non-polar components of Indigotin, Indirubin, and the nonspecific amino acid constituent as quality control markers of *Radix Isatidis* is unsuitable [22–27].

Glucosinolates are one of the characteristic components of Cruciferae. They degrade under endogenous myrosinase effects to work on multi-bioactivities including anti-virus and anti-bacteria [28,29]. *R,S*-goitrin reflects bioactivities that are relevant to the *Radix Isatidis* effects with high specificity. In our previous studies, we established thin layer chromatography (TLC) identification and high-performance liquid chromatography (HPLC) assay methods for the quality control of *Radix Isatidis* using *R,S*-goitrin as the marker [30], and these methods have been adopted by the ChP 2010 Edition [1]. Progoitrin, epiproguotrin, and *R,S*-goitrin all exist in *Radix Isatidis*. Under the effects of myrosinase, some parts or all of progoitrin and epiproguotrin transfer to *R,S*-goitrin via degradation [31].

In previous studies, based on the issues related to *Radix Isatidis* such as the improper use of NBLG, glucosinolate degradation, lack of specificity using amino acid as the BLGG quality control marker, establishing proper and scientific-based methods for quality control of *Radix Isatidis* and its finished products is necessary. Current studies about quantitative and chemical fingerprinting of *Radix Isatidis* depend on non-polar extracts or spots of chemical compositions [32–35]. Chemical fingerprint and quantitative analysis has become one of the most frequently applied approaches in the quality control of traditional Chinese medicine (TCM) and its finished products [36–41]. Studies about combining chromatographic fingerprint and multi-ingredient quantification by ultra-performance liquid chromatography with photodiode array detector (UPLC-PDA) for the quality control of *Radix Isatidis* and its preparations have not been reported.

This study aims to reveal the correlation and consistency of quality control in crude herbs, prepared slices, and *Radix Isatidis* preparations. A simple, accurate, and practical UPLC-PDA method was developed for the simultaneous determination of eight bioactive components in *Radix Isatidis* and its preparations. The chemical fingerprints of *Radix Isatidis* from various sources were established and investigated by similarity analysis (SA), hierarchical clustering analysis (HCA), and principal component analysis (PCA). The combination of chromatographic fingerprint analysis and the simultaneous determination of the eight bioactive components offer a more comprehensive strategy for the quality evaluation of *Radix Isatidis* and its finished products.

RESULTS AND DISCUSSION

Optimization of UPLC Conditions

Different UPLC parameters were examined and compared, including various columns, mobile phases, detection wavelengths, and gradient elution conditions to obtain as much chemical information as possible and to determine the best separation mechanism in chromatograms.

Four kinds of reversed-phase columns, namely, Waters ACQUITY UPLC BEH C_{18} (1.7 μm, 2.1 mm × 50 mm/100 mm) and Thermo Syncronis C_{18}/Waters ACQUITY UPLC HSS T_3 (1.7 μm, 2.1 mm × 100 mm) were investigated and compared. The Waters ACQUITY UPLC BEH C_{18} (1.7 μm, 2.1 mm × 100 mm) column had good peak separation and sharp peaks.

The effect of mobile phase composition (methanol-water and acetonitrile-water with different modifiers including acetic acid, formic acid, phosphoric acid, and triethylamine) on chromatographic separation was investigated.

Adding 0.1% triethylamine in the mobile phase and adjusting the pH to 4.0 with formic acid provided a better resolution and separation of the eight bioactivity components, and resulted in high precision sensitivity and selectivity.

Based on the maximum absorption and full-scan experiment of the marker components in the UV spectra of the three-dimensional chromatograms obtained by PDA detection, the detection wavelengths at 210, 230, 254, 280 nm were selected to compare the peak number and peak resolution of all marker compounds. Finally, the wavelength was set at 254 nm (Figure 1).

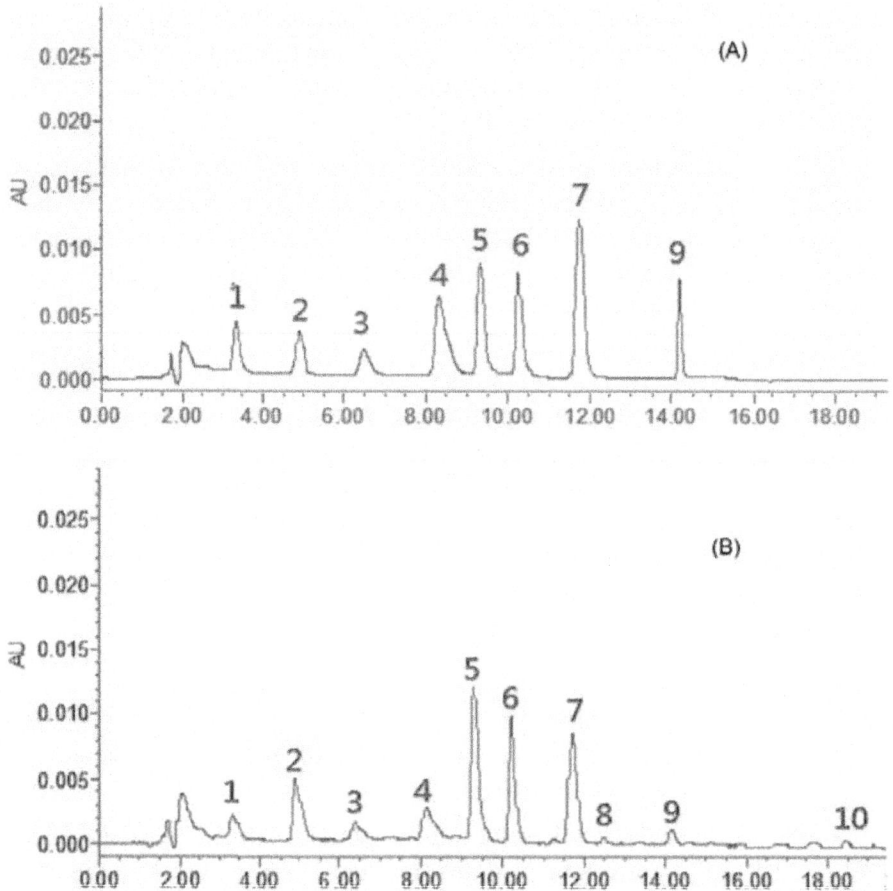

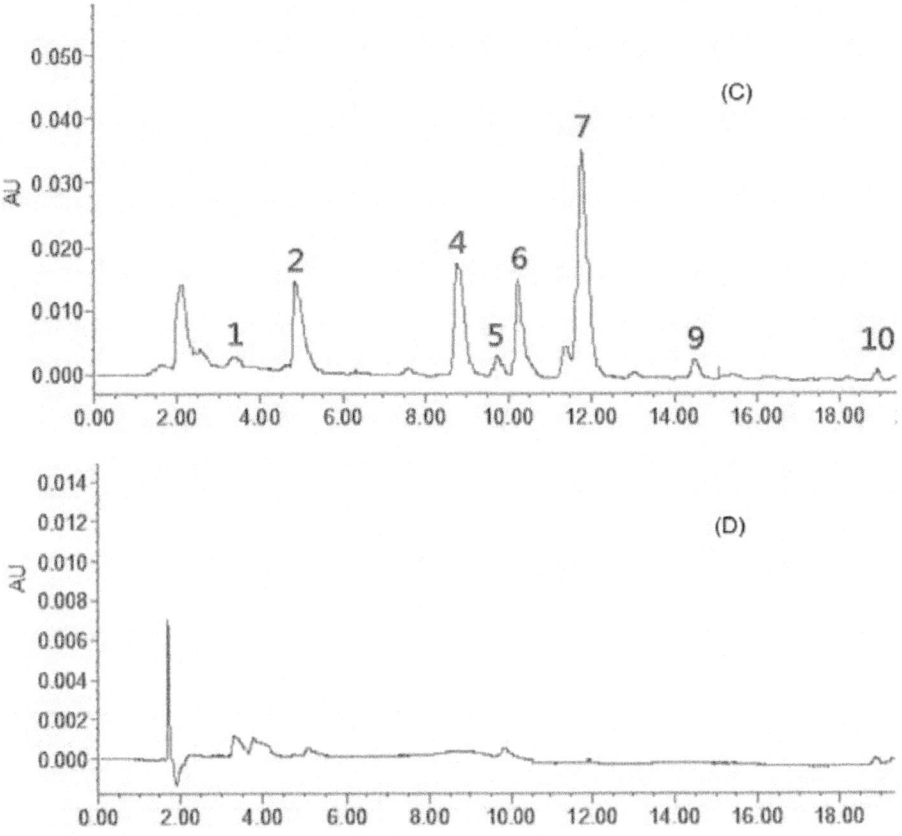

Figure 1. Typical chromatograms for determining the eight bioactive compounds in different samples. (**A**) Mixed standards; (**B**) *Radix Isatidis* (No. 20101121); (**C**) Banlangen Granules (BLGG) (No. 090921); (**D**) Nan-Banlangen (NBLG) (No. 060308-1). Peaks 1 = hypoxanthine, 2 = uridine, 3 = progoitrin, 4 = epiprogoitrin, 5 = adenosine, 6 = guanosine, 7 = *R,S*-goitrin, and 9 = luconapin.

Optimization of Extraction Methods

Satisfactory extraction efficiency was obtained by comparing water-refluxing, ultrasonic, and soxhlet extraction methods. Refluxing extraction was simpler and more effective for nucleosides and glucosinolates among the other methods. Therefore, this method was used in further experiments. In this study, different concentrations (0%, 20%, 50%, 75%, 90%, and 100%) of ethanol solutions, sample-solvent ratios (1:10, 1:20, 1:50, and 1:100, *w/v*), and extraction times (20, 30, 45, and 60 min) were used for the BLG extraction procedure (Batch No. 20101121). As a result, the best extraction condition was established as follows: the samples were extracted by refluxing extraction using 20 mL of water as the extraction solvent, and the duration was 30 min.

Method Validation of Quantitative Analysis

The method was validated in terms of linearity, limit of detection (LOD), limit of quantification (LOQ), precision, reproducibility, stability, and recovery test.

Calibration Curves, LOD, and LOQ

Methanol stock solutions containing eight analytes were diluted to appropriate concentrations for calibration curve construction. The analyte solutions at six different concentrations were injected in triplicate, and the calibration curves were established by plotting the peak area (Y) versus the concentration (x) of each component. LOD and LOQ, which were expressed by 3-and 10-fold of the signal-to-noise ratio (S/N), were also determined. The detailed information regarding the calibration curves, linear ranges, LODs, and LOQs of the eight bioactive compounds are listed in Table 1.

Table 1. Regression data, limits of detection (LODs), and limits of quantification (LOQs) for the eight bioactive constituents

Compound	Regression equation ($Y = ax + b$) a	R b	Linear range (μg mL^{-1})	LOD a (μg mL^{-1})	LOQ b (μg mL^{-1})
Hypoxanthine	$Y = 914.32x - 684.64$	0.9999	6.50–130.00	0.015	0.049
Uridine	$Y = 1945.9x + 263.11$	0.9999	4.12–82.40	0.012	0.041
Adenosine	$Y = 6870.2x - 3669.6$	0.9999	4.26–85.20	0.005	0.017
Guanosine	$Y = 12618x - 1461.7$	1.0000	3.90–78.00	0.017	0.045
Progoitrin	$Y = 2579.2x - 886.87$	0.9997	2.30–34.50	0.563	1.875
Epiprogoitrin	$Y = 4336.8x - 68.733$	0.9998	2.30–34.50	0.708	2.100
R,S-goitrin	$Y = 1756.6x - 757.89$	1.0000	4.50–90.00	0.014	0.045
Gluconapin	$Y = 1865.8x - 240.44$	0.9999	2.05–30.75	0.456	1.640

[a]Y and x stand for the peak area and the injection quantity (μg) of each standard substance, respectively;

[b]R = correlation coefficient, $n = 6$.

Precision, Reproducibility, Stability, and Recovery

As shown in Table 2, the precision based on the peak area measurements of the eight bioactive components were higher than 0.48% (RSD, $n = 6$, S-01).

The reproducibility (RSD, $n = 6$, S-01) of the proposed method based on six replicate injections was in the range of 0.10% to 1.24%. The stability (RSD, $n = 6$, S-01) of the measurements over 3 days for the eight compounds was 0.22% to 1.44%. The recovery test was performed by the standard addition method. Low, medium, and large high amounts of the standards were added to the known sample (S-01). Extraction and analysis were performed as described in Section 2.4. The mean recovery was calculated according to the following formula: recovery (%) = (amount found–original amount)/amount spiked × 100%, and RSD (%) = (SD/mean) × 100%. The mean recovery of the eight bioactive compounds was 99.5%–103.0%, and the RSD value was 0.73%–1.81%.

Table 2. Precision, reproducibility, stability, and recovery of the eight bioactive constituents

Compound	Precision RSD (%) ($n = 6$)	Reproduc-ibility RSD (%) ($n = 6$)	Stability RSD (%) ($n = 6$)	Recovery (%) ($n = 9$) Mean ± RSD (%)
Hypoxanthine	0.15	0.46	0.22	101.5 ± 1.63
Uridine	0.18	0.10	0.39	101.2 ± 1.33
Adenosine	0.07	0.61	0.49	101.1 ± 1.11
Guanosine	0.12	1.24	1.30	100.9 ± 1.72
Progoitrin	0.11	0.19	0.24	99.5 ± 1.81
Epiprogoitrin	0.11	1.92	1.44	100.2 ± 1.73
R,S-goitrin	0.48	0.72	0.91	100.4 ± 0.73
Gluconapin	0.31	0.33	0.48	103.0 ± 1.14

Sample Analysis

The newly established analytical method was subsequently applied to determine simultaneously the eight bioactive components in 21 commercial samples of *Radix Isatidis* and its finished products from different provinces or manufacturers in China. All samples were analyzed using the optimized extraction method in optimized UPLC conditions. Each sample was analyzed in triplicate to determine the mean content (mg·g^{-1}), and the results are tabulated in Table 3.

Table 3. The contents (mg g^{-1}) of eight targets in 21 commercial samples ($n = 3$)

Name	No.a	Content b (mg g^{-1})								
		1 c	2	3	4	5	6	7	9	Total
Prepared slices	S-01	0.137	0.414	5.938	6.605	0.492	0.525	0.776	4.051	18.937
	S-02	0.022	0.279	1.529	1.500	0.293	0.377	0.899	0.870	5.768
	S-03	0.042	0.436	2.389	2.094	0.343	0.465	0.789	1.210	7.767
	S-04	0.024	0.283	2.465	2.293	0.342	0.336	0.414	1.190	7.347
	S-05	0.019	0.278	3.331	3.355	0.347	0.345	0.551	1.977	10.203
	S-06	0.037	0.416	1.178	1.413	0.372	0.424	0.732	0.401	4.974
	S-07	0.021	0.384	1.416	1.750	0.336	0.347	0.483	0.574	5.310
	S-08	0.336	0.345	1.950	2.441	0.348	0.379	0.436	0.501	6.735
Crude herbs	S-09	0.056	0.083	5.929	6.605	0.114	0.095	0.057	10.020	22.959
	S-10	0.022	0.057	7.345	6.565	0.127	0.132	0.083	9.458	23.790
	S-11	0.028	0.070	5.286	6.424	0.120	0.116	0.070	10.130	22.244
	S-12	0.035	0.084	3.934	5.660	0.124	0.137	0.123	7.235	17.333
	S-13	0.048	0.057	6.396	6.534	0.105	0.135	0.062	6.747	20.084
	S-14	0.040	0.059	5.507	6.253	0.122	0.327	0.084	9.611	22.003
	S-15	0.033	0.070	6.628	5.713	0.105	0.207	0.054	5.498	18.309
Granules	S-16	N.D. d	0.045	N.D.	N.D.	0.006	0.008	0.085	0.330	0.473
	S-17	0.004	0.022	N.D.	N.D.	N.D.	N.D.	0.027	0.094	0.147
	S-18	N.D.	0.058	N.D.	N.D.	0.003	0.005	0.075	0.576	0.716
	S-19	N.D.	0.021	N.D.	N.D.	0.001	N.D.	0.033	0.100	0.154
	S-20	0.006	0.061	N.D.	N.D.	0.007	0.006	0.087	0.515	0.681
	S-21	0.036	0.053	N.D.	0.102	0.012	0.010	0.067	0.169	0.448

[a]No. means sample numbers as listed in Table 4;[b]Values in mg g^{-1} of dry raw materials; and in mean ± SD, $n = 3$, the SD was < 4% of the mean, which is not shown for clarity;[c](1–7,9) mean the bioactive constituents number as listed in Figure 1;[d]N.D.: below LOD.

Glucosinolates are the characteristic constituents in the Cruciferae plants [28,29]. The results from the quantitative analysis (Table 3) showed that the crude herbs generally contained the eight selected constituents and the glucosinolate content ranged from 6.48 mg g^{-1} to 73.63 mg g^{-1}. Table 3 and Figure 2 show that the average contents of epiprogoitrin, progoitrin, and *R,S*-goitrin in *Radix Isatidis* crude herbs are 5.86, 6.25, and 0.07 mg g^{-1}. However, the average epiprogoitrin and progoitrin contents in prepared *Radix*

Isatidis slices significantly decreased (2.53 and 2.68 mg g^{-1}, respectively), whereas the *R,S*-goitrin content clearly increased (0.64 mg g^{-1}). The traditional processing methods improve the glucosinolate biotransformation and increase the degradation product (*R,S*-goitrin) content [31].

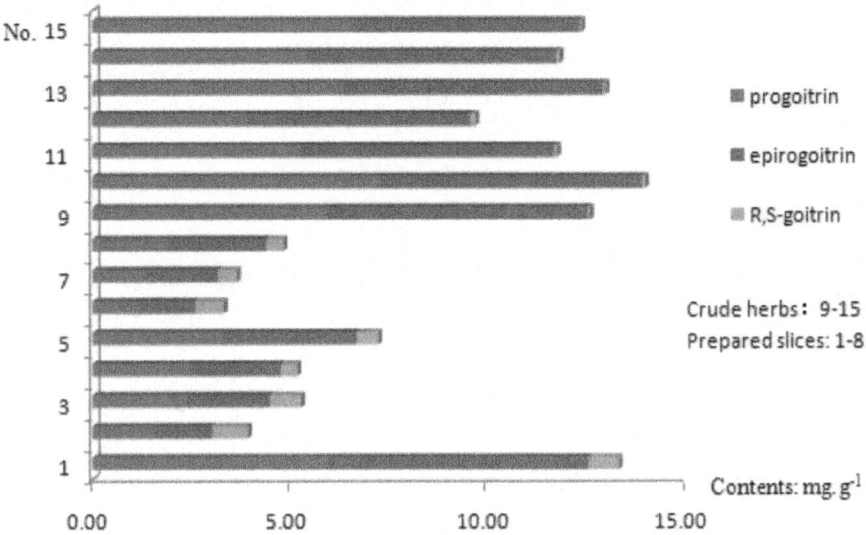

Figure 2. Bar graph of the contents of progoitrin, epiprogoitrin, and *R,S*-gotrin in 15 sample batches.

The total content of eight compounds in different samples, especially those from different sources and harvesting times, were significantly different. Moreover, epiprogoitrin and progoitrin could not be distinctly detected (below LOD) in BLGG that was prepared and precipitated with ethanol by the water extracts of *Radix Isatidis* [1].

Overall, the internal qualities of 21 batches of *Radix Isatidis* samples from different sources with different geographical sources were varied, and the qualities needed evaluation by chemical fingerprinting.

UPLC Fingerprint of Radix Isatidis

Altogether, 15 batches of samples were analyzed, and all chromatograms were introduced into the Computer-Aided Similarity Evaluation System for Chromatographic Fingerprint of TCM (China Committee of Pharmacopeia, 2004). Peaks existing in all sample chromatograms were assigned as the "common peak," and 10 common peaks were observed between 3 and 20 min in all 15 batches (Figure 1). Eight common peaks (peak 1, 2, 3, 4, 5, 6, 7, and 9) were identified as hypoxanthine, uridine, progoitrin, epiprogoitrin, adenosine, guanosine, *R,S*-goitrin, and gluconapin with reference substances (Figure 3). Peak 7 (*R,S*-goitrin, RT = 11 min), which was one of the most important active constituents of *Radix Isatidis* (China Pharmacopoeia Committee 2010), was chosen as the internal reference peak to calculate the relative retention time (RRT) and relative peak area (RPA) of the other peaks.

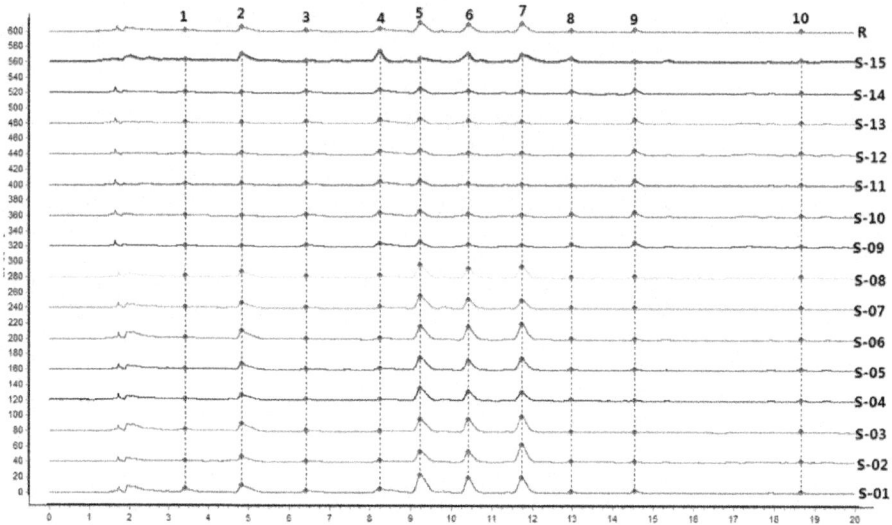

Figure 3. Ultra-performance liquid chromatography (UPLC) fingerprint chromatograms of 15 batches of *Radix Isatidis* (R: digital standard fingerprint).

Figure 1 shows that the investigated nucleosides, glucosinolates, and other compounds in *Radix Isatidis* were separated and determined using the developed UPLC-PDA method. The differences of the water-soluble bioactive components from BLG and NBLG were recognized, and the components could be rapidly and efficiently differentiated by the chromatographic method (Figure 1B,D). Moreover, the bioactive constituents of glucosinolates and nucleosides were simultaneously shown by the same chromatographic method, and the high correlation between BLG and BLGG (Figure 1B,C) provided a basis to investigate the relationship between *Radix Isatidis* and its preparations.

Similarity Analysis (SA)

The State Food and Drug Administration (SFDA) suggested that all herbal chromatograms should be evaluated in terms of similarity by calculating the correlation coefficient and/or angle cosine value of the original data [36–38]. Therefore, SA was conducted based on the standard fingerprints, and the results are shown in Table 4. The same samples showed similarities (crude herbs: 0.752 to 0.820; prepared slices: 0.933 to 0.991). The same samples had similar constituents with the slight difference resulting from the environmental conditions and the planting techniques. However, the crude herbs and the prepared slices were significantly different because of the production process and the contents of main bioactive constituents. The biotransformation, the glucosinolate contents, and the degradation products have the most important influence on the similarities.

Table 4. All samples used in this work and their corresponding similarities

Name	No.	Batch No.	Resource	Origins	Similarity
Prepared Slices	S-01	20101121	Anhui	*Isatis indigotica* Fort.	0.991
	S-02	101001	Anhui	*Isatis indigotica* Fort.	0.933
	S-03	8100182	Anhui	*Isatis indigotica* Fort.	0.948
	S-04	100709	Anhui	*Isatis indigotica* Fort.	0.947
	S-05	20110213	Jiangsu	*Isatis indigotica* Fort.	0.967
	S-06	100713	Mongolia	*Isatis indigotica* Fort.	0.954
	S-07	100602	Mongolia	*Isatis indigotica* Fort.	0.954
	S-08	101120	Zhejiang	*Isatis indigotica* Fort.	0.967
Crude herbs	S-09	20110224-1	Shanghai	*Isatis indigotica* Fort.	0.777
	S-10	20110224-2	Shanghai	*Isatis indigotica* Fort.	0.774
	S-11	20110224-3	Shanghai	*Isatis indigotica* Fort.	0.752
	S-12	20080902	Harbin	*Isatis indigotica* Fort.	0.759
	S-13	blg-081020	Henan	*Isatis indigotica* Fort.	0.818
	S-14	20110205	Anhui	*Isatis indigotica* Fort.	0.820
	S-15	20081017	Hebei	*Isatis indigotica* Fort.	0.811
Granule	S-16	090921	Shanghai	-	-
	S-17	100302	Shanghai	-	-
	S-18	A0F090	Guang-zhou	-	-
	S-19	L9F043	Guang-zhou	-	-
	S-20	100303	Sichuan	-	-
	S-21	100702	Sichuan	-	-
Nan-Ban-langen	S-22	060308-1	Jiangxi	*Baphicacanthus eusia* (Nees) Bremek.	-

	S-23	1678-3	Guizhou	*Baphicacanthus eusia* (Nees) Bremek.	-
	S-24	20110923	Fujian	*Baphicacanthus eusia* (Nees) Bremek.	-
	S-25	20111025	Fujian	*Baphicacanthus eusia* (Nees) Bremek.	-

Hierarchical Cluster Analysis (HCA)

The HCA results clearly showed that the samples were appropriately divided into two main clusters related to the *Radix Isatidis* type (Figure 4). Cluster-I was S01 to S08, which were the prepared slices of *Radix Isatidis*, whereas Cluster-II was S09 to S15, which were the crude herb samples. The HCA result was fully consistent with the actual situation of samples.

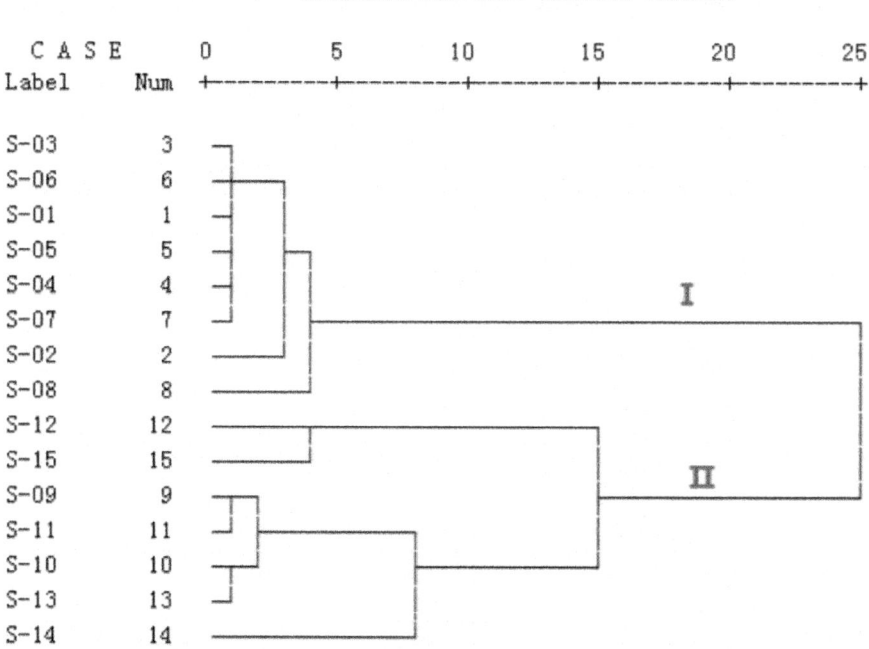

Figure 4. Dendrogram of hierarchical clustering analysis (HCA) for the 15 tested samples. The hierarchical clustering was done by the SPSS software (version 15.0; IBM: Chicago, IL, USA, 2006). Ward's method was performed, and Squared Euclidean distance was selected as a measurement.

Principal Component Analysis (PCA)

PCA, which is an unsupervised multivariate data analysis approach, is appropriate when a function of many attributes is believed to be involved in different samples [40,41]. PCA was employed to analyze the relationship of the 15 *Radix Isatidis* samples (seven crude herbs and eight prepared slices) from different sources, and the score plot derived from PCA is shown in Figure 5A.

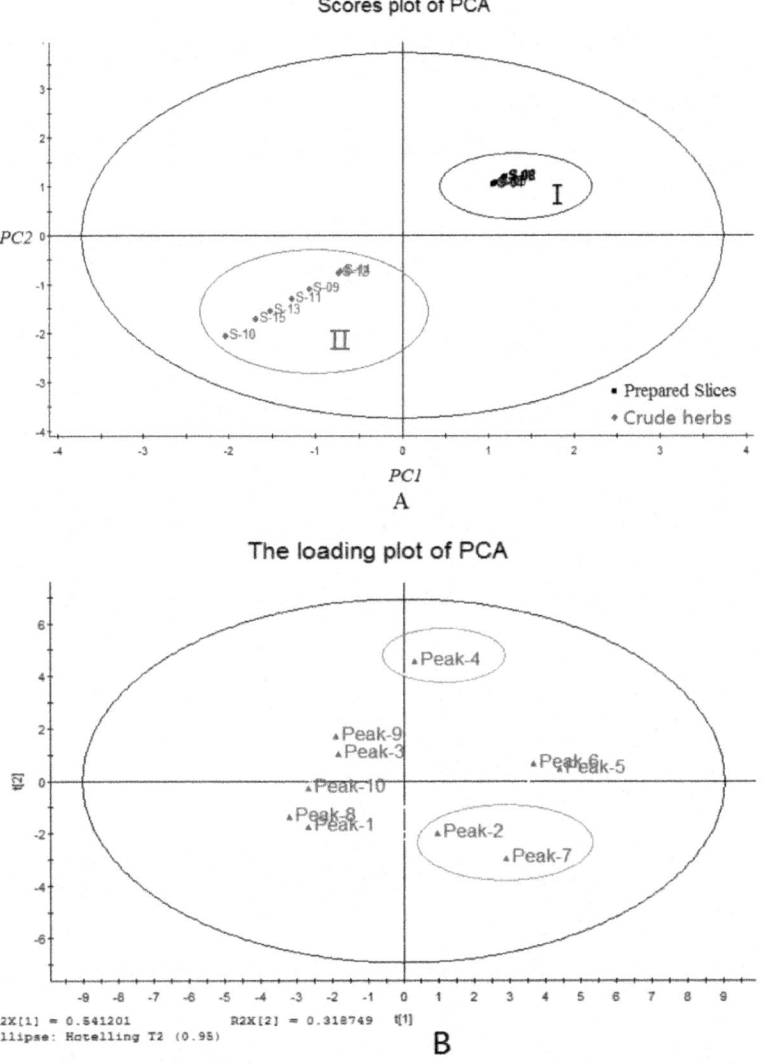

Figure 5. Score plot of principal component analysis (PCA) (**A**) and PCA loading plot (**B**) of *Radix Isatidis*.

On the basis of eigenvalues > 1, the first two principal components, PC1 and PC2, are often used to provide a convenient visual aid for identifying inhomogeneity in the data sets. The samples were clustered into two main groups. The PCA loading plot (Figure 5B) indicated that peak 4 (progoitrin) and peak 7 (*R,S*-goitrin) showed the greatest influence on the scores. Peak 2 (uridine) and other compounds also affected the scores. Progoitrin content in the crude herbs was obviously higher than that in the prepared slices, whereas *R,S*-goitrin content had a reverse trend. The difference between crude herbs and prepared *Radix Isatidis* slices could be generally tracked to the transformation of glucosinolates to their degradation products.

EXPERIMENTAL SECTION

Materials, Reagents, and Chemicals

Eight BLG (prepared slices) batches, seven BLG (crude herbs) batches, six BLGG batches, and four NBLG batches were collected from different provinces or manufacturers in China, and were numbered from S-01 to S-25 (Table 4). The BLG and NBLG samples were identified by Dr. Li-Hong Wu, and the voucher specimens were deposited in the Herbarium of Shanghai University of Traditional Chinese Medicine (Shanghai, China).

R,S-goitrin was obtained from Shanghai Research and Development Center for Standardization of Traditional Chinese Medicine (Shanghai, China). Hypoxanthine, uridine, guanosine, and adenosine were purchased from the National Institute for Food and Drug control (Beijing, China). Progoitrin, epiprogoitrin, and gluconapin were isolated from the root of *Isatis indigotica* Fort. and their structures were elucidated by mass spectrometry, ^{1}H nuclear magnetic resonance spectrometry, and ^{13}C nuclear magnetic resonance spectroscopy, and were confirmed by comparing the data with those of previous studies. The compound purity was over 98% based on the HPLC area normalization method. The standard structures are illustrated in Figure 6.

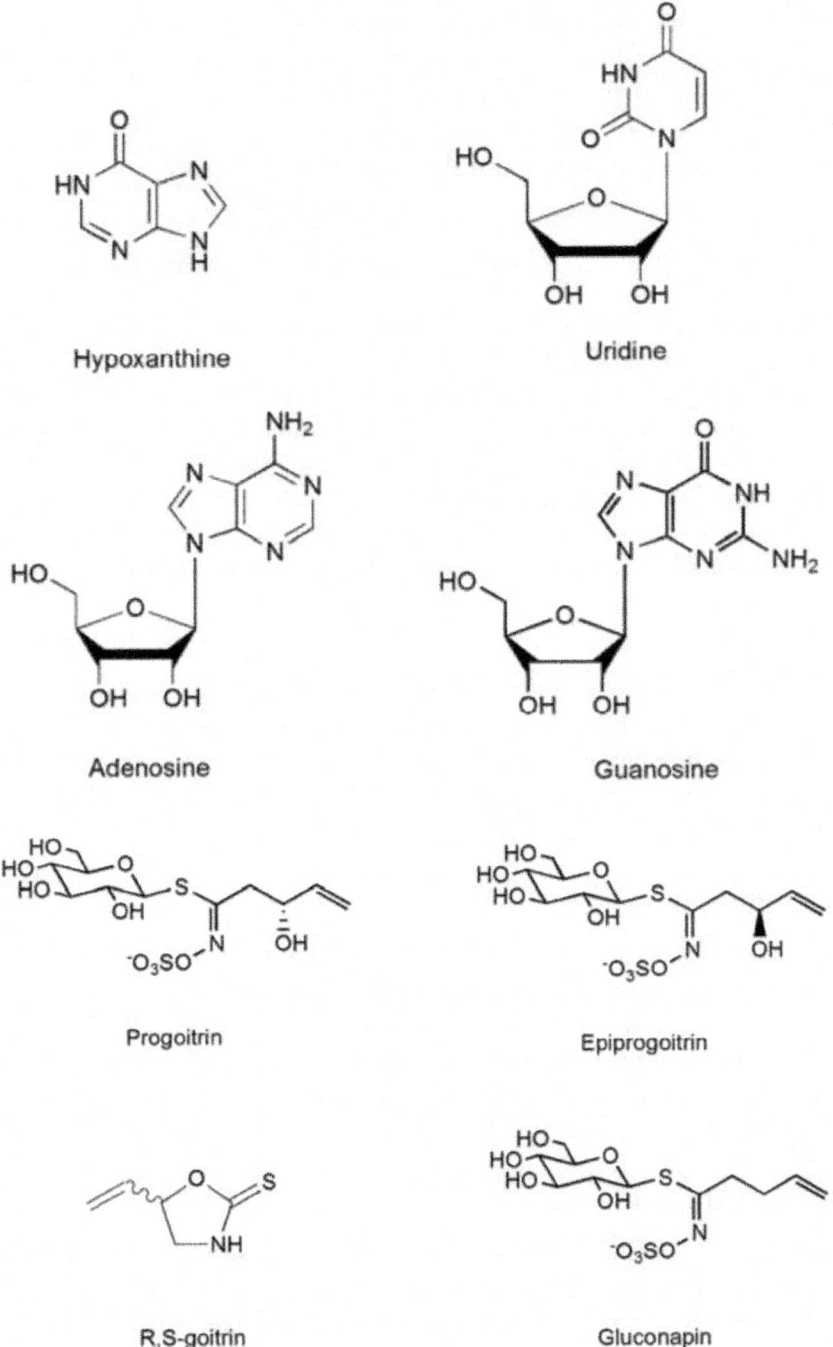

Figure 6. The structures of eight bioactive components in *Radix Isatidis.*

HPLC-grade acetonitrile, triethylamine, and formic acid (Tedia, Fairfield, OH, USA) were used for mobile phase preparation. Purified water was prepared with a Mili-Q water-purification system (Millipore, Bedford, MA, USA). Other solvents used for analyses were of analytical grade.

Instrumentation and Chromatographic Condition

UPLC analysis was performed on a Waters ACQUITY UPLC H-Class system (Waters, Milford, MA, USA) equipped with a binary solvent delivery pump, an auto sampler, and a photodiode array detector (PDA) and was controlled by the Empower-II software. UPLC fingerprinting analysis was carried out at 25 °C on a Waters ACQUITY UPLC BEH C_{18} (1.7 μm, 2.1 mm × 100 mm). A binary gradient elution system composed of acetonitrile (phase A) and 0.1% triethylamine in water (phase B, the pH was adjusted to 4.0 with formic acid) was applied to the fingerprint analysis with the gradient elution as follows: 0 min to 5 min, 0% A; 5 min to 7 min, 0% to 3% A; 7 min to 10 min, 3% A; 10 min to 20 min, 3% to 10% A. The wavelength was set at 254 nm, the mobile flow rate at 0.3 mL min^{-1}, and the on-line UV spectra was recorded in the range of 190 nm to 500 nm.

Preparation of Standard Solution

R,S-goitrin, progoitrin, epiprogoitrin, gluconapin, adenosine, uridine, guanosine, and hypoxanthine CRS were accurately weighed, and dissolved in methanol to produce a solution containing 1.0 mg mL^{-1}, which was used as the reference solution.

Preparation of Sample Solution

The crude herbs and the prepared BLG and NBLG slices were oven-dried at 50 °C until the weight remained constant. Each sample (1.0 g powder) was extracted with 20 mL of water, which was subjected to reflux for 30 min. The final solution was filtered through a 0.22 μm membrane prior to use. An aliquot of 2.0 μL of each sample solution was injected into the UPLC system for analysis.

Method Validation

The method was validated for linearity, LOD, LOQ, precision, reproducibility, stability, and accuracy following the International Conference on Harmonization (ICH) guideline and several reports on determination analysis [36,38].

Data Analysis

Data analysis was performed by the professional software of the Computer-Aided Similarity Evaluation System for Chromatographic Fingerprint of TCM (China Committee of Pharmacopeia, 2004A), which was recommended by the SFDA. The software was employed for synchronization and quantitative comparison between different samples [37].

In addition, the RRT and RPA of each characteristic peak related to the reference peak (*R,S*-goitrin) were also statistically analyzed using the SPSS software package (version 15.0; IBM: Chicago, IL, USA, 2006) and Soft Independent Modeling of Class Analogy (SIMCA)-P (version 13.0; Umetrics: Basel, Switzerland, 2012) [39–41].

CONCLUSIONS

In conclusion, the UPLC fingerprint and quantitative analysis of the water-soluble extract of *Radix Isatidis* were first established in this study to determine the eight bioactive constituents in *Radix Isatidis* and its finished products.

Meanwhile, the results from the three chemometric techniques (SA, HCA, and PCA) showed good consistency with one another. The samples from different locations of the same type (crude herbs or prepared slices) were still clustered into one group, and the eight bioactive components were mostly close, which indicates the similarity of internal quality of the samples in the same type. Furthermore, eight marker constituents were found to be specific variables, which could provide the most discrimination and quality control of *Radix Isatidis* and its finished products by quantitative analysis. The PCA loading plot identified the greatest impact factor of characteristic constituents to the classification. In addition, this chromatographic method was efficient and rapid in distinguishing *Radix Isatidis* from other species such as NBLG.

Based on this study, the UPLC fingerprint and quantitative analysis methods provided reliable assurance on systematic and complete quality control on *Radix Isatidis* and its finished products, which would be helpful in improving reasonable developments in *Radix Isatidis* and its finished products.

ACKNOWLEDGMENTS

This research project was supported by the grant from the National Natural Science Foundation of China (No. 81173518, 81172951), the Innovation Program of Shanghai Municipal Education Commission (11YZ70), and the SHUTCM Program for the Establishment of TCM Chemistry.

REFERENCES

1. Pharmacopoeia of People's Republic of China, *The State Pharmacopoeia Committee of People's Republic of China*; Chemical Industry Press: Beijing, China, 2010; Volume 1, pp. 20–21.

2. State Administration of Traditional Chinese Medicine of the People's Republic of China, *Chinese Materia Medica (No.21)*; Shanghai Scientific & Technical Publishers: Shanghai, China, 1999.

3. Wu, P.Y.; Wang, J.M. Botanical study on "Da Qing Ye", "Ban Lan Gen" and "Qing Dai". *Acta Univer. Tradit. Med. Sin. Pharmacol. Shanghai* **1996**, *10*, 50–52.

4. Wu, Q.Z.; Wang, Y.L. Comparative analysis of chemical constituents in north-banlangen and south-banlangen. *Strait Pharm. J* **2007**, *19*, 82–83.

5. Qin, H.Z.; Shi, B.; Li, S.Y.; Fan, X.J.; Song, D.Z.; Xian, H.M. Comparing research in Antiviral action of Rhizoma et *Radix Baphicacanthis Cusiae* and *Radix Isatidis* against a virus. *Chin. Arch. Tradit. Chin. Med* **2009**, *27*, 168–169.

6. Xu, Y.S.; Sun, J.; He, S.Q. Effect of three kinds of *Radix isatidis* preparation on the expression of nucleoprotein of influenza virus. *Shandong Med. J* **2010**, *50*, 8.

7. Liu, S.; Chen, W.S.; Qiao, C.Z.; Zheng, S.Q.; Zeng, M.; Zhang, H.M. Antiviral action of *Radix Isatidis* and *Folium Isatidis* from different germplasm against influenza A virus. *Acad. J. Sec. Mil. Med. Univ* **2000**, *21*, 204–206.

8. Lin, C.W.; Tsai, F.J.; Tsai, C.H.; Lai, C.C.; Wan, L.; Ho, T.Y.; Hsieh, C.C.; Chao, P.D. Anti-SARS coronavirus 3C-like protease effects of *Isatis indigotica* root and plant-derived phenolic compounds. *Antiviral Res* **2005**, *68*, 36–42.

9. Sun, H.H. *Establishment of a Mouse Model for Pandemic H1N1 Influenza Virus and Study on Effect of Banlangen Granules on Mice Challenged with Pandemic H1N1 Influenza Virus*; Peaking Union Medical College: Beijing, China, 2010; pp. 39–49.

10. Wang, Y.T.; Yang, Z.F.; Zhao, H.S.; Qin, S.; Guan, W.D. Screening of anti-H1N1 active constituents from *Radix Isatidis.J. Guangzhou Uni. Tradit. Chin. Med* **2011**, *28*, 419–422.

11. Sun, H.H.; Deng, W.; Zhan, L.J.; Xu, L.L.; Li, F.D.; Nv, Q.; Zhu, H. Effect of Banlangen Granules on mice challenged with a/California/7/2009. *Chin. J. Comp. Med* **2010**, *20*, 53–56.

12. Cheng, Y.; Li, X.; Chen, J.W.; He, L.W.; Wang, Y.Y. A study on antiviral effects of the active parts from *Radix Isatidis*. *J. Nanjing TCM Univ* **2011**, *27*, 155–157.

13. Ye, W.Y.; Li, Y.; Guo, J.W. Screening of 11 antiviral effect constituents from *Radix Isatidis*. *J. Emerg. Tradit. Chin. Med***2011**, *20*, 1172–1774.

14. Li, X.M. Remarkable progresses in studies on antiviral mechanism researches of *Radix Isatidis* (Baiyunshan, Guangdong). *China News of TCM* **2011**, 8.

15. Liu, S.Y. Tucking up veil the Antiviral causes of *Radix Isatidis*. *China News of TCM* **2011**, A13.

16. Liu, Y.H.; Wu, X.Y.; Fang, J.G.; Xie, W. Chemical constituents from *Radix Isatidis*. *Cent. South Pharm* **2003**, *5*, 302–305.

17. Fang, J.G.; Wang, S.B.; Xu, H.; Liu, Y.H.; Liu, Y.W. Chemical constituents from *Radix Isatidis*. *Chin. Tradit. Herb. Drugs***2004**, *35*, 845–846.

18. He, L.W.; Li, X.; Chen, J.W.; Sun, D.D.; Ju, W.Z.; Wang, C.H. Chemical constituents from water extract of *Radix isatidis*.*Acta Pharm. Sin* **2006**, *41*, 1193–1196.

19. Wang, Y.Y.; Li, X.; Chen, J.W.; Chen, Y.; Wang, S. Study on Chemical constituents from water extract of *Radix isatidis*.*Res. Pract. Chin. Med* **2009**, *23*, 54–56.

20. Lee, K.C.; Cheuk, M.W.; Chan, W. Determination of glucosinolates in traditional Chinese herbs by high-performance liquid chromatography and electrospray ionization mass spectrometry. *Anal. Bioanal. Chem* **2006**, *386*, 2225–2232.

21. Xu, L.H.; Huang, F.; Chen, T.; Wu, J. Antivirus constituents of radix of *Isatidis indigotica*. *Chin. J. Nat. Med* **2005**, *3*, 359–360.

22. Deng, X.Y.; Gao, G.H.; Zheng, S.N.; Li, F.M. Qualitative and quantitative of flavonoids in the leaves of *Isatis indigatica*Fort. by ultra-performance liquid chromatography with PDA and electrospray ionization tandem mass spectrometry detection. *J. Pharm. Biomed. Anal* **2008**, *48*, 562–567.

23. Ye, W.T.; Ou-Yang, H.F.; Li, X.L. Study on quality standard of CO Banlangen Granules. *Clin. Med. Eng* **2010**, *4*, 51–52.

24. Kong, W.J.; Zhao, Y.L.; Shan, L.M.; Xiao, X.H.; Guo, W.Y. Investigation on the spectrum-effect relationships of ETOAc extract from *Radix isatidis* based on HPLC fingerprints and microcalorimetry. *J. Chromatogr. B* **2008**, *1*, 109–144.

25. Kong, W.J.; Zhao, Y.L.; Shan, L.M.; Xiao, X.H.; Guo, W.Y. Thermochemical studies on the quantity-antibacterial effect relationship of four organic

Accids from *Radix Isatidis* on *Escherichia coli* growth. *Biol. Pharm. Bull* **2008**, *31*, 1301–1305.

26. Kong, W.J.; Zhao, Y.L.; Shan, L.M.; Xiao, X.H.; Guo, W.Y.; Wang, J.B. Determination of salicylic acid in total acid extraction of *Radix Isatidis* by HPLC. *Cent. South Pharm* **2008**, *2*, 137–140.

27. Fan, L.F.; Zhang, L.T.; Yuan, Z.F.; Xu, H.J.; He, W. HPLC determination the contents of indigo and indirubin in *Radix Isatidis*. *Chin. J. Pharm. Anal* **2008**, *4*, 540–543.

28. Na, G.X.; Fang, P. Research overview in degradation of glucosinolates. *Food Sci* **2008**, *29*, 350–354.

29. Hashem, F.A.; Saleh, M.M. Antimicrobial components of some cruciferae plants (*Diplotaxis harra* Forsk. and *Erucaria microcarpa Boiss.*). *Phytother. Res* **1999**, *13*, 329–332.

30. Wang, R.; Yang, H.Y.; Yang, Q.W. Study on quality standard of *Isatidis Radix*. *Chin. Tradit. Herb. Drug* **2010**, *41*, 478–480.

31. Xie, Z.Y.; Shi, Y.H.; Wang, Z.T.; Wang, R.; Li, Y.M. Biotransformation of glucosinolates epiprogoitrin and progoitrin to (*R*)- and (*S*)-goitrin in *Radix isatidis*. *J. Agric. Food. Chem* **2011**, *59*, 12467–12472.

32. Zhou, W.; Xie, M.F.; Zhang, X.Y.; Liu, T.T.; Duan, G.L. Improved liquid chromatography fingerprint of fat-soluble*Radix Isatidis* extract using multi-wavelength combination technique. *J. Sep. Sci* **2011**, *34*, 1123–1132.

33. Xiao, X.; Luo, G.A.; Wang, Y.M. Fingerprints of *Isatidis Radix* by HPLC. *J. Jiangxi Univ. TCM* **2010**, *22*, 67–69.

34. Zou, P.; Hong, Y.; Koh, H.L. Chemical fingerprinting of *Isatis Indigatica* root by RP-HPLC and Hierarchical clustering analysis. *J. Pharm. Biomed. Anal* **2005**, *38*, 514–520.

35. Shi, Y.H.; Xie, Z.Y.; Wu, Y.C.; Li, Y.M.; Wang, R.; Wang, Z.T. Determination of R,S-goitrin of Banlangen preparations by RP-HPLC. *Chin. J. Exp. Tradit. Med. Form* **2010**, *17*, 144–146.

36. Tang, D.Q.; Yang, D.Z.; Tang, A.B.; Gao, Y.Y.; Jiang, X.G.; Mou, J.; Yin, X.X. Simultaneous chemical fingerprint and quantitative analysis of *Ginkgo biloba* extract by HPLC-DAD. *Anal. Bioanal. Chem* **2010**, *396*, 3087–3095.

37. Wei, H.; Sun, L.N.; Tai, Z.G.; Gao, S.H.; Xu, W.; Chen, W.S. A simple and sensitive HPLC method for the simultaneous determination of 8 bioactive components and fingerprint analysis of *Schisandra sphenanthera*. *Anal. Chim. Acta* **2010**,*662*, 97–104.

38. Jin, X.F.; Lu, Y.H.; Wei, D.Z.; Wang, Z.T. Chemical fingerprint and quantitative analysis of *Salvia plebeian* R.Br. by high-performance liquid chromatography. *J. Pharm. Biomed. Anal* **2008**, *48*, 100–104.

39. Li, Y.; Wu, T.; Zhu, J.J.; Wan, L.L.; Yu, Q.; Li, X.X.; Cheng, Z.Z.; Guo, C. Combinative method using HPLC fingerprint and quantitative analyses for quality consistency evaluation of an herbal medicinal preparation produced by different manufactures. *J. Pharm. Biomed. Anal* **2010**, *52*, 597–602.

40. Wang, Y.X.; Li, Q.; Wang, Q.; Li, Y.J.; Ling, J.H.; Liu, L.L.; Chen, X.H.; Bi, K.S. Simultaneous determination of seven bioactive components in Oolong Tea *Camellia sinensis*: Quality control by chemical composition and HPLC fingerprints. *J. Agric. Food Chem* **2012**, *60*, 256–260.

Chapter 2

QUANTITATIVE AND CHEMICAL FINGERPRINT ANALYSIS FOR THE QUALITY EVALUATION OF RECEPTACULUM NELUMBINIS BY RP-HPLC COUPLED WITH HIERARCHICAL CLUSTERING ANALYSIS

Yan-Bin Wu [1], Li-Jun Zheng [1], Jun Yi [2], Jian-Guo Wu [1], Ti-Qiang Chen [3] and Jin-Zhong Wu [1]

[1]Academy of Integrative Medicine, Fujian University of Traditional Chinese Medicine, Fuzhou 350122, China

[2]Department of Science, Fujian Institute of Education, Fuzhou 350001, China

[3]Fujian Academy of Agricultural Sciences, Fuzhou 350013, China

ABSTRACT

A simple and reliable method of high-performance liquid chromatography with photodiode array detection (HPLC-DAD) was developed to evaluate the quality of Receptaculum Nelumbinis (dried receptacle of *Nelumbo nucifera*) through establishing chromatographic fingerprint and simultaneous determination of five flavonol glycosides, including hyperoside, isoquercitrin, quercetin-3-O-β-d-glucuronide, isorhamnetin-3-O-β-d-galactoside and syringetin-3-O-β-d-glucoside. In quantitative analysis, the five components showed good regression ($R > 0.9998$) within linear ranges, and their recoveries were in the range of 98.31%–100.32%. In the chromatographic fingerprint, twelve peaks were selected as the characteristic peaks to assess the similarities of different samples collected from different origins in China according to the State Food and Drug Administration (SFDA) requirements. Furthermore, hierarchical cluster analysis (HCA) was also applied to evaluate the variation of chemical components among different sources of Receptaculum Nelumbinis in China. This study indicated that the combination of quantitative and chromatographic fingerprint analysis can be readily utilized as a quality control method for Receptaculum Nelumbinis and its related traditional Chinese medicinal preparations.

INTRODUCTION

Lotus (*Nelumbo nucifera* Gaertn.), belonging to the family Nymphaeaceae, is a kind of perennial aquatic herbage plant, which is one of the most important aquatic vegetables widely grown in China, due to its pleasant flavor and high nutritional value, especially its seeds, rhizomes and leaves. It's easy to be cultivated and distributed in wetlands throughout temperate and tropical Asia from Iran to Japan and from China to Queensland [1]. Up to the year 2002, a total of 572 lotus accessions (including landraces, cultivars and breeding lines) with different germplasm resources were conserved in the National Garden of Aquatic Vegetable in Wuhan, Hubei province, China, including those collections from 153 counties in 18 provinces and lines bred by breeders. According to different purposes or morphological differences, the lotus is usually classified into three types: rhizome lotus, seed lotus and flower lotus. Rhizome lotus is mainly cultivated in Hubei, Jiangsu, Anhui and Zhejiang provinces, seed lotus in Jiangxi, Fujian and Hunan and flower lotus in Wuhan city, Hubei province and Beijing [2].

The previously study reported that different types of lotus have show dissimilar characteristics, which is indicative of their distinct genetic differentiations [3]. There are inextricable links between medicinal plants and their ecological environment in the process of long-term survival competition and natural selection. The genetic variation of active ingredients in germplasm resources is an important factor affecting yield and quality of drugs. To some extent, the formulation of "authentic ingredients" with excellent efficacy is attributed to the action of the "local variety". In addition, herbs collected at different times and planted in different regions may affect the quality of their chemical composition and the amounts of major bioactive constituents [4].

Receptaculum Nelumbinis, commonly used traditional Chinese medicine (TCM), called Lianfang in Chinese, is derived from the dried receptacle of *N. nucifera*. It is used as an antihemorrhagic agent, especially for excess menstrual bleeding and irregular genital bleeding and also as a remedy for dehydration caused by diarrhea in summer and for prevention of miscarriage in traditional Chinese medicine [5]. In previous bioactivity research on this herb, Receptaculum Nelumbinis have exhibited a wide spectrum of biopharmacological effects, including antioxidation, improving learning and memory abilities, protective effects against experimental myocardial injury and ischemia, radioprotective activity and anti-tumor effects [6–10]. Our previous phytochemical investigations of Receptaculum Nelumbinis have revealed that its main components are phenolic constituents [11]. In the official Chinese pharmacopoeia (China Pharmacopoeia Committee, 2010) [12], only microscopic identification methods were used to identify this medicinal material, and the content determination of marker compounds were not recorded [12]. In the aspects of literature, only the assay of hyperoside and quercetin in Receptaculum Nelumbinis was reported [13,14]. This can't fully account for all the activities of Receptaculum Nelumbinis and does not meet the need to effectively control the quality for Receptaculum Nelumbinis.

Therefore, in order to control the quality and to clarify the differentiation of chemical constituents in Receptaculum Nelumbinis, a HPLC-DAD method of multiple compounds determination in combination with chemical fingerprinting methodology was developed for the quality evaluation of Receptaculum Nelumbinis. In consideration of the complexity of herb medicine, the HPLC chromatograms are complex multivariate data sets, so minor differences between very similar chromatograms might be missed; the chemical pattern recognition methods, such as similarity analysis (SA) and hierarchical clustering analysis (HCA), were used to reasonably define the class of the herbal medicine and to efficiently evaluate the differentiation of the Receptaculum Nelumbinis samples. We expected that this HPLC method would be helpful for the quality control of Receptaculum Nelumbinis in the future.

RESULTS AND DISCUSSION

Optimization of HPLC Conditions

In order to obtain the chromatograms with better separation of adjacent peaks within a short time, the column, mobile phase and detection wavelength were investigated. Different HPLC columns were tested for better resolution, and then baseline separation of the five constituents was obtained on an Agilent HC-C_{18} column. Acetonitrile-Water system was used as the mobile phase. It could give rise to more peaks, but separation was not satisfactory. According to the literature, acid could achieve better separation for dihydrochalcones [15], thus, 0.2% acetic acid was added to the acetonitrile-water system to further improve the peak shape. Due to a full-scan experiment of the five active components from 200 to 400 nm, 360 nm was selected as the detection wavelength, so that more characteristic peaks could be obtained, and the baseline was well improved on the chromatographic profiles.

Method Validation of Quantitative Analysis

The method was validated in terms of linearity, precision, repeatability, stability and recovery test. The Receptaculum Nelumbinis for method validation was collected from Fujian, China, and the variety of *Nelumbo nucifera* Gaertn was named Taikong 36.

The calibration curve was generated to confirm the linear relationship between the peak area and the concentrations of each reference compound in the test samples. The five standards of hyperoside, isoquercitrin, quercetin-3-O-β-d-glucuronide, isorhamnetin-3-O-β-d-galactoside and syringetin-3-O-β-d-glucoside were accurately weighed, dissolved and diluted with 50% methanol in a volumetric flask to obtain standard solutions for the calibration curves. Calibration curves were peak area *versus* concentration for each analyte. The linear regression equations, correlation coefficients and ranges of calibration curves for the listed flavonoid derivatives are shown in Table 1. The calibration curves showed good linear regression, with correlation coefficience over 0.9998 within test ranges.

Table 1. Regression equation and correlation coefficient of calibration curves for the five compounds

Compound	Regression equationa	Rb	Linearity range (µg/mL)
Hyperoside	$Y = 25.303x + 13.352$	0.9998	8–104 µg/mL
Isoquercitrin	$Y = 16.058x + 22.746$	0.9991	2–64 µg/mL
Quercetin-3-O-β-d-glucuronide	$Y = 9.6542x - 26.718$	0.9998	30–960 µg/mL
Isorhamnetin-3-O-β-d-galactoside	$Y = 19.097x - 1.1527$	0.9999	1.4–42.6 µg/mL
Syringetin-3-O-β-d-glucoside	$Y = 13.305x + 0.9321$	0.9997	0.7–22.8 µg/mL

[a]Y peak area, x concentration of compound (µg/mL);

[b]R = correlation coefficient, $n = 6$.

A sample of the medicinal material was prepared as described above and was subjected to HPLC analysis six times in the same day to evaluate the precision. The repeatability was examined by the injection of six different samples, which were prepared with the same sample preparation procedure. Variations were expressed as relative standard deviations (RSD).Table 2 showed the results of the tests of precision and repeatability. The stability was analyzed in 0, 4, 8, 12, 24 and 48 h within 2 days. Stability was expressed as the RSD, and the values were less than 0.69% for the five compounds (Table 2). The recovery test was determined by the standard addition method. Five flavonol glycosides were added to the samples, and then, the extraction and analysis were performed according to the above sample preparation procedure. The mean recovery was calculated according to the following formula: recovery (%) = [(found amount − original amount)/spiked amount] × 100% and RSD (%) = (SD/mean) × 100%. The mean recovery of the five flavonoids compounds was 98.31%–100.32%, and their RSD values were less than 3.00% (Table 2).

Table 2. Precision, reproducibility, stability and recovery of the five compounds

Compound	Precision RSD (%) ($n = 6$)	Reproducibility RSD (%) ($n = 6$)	Stability RSD (%) ($n = 6$)	Recovery (%) ($n = 6$) Mean ± RSD (%)
Hyperoside	0.07	1.99	0.09	99.23 ± 2.61
Isoquercitrin	0.14	2.37	0.17	99.72 ± 2.84
Quercetin-3-O-β-d-glucuronide	0.48	2.88	0.68	99.54 ± 2.91
Isorhamnetin-3-O-β-d-galactoside	0.08	2.44	0.18	98.31 ± 3.00
Syringetin-3-O-β-d-glucoside	0.49	2.96	0.69	100.32 ± 2.71

Establishment of Chromatographic Fingerprint of Receptaculum Nelumbinis and Similarity Analysis (SA)

To standardize the HPLC profile, 20 samples of Receptaculum Nelumbinis were analyzed, and all chromatograms were introduced into the Computer-Aided Similarity Evaluation System for Chromatographic Fingerprint of TCM (China Committee of Pharmacopeia, 2004). Peaks that existed in all chromatograms of samples with reasonable heights and good resolutions were assigned as "common peak" for Receptaculum Nelumbinis. As shown in Figure 1, there are 12 distinct common peaks (from peak 1 to peak 12) in the HPLC fingerprint common patterns from the 20 samples of Receptaculum Nelumbinis, and the representative standard fingerprints of the investigated samples is shown in Figure 2. Five common peaks (peak 4, 5, 6, 8 and 9) were identified as hyperoside, isoquercitrin, quercetin-3-O-β-d-glucuronide, isorhamnetin-3-O-β-d-galactoside and syringetin-3-O-β-d-glucoside.

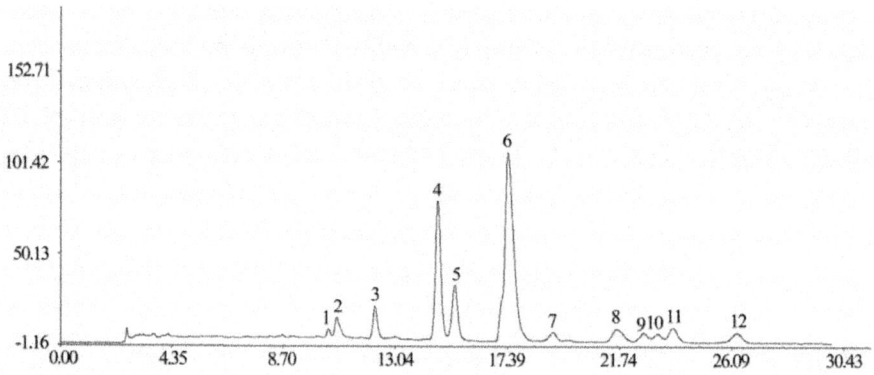

Figure 1. Average artificial HPLC fingerprint common pattern of Receptaculum Nelumbinis based on 20 samples.

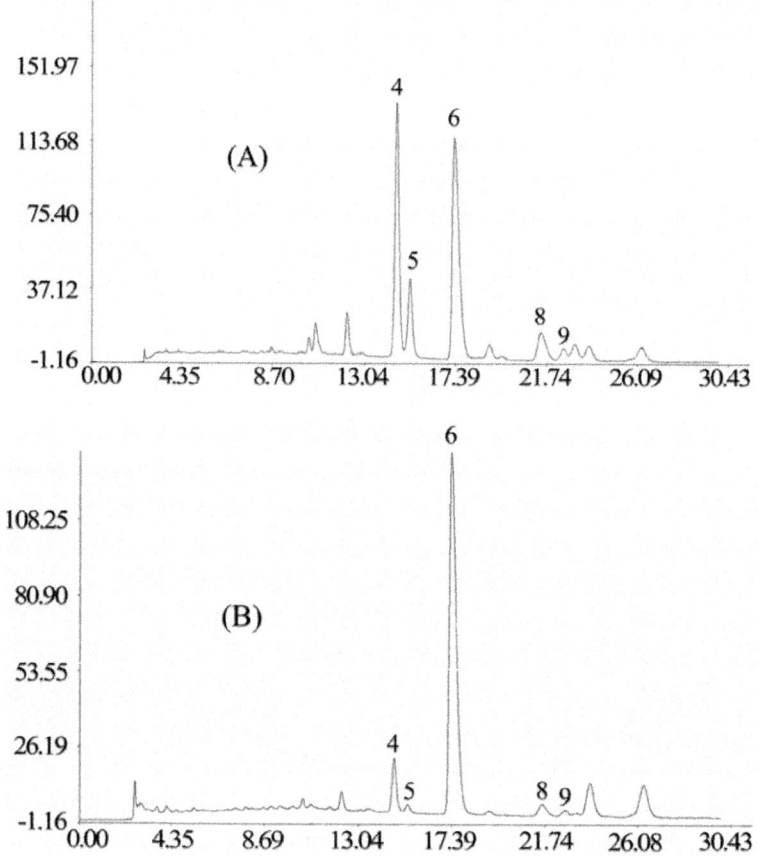

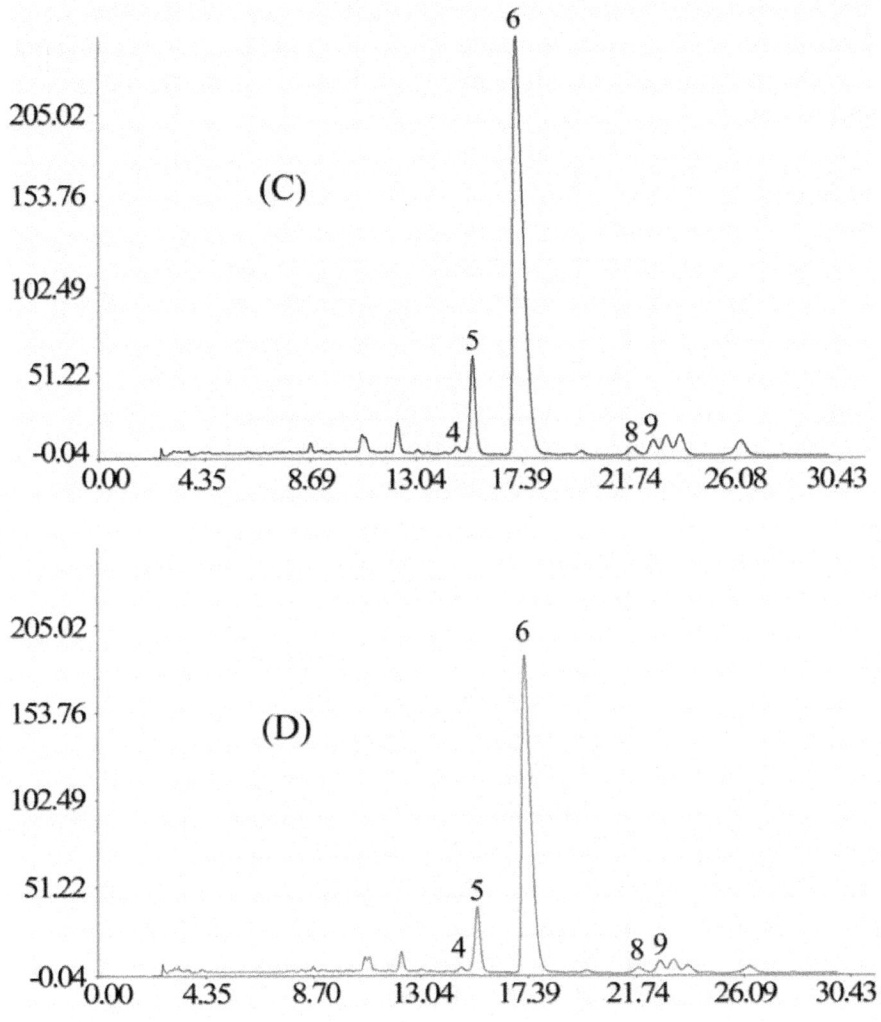

Figure 2. The representative standard fingerprints of Receptaculum Nelumbinis obtained by Similarity Evaluation System: **(A)** L1; **(B)** L11; **(C)** L13; **(D)** L17.

The similarities of chromatograms for the varieties of the receptacles of *N. nucifera* were calculated using the similarity evaluation system recommended by SFDA. Similarity comparison of the standard fingerprints of different samples showed that the similarity ranged from 0.282 to 0.966 (Table 3). The results (Table 3) showed that sample L11, L13 and L17 have a small similarity with its similarity, respectively, as 0.867, 0.283 and 0.282. The similarity values of the other samples were more than 0.918. These results indicated that the chemical composition and content in the Receptaculum Nelumbinis varied significantly.

Table 3. The similarities of the chromatograms of twenty varieties of the receptacles of *N. Nucifera*

Samples	Similarities
L1	0.956
L2	0.962
L3	0.934
L4	0.957
L5	0.954
L6	0.941
L7	0.918
L8	0.959
L9	0.966
L10	0.959
L11	0.867
L12	0.961
L13	0.283
L14	0.959
L15	0.965
L16	0.966
L17	0.282
L18	0.965
L19	0.956
L20	0.956

Quantitative Determination of Five Compounds in Receptaculum Nelumbinis

According to the contents and pharmacological properties of major constituents in Receptaculum Nelumbinis, the peak of hyperoside, isoquercitrin, quercetin-3-*O*-β-d-glucuronide, isorhamnetin-3-*O*-β-d-galactoside and syringetin-3-*O*-β-d-glucoside were chosen as reference peaks. The contents of the five compounds in the twenty varieties of the receptacles of *N. nucifera* from different sources in China were determined by the establishing HPLC method. The representative

HPLC chromatogram of the five compounds is shown in Figure 3. Each sample was analyzed in triplicate to determine the mean content (mg/g), and the results are listed in Table 4. The quantitative analysis results showed that the twenty varieties of the receptacles of *N. nucifera* generally contained the five flavonol glycosides, and the content of the five compounds were significantly different. The content ranges for hyperoside, isoquercitrin, quercetin-3-*O*-β-d-glucuronide, isorhamnetin-3-*O*-β-d-galactoside and syringetin-3-*O*-β-d-glucoside were 0.1–9.1, 0.1–6.0, 9.3–72.2, 0.3–3.3 and 0.1–1.9 mg/mL, respectively. The results in Table 4 showed that the content of each flavonol glycoside in different samples were significantly different, which is consistent with that of HPLC fingerprint analysis. Moreover, hyperoside, isoquercitrin and quercetin-3-*O*-β-d-glucuronide were found to be predominant among the five determined analytes. Many studies have shown that hyperoside, isoquercitrin and quercetin-3-*O*-β-d-glucuronide have exhibited a wide spectrum of biopharmacological effects, including beneficial cardiovascular effect, antioxidation, anti-hypertrophic effect on vascular smooth muscle cell, antiviral, anti-inflammatory and anti-tumor effect [11,16–20]. The high yield of hyperoside, isoquercitrin and quercetin-3-*O*-β-d-glucuronide in 50% ethanol extract may contribute to the curative effect of Receptaculum Nelumbinis.

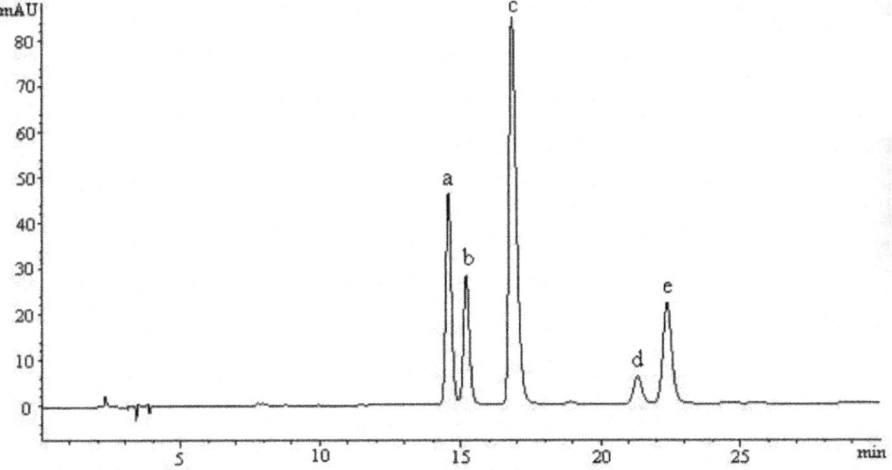

Figure 3. HPLC chromatogram of mixture standard of the five compounds: (a) Hyperoside; (b) Isoquercitrin; (c) Quercetin-3-*O*-β-d-glucuronide; (d) Isorhamnetin-3-*O*-β-d-galactoside; (e) Syringetin-3-*O*-β-d-glucoside.

Table 4. Contents (mg/g) of five compounds in twenty varieties of the receptacles of *N. nucifera* (*n* = 3)

Samples	Cultivated varieties	Collected location	Collection time	Contents (mg/g)				
				1a	2	3	4	5
L1	Taikong 36	Jianou, Fujian	October 2010	7.4	3.9	27.0	2.1	1.0
L2	Zajiao 8236	Jianou, Fujian	October 2010	6.2	2.9	29.1	1.8	0.9
L3	Jianjibaihualian	Jianou, Fujian	October 2010	5.5	4.9	18.4	1.3	0.6
L4	Guangchang Taikong 3	Guangchang, Jiangxi	September 2010	2.5	1.4	14.7	1.2	0.7
L5	Baihualian	Guangchang, Jiangxi	September 2010	1.4	0.5	8.5	0.6	0.3
L6	Shilihe 1	Hangzhou, Zhejiang	September 2010	2.9	1.8	9.3	0.3	0.4
L7	Liyebailian	Hangzhou, Zhejiang	September 2010	0.8	0.2	9.4	0.3	0.1
L8	Jianxuan 17	Hangzhou, Zhejiang	August 2010	6.8	6.0	41.8	1.4	1.5
L9	Taikong 3	Hangzhou, Zhejiang	August 2010	7.9	3.8	33.7	1.9	1.1
L10	Ganxuan 62	Hangzhou, Zhejiang	August 2010	2.6	1.7	14.4	0.4	0.4
L11	Jinfunong	Hangzhou, Zhejiang	August 2010	1.1	0.1	30.6	0.6	0.4
L12	Dahonglian	Hangzhou, Zhejiang	August 2010	7.3	3.6	29.7	0.9	1.0
L13	Guanshanglian	Yuanmingyuan, Beijing	August 2010	0.2	5.5	76.2	0.6	1.4
L14	Baoyingmeirenhon-glian	Baoying, Jiangsu	August 2010	4.4	3.2	19.3	0.8	0.5
L15	Honghelian	Honghu, Hubei	August 2010	4.1	2.1	21.3	1.2	0.8
L16	Jianlian	Jianning, Fujian	August 2010	4.1	2.1	21.3	1.2	0.8
L17	Baiyangdingyesh-englian	Baoding, Hebei	August 2010	0.1	3.6	49.2	0.4	1.1
L18	Xingkongmudan	Guangchang, Jiangxi	August 2010	7.5	4.2	34.8	1.5	0.9
L19	Cunsanlian	Xiangtan, Hunan	August 2010	4.8	2.8	19.8	0.6	0.7
L20	Taikong 1	Hangzhou, Zhejiang	August 2010	9.1	3.2	36.5	3.3	1.9

[a]The data was present at average of duplicates. 1: Hyperoside; 2: Isoquercitrin; 3: Quercetin-3-*O*-β-d-glucuronide; 4: Isorhamnetin-3-*O*-β-d-galactoside; 5: Syringetin-3-*O*-β-d-glucoside.

According to the quantitative analysis results, it was suggested that the genetic variation was one of the key factors affecting the contents of bioactive constituents. The results also indicated that the internal qualities of 20 batches of Receptaculum Nelumbinis samples from different sources with different varieties had marked variations, and the quality control needed evaluation by chemical fingerprinting. Multiple factors for Receptaculum Nelumbinis, such as various regions, source and different harvesting time various, would accordingly result in the differences in their qualities. Thus, the selection of the stable source of the Receptaculum Nelumbinis is quite important and meaningful for the clinical effect and quality evaluation of this medicine.

Hierarchical Clustering Analysis (HCA)

HCA is a multivariate analysis technique that is used to sort samples into groups [21]. The HCA method is well known and has been applied for fingerprint analysis, because it is a nonparametric data interpretation method and simple to use [21–24]. HCA provides a visual representation of complex data. A method called average linkage between groups was applied, and Pearson correlation was selected as a measurement. The method can classify different herbs by measuring the peak areas from their corresponding LC fingerprints. The common characteristic peaks, which were calculated by the Similarity Evaluation System, were selected for the hierarchical cluster analysis [4].

In order to assess the resemblance and differences of these samples, a hierarchical agglomerative clustering analysis of Receptaculum Nelumbinis samples was performed based on the relative peak areas of all the 12 characteristics chromatographic peaks. The peak areas of characteristics constituents in 20 batches of Receptaculum Nelumbinis samples from various sources formed a matrix of 12×20. The results of HCA were shown in Figure 4 from which the quality characteristics were revealed more clearly. The results of the hierarchical cluster analysis showed that the samples could be divided into two quality clusters. Among them, Cluster I includes the samples L7, L11, L13 and L17 and the other in Cluster II. Cluster I was distinguished as hyperoside—poor chemotype—which contains less hyperoside than the Cluster II. These results were in correspondence to the SA. The low concentration of hyperoside in the Cluster I may be due to the poor herb quality of Receptaculum Nelumbinis. This indicated that hyperoside could be used as a marker compound to distinguish the Receptaculum Nelumbinis with different quality. The results of HCA could be validated each other and provided more references for the quality evaluation of Receptaculum Nelumbinis.

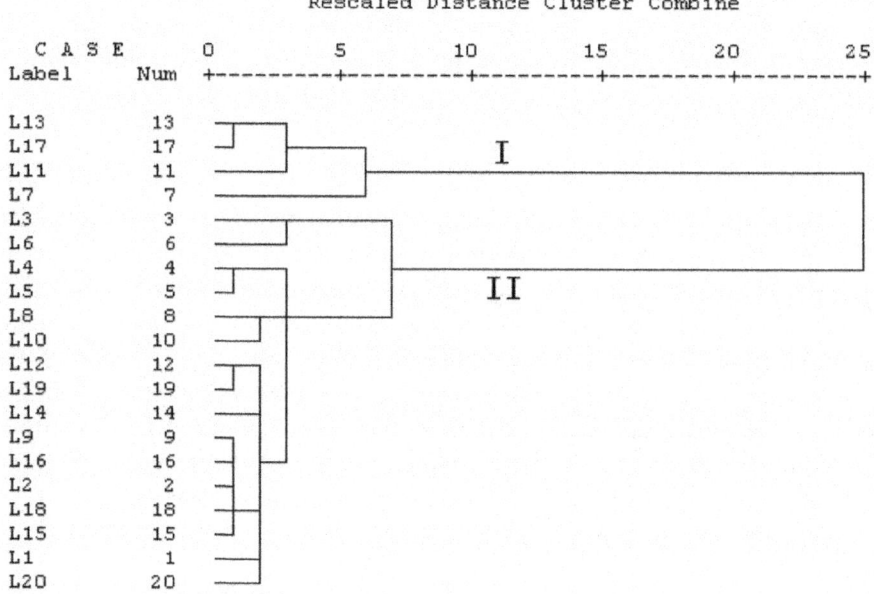

Figure 4. Results of hierarchical cluster analysis of 20 Receptaculum Nelumbinis samples (dendrogram using average linkage between groups).

EXPERIMENTAL SECTION

Plant Materials and Reagents

Twenty Receptaculum Nelumbinis populations were collected from different regions of China, and all voucher specimens were taxonomically identified based on morphological characteristics by Professor J. Z. Wu and deposited at Herbarium of Academy of Integrative Medicine, Fujian University of Traditional Chinese Medicine in Fuzhou 350108, China.

HPLC grade acetonitrile and methanol were purchased from Fisher Scientific (Pittsburgh, PA, USA). HPLC grade water was prepared using a Milli-Q water purification system (Millipore, Bedford, MA, USA). Analytical grade methanol, ethanol and acetic acid were purchased from Sinopharm Chemical Reagent Co. Ltd, Shanghai, China. All the solutions were filtered through 0.45 μm membranes (Schleicher & Schuell, Dassel, Germany) and degassed by ultrasonic bath before use.

Instrument and Chromatographic Conditions

The chromatographic separation was performed on an Angilent 1200 HPLC system (Agilent Technologies Inc., Santa Clara, CA, USA), equipped with a quaternary pump, an autosampler, a degasser, an automatic thermostatic column compartment, a DAD detector and a computer with a Chemstation software program for analysis of the HPLC data. Agilent HC-C$_{18}$ reversed-phase column (250 mm × 4.6, 5 μm) together with an Agilent HC-C$_{18}$ guard column (12.5 mm × 4.6, 5 μm) were used with column temperature set at 25 °C. HPLC-DAD detection was used for purity assay of reference compounds. The mobile phase consisted of acetonitrile (A) and 0.2% acetic acid (v/v, B) using a gradient program of 16%–19.6% A in 0–6 min, 19.6%–20.7% A in 6–30 min. This was followed by a 10 min equilibration period to the injection of each sample. The flow rate was 1 mL/min, and detection wavelength was set at 360 nm; an aliquot of 10 μL solution was injected for acquiring chromatograms.

Preparation of Standard Solutions

Hyperoside, isoquercitrin, quercetin-3-*O*-β-d-glucuronide, isorhamnetin-3-*O*-β-d-galactoside and syringetin-3-*O*-β-d-glucoside were extracted, isolated and purified from Receptaculum Nelumbinis in our laboratory. The chromatogram of the five mixture compounds was shown in Figure 4. All were identified using ESI-MS, ^{1}H-NMR and ^{13}C-NMR spectrometric techniques. The purity of each compound was determined to be higher than 98% by HPLC. Their structures and the detailed procedures for isolation and spectrometric identification have been reported in another paper [11].

Five reference compounds were accurately weighed and dissolved in 20% acetonitrile, then diluted to appropriate concentration ranges for the establishment of calibration curves. All stock and working standard solutions were stored at 4 °C until used for analysis.

Preparation of Sample Solutions

The open-air dried lotus receptacles were cut into small pieces and ground into powder, and then 1.0 g of sample fine powder was extracted twice with 25 mL of 50% ethanol by ultrasonic for 30 min. The extracts was filtered and evaporated under vacuum, the residues were dissolved with 10 mL 20% acetonitrile solution and sonicated for 10 min. The sample solution was filtered through a 0.45 μm membrane filter prior to HPLC analysis, and the injection volume was 10 μL.

Data Analysis

The chromatographic profiles of all extracts were performed by professional software named Similarity Evaluation System for Chromatographic Fingerprint of Traditional Chinese Medicine (Version 2004 A), which was recommended by the State Food and Drug Administration of China (SFDA) for evaluating similarities of chromatographic profiles of TCM [19]. The hierarchical cluster analysis (HCA) of samples was performed using SPSS software (SPSS 16.0 for Windows Vista™, SPSS Inc., Chicago, IL, USA).

CONCLUSIONS

In this paper, a HPLC fingerprint and quantitative analysis method was developed to evaluate the quality of Receptaculum Nelumbinis from different sources. The method was well validated by systematically comparing chromatograms of all samples from different sources and certified helpfully to improve the quality control. Meanwhile, the chemometrics methods were applied with the HPLC fingerprint techniques for analysis of chemical variation of Receptaculum Nelumbinis samples. Chemometrics analysis indicated that the quality of Receptaculum Nelumbinis have no significant relativity with geographic location and germplasm resources. HCA could distinguish these samples as different chemical-types, but not different geographic population and germplasm resources. In the view of results of content analysis, the samples L7, L11, L13 and L17 have a lower content of hyperoside, with its content, respectively, as 0.8, 1.1, 0.1 and 0.2 mg/g. This also explained why samples L7, L11, L13 and L17 were grouped as the same type in HCA analysis and have a small similarity. Furthermore, five marker constituents were found to be specific variables, which could provide the most discrimination and quality control of Receptaculum Nelumbinis by quantitative analysis. The results demonstrated that the chemometrics techniques, such as SA and HCA, were able to classify samples objectively and successfully in accordance with their chemical constituents and content. Further, the method is a powerful, practical tool for quality control of Receptaculum Nelumbinis samples or other related traditional Chinese medicinal preparations.

ACKNOWLEDGMENTS

This work was conducted in the Class III Laboratory of Traditional Chinese Medicine on Pharmacognosy of State Administration of Traditional Chinese Medicine of People's Republic of China and was financially supported by Key Project of Fujian Provincial Universities for Haixi Development (Grant No. 5), the Research Foundation of Health Bureau of Fujian Province of China

(Grant No. wzzd0901) and the Natural Science Foundation of Fujian province of China (Grant No. 2011J01212).

REFERENCES

1. Woranuch, L.; Chusie, T.; Henrik, B. Management and use of *Nelumbo nucifera* Gaertn. In Thai wetlands. *Wetl. Ecol. Manag* **2009**, *17*, 279–289.

2. Guo, H.B. Cultivation of lotus (*Nelumbo nucifera* Gaertn. ssp. nucifera) and its utilization in China. *Genet. Resour. Crop Evol* **2009**, *56*, 323–330.

3. Han, Y.C.; Teng, C.Z.; Chang, F.H.; Robert, G.W.; Zhou, M.Q.; Hu, Z.L.; Song, Y.H. Analyses of genetic relationships in*Nelumbo nucifera* using nuclear ribosomal ITS sequence data, ISSR and RAPD markers. *Aquat. Botan* **2007**, *87*, 141–146.

4. Liu, J.; Chen, X.; Yang, W.; Zhang, S.; Wang, F.; Tang, Z. Chemical fingerprinting of wild germplasm resource of*Ophiopogon* japonicus from Sichuan basin, China by RP-HPLC coupled with hierarchical cluster analysis. *Anal. Lett***2010**, *43*, 2411–2423.

5. Ishida, H.; Umino, T.; Tsuji, K.; Kosuge, T. Studies on the antihemorrhagic substances in herbs classified as hemostatics in Chinese medicine. VIII. On the antihemorrhagic principle in Nelumbins Receptaculum. *Chem. Pharm. Bull* **1988**, *36*, 4585–4587.

6. Wu, Y.B.; Zheng, L.J.; Yi, J.; Wu, J.G.; Tan, C.J.; Chen, T.Q.; Wu, J.Z.; Wong, K.H. A comparative study on antioxidant activity of ten different parts of *Nelumbo nucifera* Gaertn. *Afr. J. Pharm. Pharmacol* **2011**, *5*, 2454–2461.

7. Gong, Y.S.; Liu, L.G.; Xie, B.J.; Liao, Y.C.; Yang, E.L.; Sun, Z.D. Ameliorative effects of lotus seedpod proanthocyanidins on cognitive deficits and oxidative damage in senescence-accelerated mice. *Behav. Brain Res* **2008**,*194*, 100–107.

8. Zhang, X.H.; Zhang, B.; Gong, P.L.; Zeng, F.D. Protective effect of procyanidins from the seedpod of the lotus on myocardial ischemia and its mechanisms. *Yao Xue Xue Bao* **2004**, *39*, 747–750.

9. Duan, Y.; Zhang, H.; Xie, B.; Yan, Y.; Li, J.; Xu, F.; Qin, Y. Whole body radioprotective activity of an acetone-water extract from the seedpod of *Nelumbo nucifera* Gaertn. *Food Chem. Toxicol* **2010**, *48*, 3374–3384.

10. Duan, Y.; Zhang, H.; Xu, F.; Xie, B.; Yang, X.; Wang, Y.; Yan, Y. Inhibition effect of procyanidins from lotus seedpod on mouse B16 melanoma *in vivo* and *in vitro*. *Food Chem* **2010**, *122*, 84–91.

11. Wu, Y.B.; Zheng, L.J.; Wu, J.G.; Chen, T.Q.; Yi, J.; Wu, J.Z. Antioxidant

activities of extract and fractions from Receptaculum Nelumbinis and related flavonol glycosides. *Int. J. Mol. Sci* **2012**, *13*, 7163–7173.

12. Chinese Pharmacopoeia Committee, *Pharmacopoeia of the People's Republic of China*; Chemical Industry Press: Beijing, China, 2010; Volume 1, pp. 193–195.

13. Wang, J.; Liu, Y.; Cheng, B.; Shi, Y.; Chen, H. The research on the isolation and identification of hyperin in Receptaculum Nelumbinis and quality standard of Receptaculum Nelumbinis. *J. Chengdu Med. Coll* **2008**, *3*, 35–37.

14. Wang, C.; Zhang, X. Determination of hyperin and quercetin in different processed products of Nelumbins Receptaculum by HPLC. *Chin. Tradit. Patent Med* **2010**, *32*, 1729–1732.

15. Fan, L.L.; Tu, P.F.; Chen, H.B.; Cai, S.Q. Simultaneous quantification of five major constituents in stems of *Dracaena*plants and related medicinal preparations from China and Vietnam by HPLC-DAD. *Biomed. Chromatogr* **2009**, *23*, 1191–1200.

16. Li, Z.; Liu, J.; Hu, J.; Li, X.; Wang, S.; Yi, D.; Zhao, M. Protective effects of hyperoside against human umbilical vein endothelial cell damage induced by hydrogen peroxide. *J. Ethnopharmacol* **2012**, *139*, 388–394.

17. Silva, C.G.; Raulinoa, R.J.; Cerqueirab, D.M.; Mannarinoa, S.C.; Pereira, M.D.; Paneka, A.D.; Silvac, J.F.M.; Menezes, F.S.; Eleutherio, E.C.A. *In vitro* and *in vivo* determination of antioxidant activity and mode of action of isoquercitrin and *Hyptis fasciculata*. *Phytomedicine* **2009**, *16*, 761–767.

18. Yoshizumi, M.; Tsuchiya, K.; Suzaki, Y.; Kirima, K.; Kyaw, M.; Moon, J.; Terao, J.; Tamaki, T. Quercetin glucuronide prevents VSMC hypertrophy by angiotensin II via the inhibition of JNK and AP-1 signaling pathway. *Biochem. Bioph. Res. Commun* **2002**, *293*, 1458–1465.

19. Fan, D.; Zhou, X.; Zhao, C.; Chen, H.; Zhao, Y.; Gong, X. Anti-Inflammatory, antiviral and quantitative study of quercetin-3-*O*-β-d-glucuronide in *Polygonum perfoliatum* L. *Fitoterapia* **2011**, *82*, 805–810.

20. Yang, J.H.; Hsia, T.C.; Kuo, H.M.; Chao, P.D.L.; Chou, C.C.; Wei, Y.H.; Chung, J.G. Inhibition of lung cancer cell growth by quercetin glucuronides via G2/M arrest and induction of apoptosis. *Drug Metab. Dispos* **2006**, *34*, 296–304.

21. Kong, W.; Zhao, Y.; Xiao, X.; Jin, C.; Li, Z. Quantitative and chemical fingerprint analysis for quality control of Rhizoma Coptidischinensis based on UPLC-PAD combined with chemometrics methods. *Phytomedicine* **2009**, *16*, 950–959.

22. Zou, P.; Hong, Y.; Koh, H.L. Chemical fingerprinting of *Isatis indigotica* root by RP-HPLC and hierarchical clustering analysis. *J. Pharmaceut. Biomed. Anal* **2005**, *38*, 514–520.

23. Lu, F.; Li, D.; Fu, C.; Liu, J.; Huang, Y.; Chen, Y.; Wen, Y.; Nohara, T. Studies on chemical fingerprints of *Siraitia grosvenorii* fruits (Luo Han Guo) by HPLC. *J. Nat. Med* **2012**, *66*, 70–76.

24. Liang, Y.Z.; Xie, P.; Chan, K. Quality control of herbal medicines. *J. Chromatogr. B* **2004**, *812*, 53–70.

Chapter 3

APPLICATION OF ULTRA-HIGH-PERFOR-MANCE LIQUID CHROMATOGRAPHY COUPLED WITH LTQ-ORBITRAP MASS SPECTROMETRY FOR THE QUALITATIVE AND QUANTITATIVE ANALYSIS OF POLYGONUM MULTIFLORUM THUMB. AND ITS PROCESSED PRODUCTS

Teng-Hua Wang [1,2], Jing Zhang [1], Xiao-Hui Qiu [1], Jun-Qi Bai [1], You-Heng Gao [2] and Wen Xu [1]

[1]Lab of Chinese Materia Medica Preparation, the Second College of Clinic Medicine, Guangzhou University of Chinese Medicine; Guangdong Province Institute of TCM, Guangzhou 510006, China

[2]School of Chinese Materia Medica, Guangzhou University of Chinese Medicine, Guangzhou 510006, China

ABSTRACT

In order to quickly and simultaneously obtain the chemical profiles and control the quality of the root of *Polygonum multiflorum* Thumb. and its processed form, a rapid qualitative and quantitative method, using ultra-high-performance liquid chromatography coupled with electrospray ionization-linear ion trap-Orbitrap hybrid mass spectrometry (UHPLC-LTQ-Orbitrap MS^n) has been developed. The analysis was performed within 10 min on an AcQuity UPLC™ BEH C_{18} column with a gradient elution of 0.1% formic acid-acetonitrile at flow rate of 400 μL/min. According to the fragmentation mechanism and high resolution MS^n data, a diagnostic ion searching strategy was used for rapid and tentative identification of main phenolic components and 23 compounds were simultaneously identified or tentatively characterized. The difference in chemical profiles between *P. multiflorum* and its processed preparation were observed by comparing the ions abundances of main constituents in the MS spectra and significant changes of eight metabolite biomarkers were detected in the *P. multiflorum* samples and their preparations. In addition, four of the representative phenols, namely gallic acid, *trans*-2,3,5,4'-tetra-hydroxystilbene-2-*O*-β-d-glucopyranoside, emodin and emodin-8-*O*-β-d-glucopyranoside were quantified by the validated UHPLC-MS/MS method. These phenols are considered to be major bioactive constituents in *P. multiflorum*, and are generally regarded as the index for quality assessment of this herb. The method was successfully used to quantify 10 batches of *P. multiflorum* and 10 batches of processed *P. multiflorum*. The results demonstrated that the method is simple, rapid, and suitable for the discrimination and quality control of this traditional Chinese herb.

INTRODUCTION

The root of *Polygonum multiflorum* Thumb. (*Fallopia multiflora*), well known as He-shou-wu in China, has been widely used as a tonic and purgative in traditional Chinese medicine for thousands of years [1]. Previous phytochemical studies revealed that the constituents of *P. multiflorum* are mainly two kinds of active constituents—anthraquinones and stilbenes—as well as other compounds such as flavonoids, tannins and phospholipids [2,3,4]. Recently, four novel stilbene derivatives, named polygonumosides A–D, were isolated from the processed root of *P. multiflorum* [5].

The anthraquinones, such as emodin (EM) and emodin-8-*O*-β-d-glucopyranoside (EMG), provide immunomodulating, anti-inflammatory, anticancer, antimutation, antibacterial, gastrointestinal smooth muscle prokinetic actions, dose-dependent protection against myocardial ischemia-reperfusion injury and cerebral ischemia-induced infarct volume reduction biological effects [6,7,8,9]. Some research supports the notion that the stilbenes in PM possess anti-inflammatory [10,11,12], antioxidant activity [13,14,15,16], anti-hyperlipidaemia [17], anti-melanogenic activity, prophylactic and therapeutic activity against Alzheimer's disease and Parkinson's disease [18,19,20], free radical scavenging activity [21], and hair growth properties [22,23,24,25]. According to Chinese medicine theory, the crude *P. multiflorum* (CPM) should be processed before use, involving steaming the crude roots with or without black soybean extract (namely *Paozhi*), which could reduce the laxative effects and strengthen the tonic effects [26]. Both the crude *P. multiflorum* (CPM) and processed *P. multiflorum* (PPM) have been officially listed in the successive editions of the Chinese Pharmacopoeia.

As a consequence of the potential medicinal value of PM, several methods have been published for identification and/or quantification of its chemical components, such as high performance liquid chromatography coupled with photodiode array (HPLC-PDA) detection [27], capillary zone electrophoresis (CZE) [28], and HPLC-DAD coupled with electrospray ionization tandem mass spectrometry (ESI-MS[n]) [29,30]. The present method of using several ingredients in evaluating PM preparations cannot reflect the overall changes of chemical composition that occur during the course of processing. Any new method should overcome this problem by using multi-target ingredient determination and fingerprinting analysis technologies simultaneously, which could control the preparation process and product quality precisely, and ensure the stability and effectiveness of products.

In our previous work, the fragmentation pathways of typical constituents and chemical profiles of PM have been studied by an on-line UHPLC-ESI-linear ion trap-Orbitrap hybrid mass spectrometry (LTQ-Orbitrap) method [31,32]. UHPLC coupled with the high resolution tandem mass spectrometric techniques have been proven to be a powerful tool for the rapid identification of unknown components in botanical extracts. In this paper, a simple and rapid method for the comprehensive qualitative and quantitative analysis of the major constituents was successfully developed for quality evaluation of CPM and PPM. Up to now, this is the first report on the simultaneous identification and determination of multiple components in botanical herb products by an UHPLC-LTQ-Orbitrap MS[n] technique.

RESULTS AND DISCUSSION

Optimization of the LC-MS Conditions

The chromatographic conditions, such as the chromatographic column and mobile phase, were optimized to achieve the best separation efficiency. Three reversed-phase chromatographic columns, including a Hypersil C_{18} (2.1 mm × 100 mm, 5 μm, Thermo Fisher Scientific, Carlsbad, CA, USA), a Kinetex XB C_{18} (2.1 mm × 100 mm, 1.7 μm, Phenomenex, Torrance, WA, USA) and an AcQuity UPLC™ BEH C_{18} column (2.1 mm × 50 mm, 1.7 μm, Waters, Milford, MA, USA), were selected to test the separation ability of the four investigated compounds. It was shown that an UHPLC system with a 1.7 μm small particle size column had more powerful separation ability with a higher peak resolution and the total analysis time was less than 10 min, which was approximately fourfold faster than that for a conventional column packed with 5 μm particles (Figure 1). Thus the AcQuity UPLC™ BEH C_{18} column (2.1 mm × 50 mm, 1.7 μm) was proved to be the best in this application.

The experimental results showed that an acetonitrile-H_2O mobile phase separated the target compounds more effectively (Figure 2B) than a methanol-H_2O system (Figure 2A). Thus acetonitrile was selected as the organic phase, so as to obtain the optimal separation of adjacent peaks and to avoid peak tailing. Formic acid (Figure 2B), acetic acid (Figure 2C), ammonium acetate

(Figure 2D) and ammonium formate (Figure 2E) were added to the mobile phase, respectively, to achieve high MS sensitivity and restrain peak tailing. Formic acid was screened as an effective mobile phase additive with which improvements of peak shape, peak width and decreased ion suppression effects could be directly and efficiently obtained, compared to others.

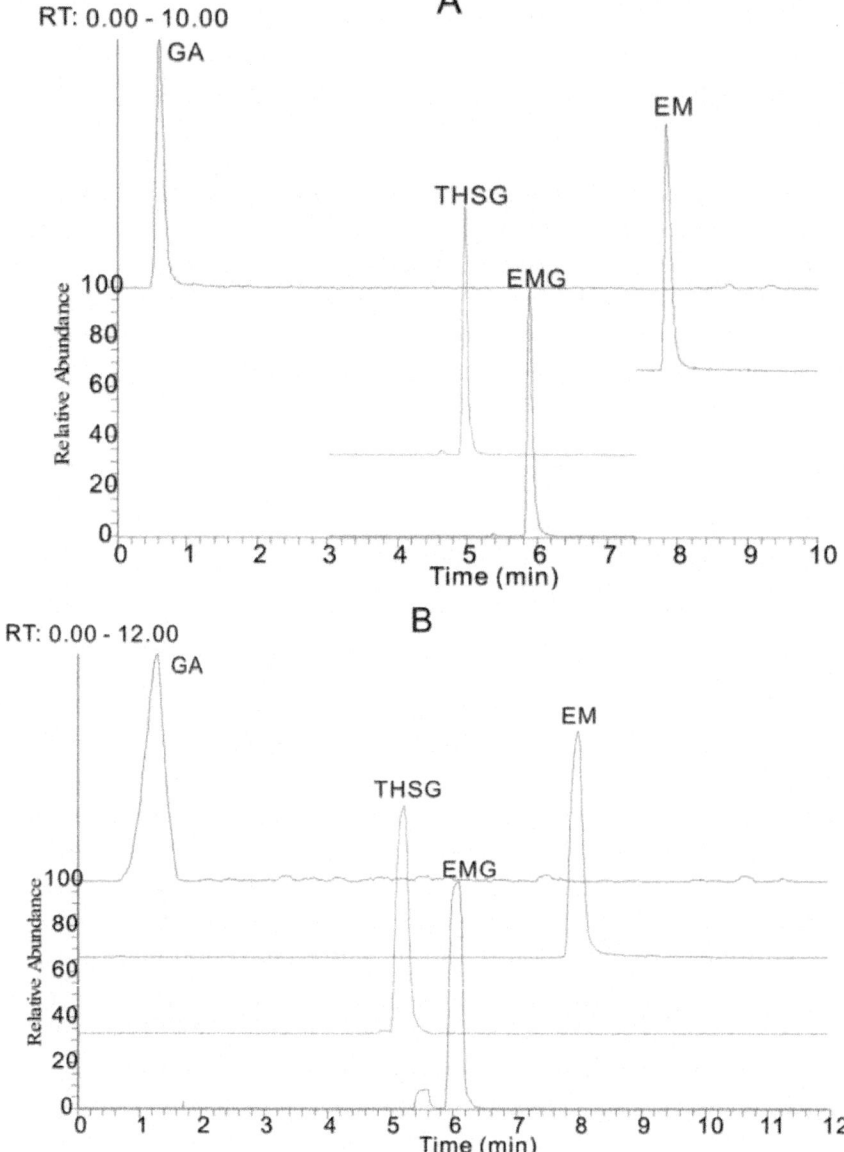

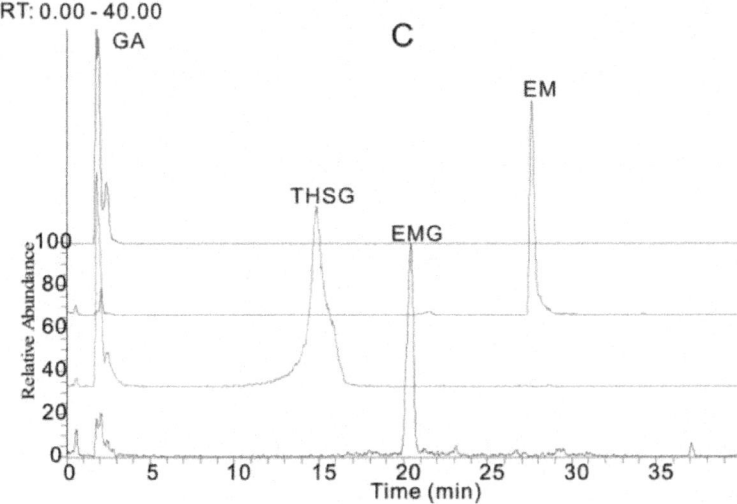

Figure 1. The extracted ion chromatograms (EIC) of gallic acid (GA), *trans*-2,3,5,4′-tetrahydroxystilbene-2-*O*-β-d-glucopyranoside (THSG), emodin-8-*O*-β-d-glucopyranoside (EMG) and emodin (EM) with three different reversed-phase columns. (**A**) AcQuity UPLC™ BEH C$_{18}$ column (2.1 mm × 50 mm, 1.7 μm); (**B**) Kinetex XB C$_{18}$ (2.1 mm × 100 mm, 1.7 μm) and (**C**) Hypersil C$_{18}$ (2.1 mm × 100 mm, 5 μm). The mobile phases consisted of acetonitrile (a) and water containing 0.1% formic acid (b), with the following elution gradient program: (**A**) used the optimized mobile phase and gradient (see Section 3.2); (**B**) 13% a (0 min), 35% a (3.5 min), 90% a (7.5 min), 95% a (8.5 min) and 95% a (12 min); (**C**). 13% a (0 min), 35% a (12 min), 90% a (32 min), 95% a (35 min) and 95% a (40 min).

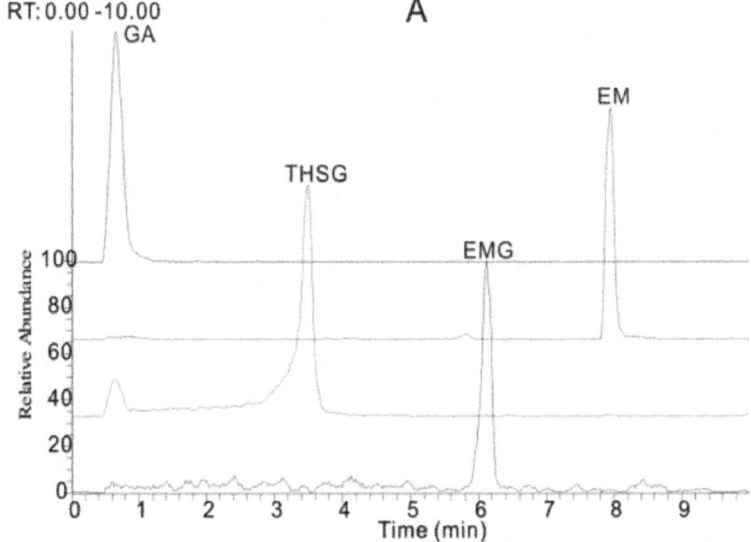

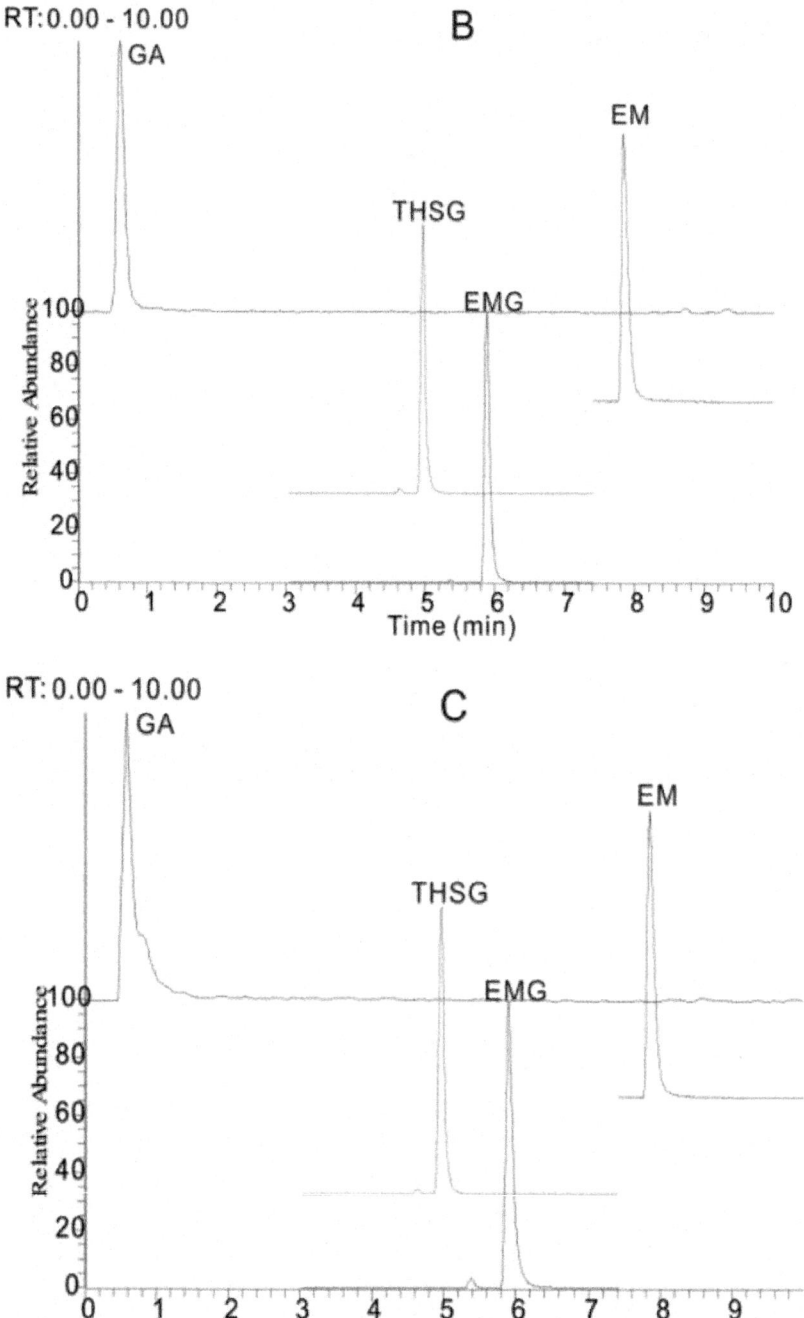

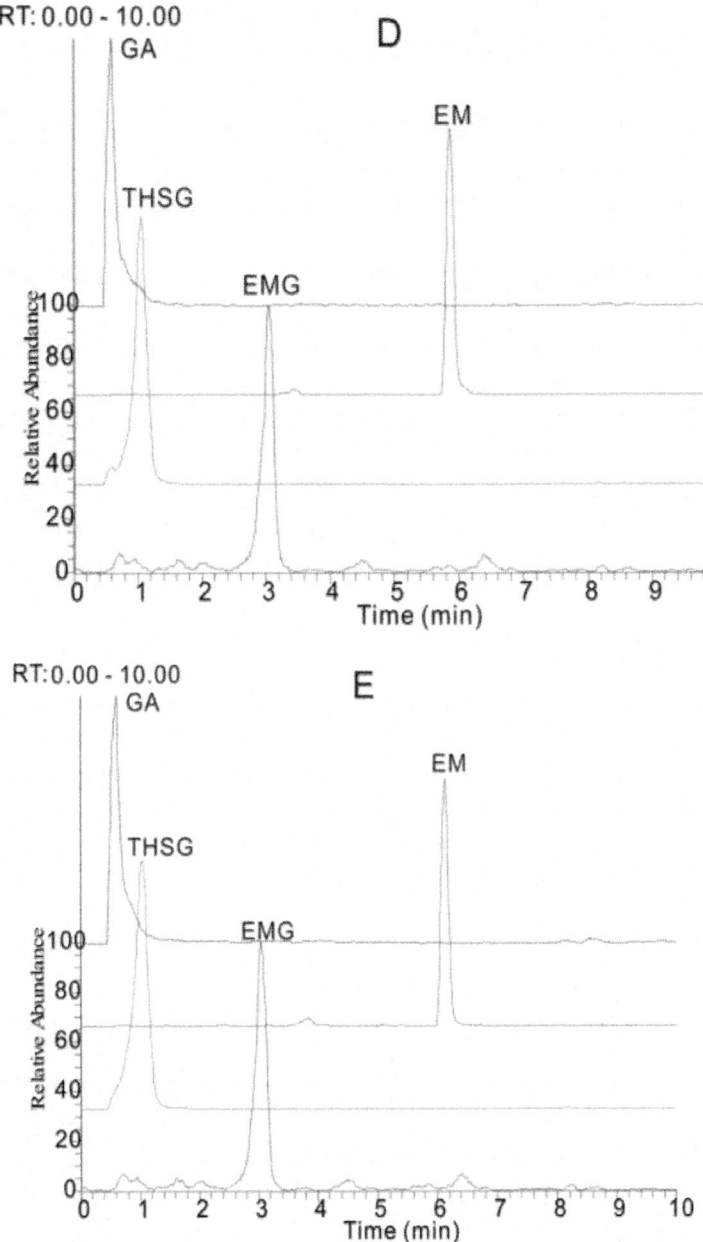

Figure 2. The extracted ion chromatograms (EIC) of gallic acid (GA), *trans*-2,3,5,4′-tetra-hydroxystilbene-2-*O*-β-d-glucopyranoside (THSG), emodin-8-*O*-β-d-glucopyranoside (EMG) and emodin (EM) under five different chromatographic conditions: (**A**) methanol/water system containing 0.1% formic acid; acetonitrile/water

system; (**B**) containing 0.1% formic acid; (**C**) containing 0.1% acetic acid; (**D**) containing 0.1% ammonium acetate and (**E**) containing 0.1% ammonium formate.

Sample Extraction Optimization

In order to optimize the sample extraction conditions, the factors affecting the extraction efficiency of the four main compounds, including extraction solvents, solid-liquid ratio and extraction times, were studied. In our research, the extraction efficiency was evaluated by using a single factor method and choosing the extraction yield (extraction yield = the extract weight × the content of target analyte/the weight of crude drug × 100%) as the index. The results suggested that single-solvent system (pure methanol) was superior to a binary solvent system in the extraction of all four compounds (Table 1). The extraction yield was as high as 5.27% (summation of the four test components) as a maximum value with methanol at 100%. Furthermore, the extraction yield increased by 75.9% (summation of the four test components) when the amount of liquid was increased from 10 to 25 mL (*i.e.*, solid to liquid ratio from 1:50 to 1:125). When the liquid amount was 50 mL (solid to liquid ratio 1:250), the average extraction yield and extraction yield of gallic acid showed no obvious increase. Thus the solid to liquid ratio of 1:125 was recommended. It was also found that when the extraction time was within 30 min, the extraction yield increased significantly with extraction time. The total extraction yield was 5.85% and the highest yield was reached at an extraction time of 30 min. After that, the yield decreased slightly with time. According to the above results, the optimum extraction conditions were: solid-liquid ratio 1:125 and extraction with 25 mL of 100% methanol at reflux for 30 min.

Table 1. Effects of methanol concentration, solid-liquid ratio and extraction time on extraction yield [a]

Compound	Methanol Concentration (%)			Solid-Liquid Ratio			Extraction Time (min)		
	50	70	100	1:50	1:125	1:250	15	30	60
Gallic acid	0.46	0.52	0.67	0.69	0.88	0.89	0.26	0.77	0.73
Emodin	1.25	1.42	1.55	1.23	1.75	1.77	0.37	1.88	1.68
trans-2,3,5,4'-Tetrahydroxy-stilbene-2-*O*-β-d-glucopyranoside	1.26	1.32	1.61	1.18	1.93	1.69	0.79	1.78	1.86
Emodin-8-*O*-β-d-glucopyranoside	0.86	1.46	1.44	0.43	1.65	1.44	0.73	1.42	1.22

[a] extraction yield (%) = the extract weight × the content of target analyte/the weight of crude drug × 100%.

Tentative Identification of the Major Compounds

The four reference compounds (gallic acid, EM, THSG and EMG) were initially analyzed to obtain the corresponding retention times and characteristic fragmentation pathway data (Figure 3). The precise quasi-molecular and fragment ions were determined within a reasonable degree of measurement error using the Orbitrap instrument (the mass error is less than 2 ppm in most cases). Other clues such as potential elemental composition, degree of unsaturation and fractional isotope abundance of the compounds were also utilized for structural conformation. In our previous work, the fragmentation rules and diagnostic fragment ions of anthraquinones, stilbenes, tannins and naphthalenes have been investigated, which provided useful information for the elucidation of the chemical structures of the prescription components. As for the untargeted compounds, the characteristic diagnostic ions could be used to filter and classify them into a particular chemical family.

Twenty five compounds, including anthraquinones, stilbenes, tannins and naphthalenes were primarily identified in both of the extracts of CPM and PPM, and three compounds were only detected in PPM, which were probably new components produced during the steaming process (Table 2).

Table 2. Identification of the chemical constituents in methanol extracts of CPM and PPM by UHPLC-LTQ-Orbitrap MS

No.	t_R(min)	Precursor ions [M − H]	Formula	Mass Error (ppm)	MS^n	Identification
1	0.89	341.1078	$C_{12}H_{21}O_{11}$	0.18	MS^2[377]: 341 (100), 215 (15)	Sucrose *
		377.0843 [M + Cl]			MS^3: 179 (100), 161 (23), 143 (22), 113 (17)	
2	1.12	179.0556 225.0609 [M + HCOO]	$C_6H_{11}O_6$	3.38	MS^2: 161 (100), 143 (90), 119 (44), 113 (40), 89 (40)	Glucose *
3	2.17	169.0136	$C_7H_5O_5$	2.66	MS^2: 125 (100)	gallic acid *
					MS^3: 81	
4	4.24	577.1330	$C_{30}H_{25}O_{12}$	−1.8	MS^2: 425 (100), 407 (48), 457 (20), 471 (17), 289 (10)	procyanidin B
					MS^3: 407 (100)	

5	4.34	289.0708	$C_{15}H_{13}O_6$	0.47	MS^2: 245 (100), 205 (42), 179 (19)	epicatechin/ catechin
					MS^3: 203 (100), 227 (23), 187 (22), 161 (20)	
6 #	4.48	531.1488	$C_{26}H_{27}O_{12}$	−2.2	MS^2: 369 (100), 351 (29), 405 (21), 243 (18)	unknown
					MS^3: 351 (100)	
7 #	4.63	549.1594	$C_{26}H_{29}O_{13}$	−0.97	MS^2: 387 (100), 459 (73), 531 (22), 297 (16)	unknown
8	4.82	577.1330	$C_{30}H_{25}O_{12}$	−1.53	MS^2: 425 (100), 407 (48), 457 (20), 471 (17), 289 (10)	procyanidin B
					MS^3: 407 (100)	
9 #	4.98	421.1123	$C_{20}H_{21}O_{10}$	−1.38	MS^2: 259 (100)	6-methoxyl-2-acetyl-3-methylju-glone-8-O-glu
					MS^3: 259 (100), 331 (50), 128 (20)	
10	5.02	613.1751 [M + HCOO]	$C_{27}H_{33}O_{16}$	−1.21	MS^2: 405 (100), 567 (36)	tetrahydroxys-tilbene-O-di-glu
					MS^3: 243 (100)	
11	5.37	405.1177	$C_{20}H_{21}O_9$	−0.74	MS^2: 243 (100)	THSG *
		811.2428 [2M − H]			MS^3: 225 (100), 149 (79), 137 (73), 215 (70), 173 (36)	
12	5.41	557.1286	$C_{27}H_{25}O_{13}$	−0.66	MS^2: 313 (100), 243 (30), 405 (20), 169 (5)	tetrahydroxys-tilbene-O-(galloyl)-glu
					MS^3: 169 (100), 125 (20), 151 (20), 295 (17)	
13	5.71	557.1285	$C_{27}H_{25}O_{13}$	−0.82	MS^2: 313 (100), 243 (80), 405 (70), 169 (10)	tetrahydroxys-tilbene-O-(galloyl)-glu
					MS^3: 169 (100), 125 (20), 151 (20), 295 (17)	
14	5.72	431.0970	$C_{21}H_{19}O_{10}$	−0.63	MS^2: 269 (100)	emodin-1-O-glu
					MS^3: 225 (100), 241 (21), 181 (4)	
15	5.75	567.1488	$C_{29}H_{27}O_{12}$	−1.59	MS^2: 243 (100)	tetrahydroxys-tilbene-O-(caffeoyl)-glu
					MS^3: 225 (100), 215 (72), 149 (67)	

16	5.90	551.1543	$C_{29}H_{27}O_{11}$	−0.89	MS²: 405 (100), 243 (31)	tetrahydroxystilbene-2-O-(coumaroyl)-glu
17	5.92	447.0919	$C_{21}H_{19}O_{11}$	−0.64	MS²: 303 (100), 285 (100)	citreorosein-O-glu
					MS³: 285 (100), 177 (11), 125 (8)	
18	6.02	407.1334	$C_{20}H_{23}O_{9}$	−0.64	MS²: 245 (100)	torachrysone-O-glu
					MS³: 230 (100)	
19	6.11	431.0973	$C_{21}H_{19}O_{10}$	0.11	MS²: 269 (100)	emodin-8-O-glu *
					MS³: 225 (100), 241 (21), 197 (5)	
20	6.26	517.0978	$C_{24}H_{21}O_{13}$	0.26	MS²: 473 (100), 431 (10)	emodin-O-(malonyl)-glu
					MS³: 269 (100), 311 (12), 225 (5)	
21	6.38	445.1127	$C_{22}H_{21}O_{10}$	−0.22	MS²: 283 (100), 445 (42)	physcion-8-O-glu
		491.1182 [M + HCOO]			MS³: 240 (100), 268 (36)	
22	6.42	313.0345	$C_{16}H_{9}O_{7}$	0.96	MS²: 269	carboxyl emodin
					MS³: 225, 241, 197	
23	7.21	269.0447	$C_{15}H_{9}O_{5}$	0.33	MS²: 225, 241	Emodin *
24	7.63	269.0446	$C_{15}H_{9}O_{5}$	0.17	MS²: 225, 241, 254	aloe-emodin
25	8.30	283.0605	$C_{16}H_{11}O_{5}$	1.41	MS²: 240	physcion

* Compared with standard compouds; # Detected only in methanol extracts of PPM.

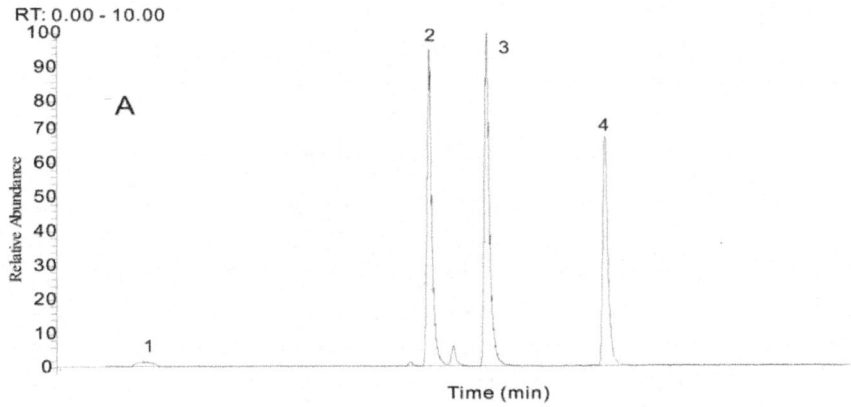

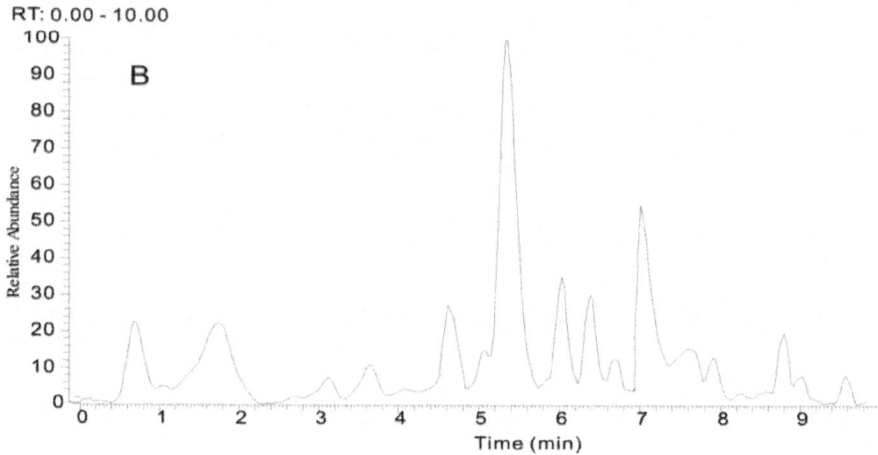

Figure 3. The selected reaction monitoring (SRM) chromatogram of four compounds (**A**) and the total ion chromatogram (TIC) of typical *P. multiflorum* sample (**B**). (1) gallic acid (GA), (2) *trans*-2,3,5,4'-tetrahydroxystilbene-2-*O*-β-d-glucopyranoside (THSG), (3) emodin-8-*O*-β-d-glucopyranoside (EMG) and (4) emodin (EM).

Identification of the Main Stilbene Glycosides

Most of the stilbene glycosides in PM showed common fragmentation pathways and two diagnostic fragment ions, 405.1177 ($C_{20}H_{21}O_9$) and 243.0661 ($C_{14}H_{11}O_4$), were used for rapidly extracting unknown tetrahydroxystilbene glucosides. For example, peaks 12 and 13 exhibited a molecular ion at m/z 557.1286 ($C_{27}H_{25}O_{13}$) in the negative ion mode. It produced a base peak at 313.0548 ($C_{13}H_{13}O_9$) and two prominent fragment ions at m/z 405.1164 ($C_{20}H_{21}O_9$) and 243.0649 ($C_{14}H_{11}O_4$) in the MS^2 spectra, indicating it was a tetrahydroxystilbene derivative. Then, the m/z 313 ion generated the characteristic ions of gallic acid, such as m/z 169 and 125. Thus, peaks 12 and 13 were tentatively identified as tetrahydroxystilbene-*O*-(galloyl)-glucoside and the proposed fragmentation pathways are shown in Figure 4.

Identification of the Main Anthraquinones

Most of the anthraquinones in PM are emodin and physcion derivatives and they generated typical adducts [M + HCOO]⁻and/or [M − H]⁻ with high intensity in negative ion mode. Diagnostic ions at m/z 269.0447, 225.0544 and 241.0492 could be used for rapid extraction and identification of emodin derivatives. For physcion derivatives, two odd-electron ions (OE⁻) at m/z 240 and 268 were produced from the simultaneous loss of free radicals on the side chains, which could be used for the identification of such derivatives.

For example, peak 20 gave a $[M - H]^-$ ion at m/z 517.0978 ($C_{24}H_{21}O_{13}$) and a prominent fragment ion at m/z 473.1063 ($C_{23}H_{21}O_{11}$), as well as a characteristic ion at m/z 431.0960 ($C_{21}H_{19}O_{10}$). Additional fragmentation yielded diagnostic ions at m/z 269 and 225 in the MS^3 spectra. High resolution MS detection indicated it was malonyl-substituted glycoside derivative, which was tentatively identified as emodin-O-(malonyl)-glucoside.

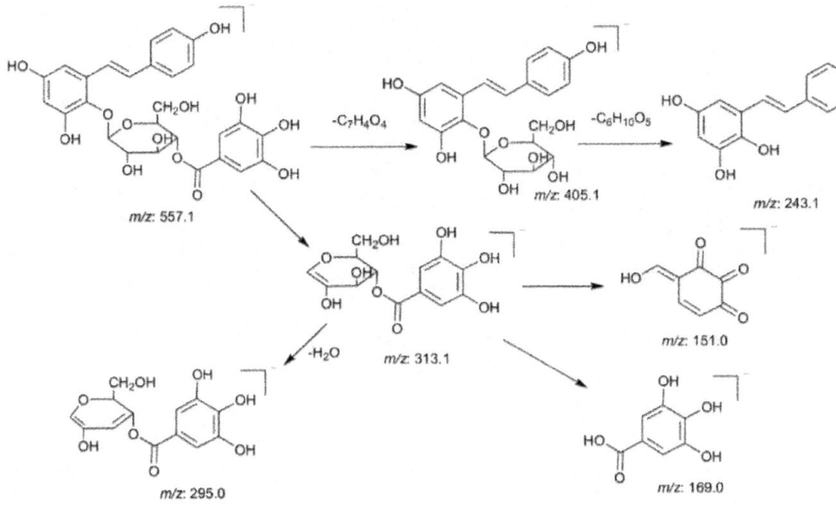

Figure 4. The proposed fragmentation pathways of peaks 12 and 13.

Changes of the Relative Intensity of the Main Chemical Components

Statistical analysis based on the main chemical metabolic components clearly showed differences among 10 batches of CPM and 10 batches of PPM by two-way ANOVA and Bonferroni correction. In examining the relative intensity of the 12 main chemical components in more detail, a histogram of compounds revealed the chemical profiles differ between PM and its processed products. In the present study, obvious differences existed in the compounds sucrose, gallic acid, procyanidin B, catechin, THSG, torachrysone-O-glu, emodin-8-O-glu and emodin-O-(malonyl)-glu between the 10 batches of CPM and 10 batches of PPM (Figure 5). The contents of those compounds were destroyed may due to long-time steaming [33,34].

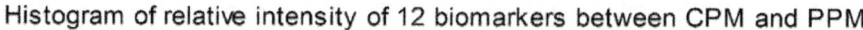

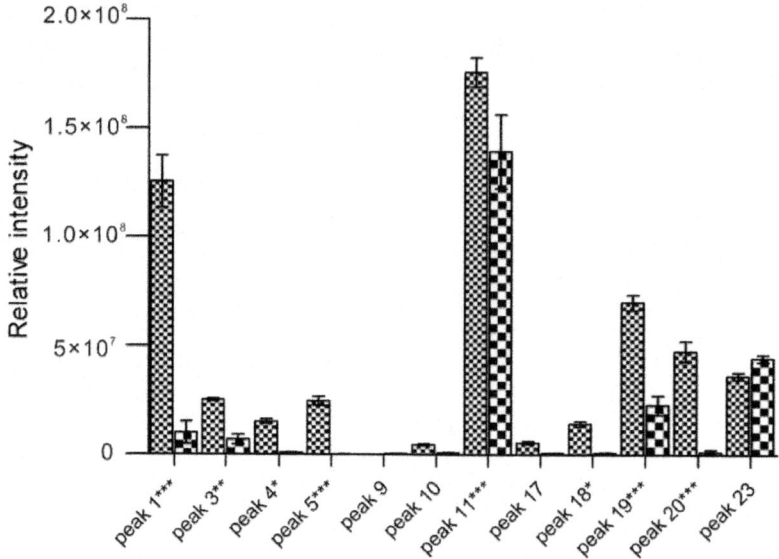

Figure 5. Relative intensity of the extracted ion chromatograms (EIC; mean ± SD) of 12 biomarkers between PM and its processed products. The peak names were originated from Table 2. Statistics are based on three independent experiments for each sample. $*p < 0.05$, $** p < 0.01$ and $*** p < 0.001$.

Quantification of Four Major Phenolic Components of the UHPLC-MS/MS System

A rapid method was developed to quantify four representative phenolic compounds (galic acid, EM, THSG and EMG) in 10 batches of CPM and 10 batches of PPM that were collected from different regions in China (Supplementary Data, Table S1). Since the four investigated bioactive compounds vary greatly in polarity, it is difficult to accomplish a satisfactory separation using a conventional HPLC-UV method. The UHPLC-ESI-MS/MS method has demonstrated higher selectivity, sensitivity and shorter running times, thus it was utilized for rapid determination of these biomarkers. The UHPLC conditions and spectrometric parameters transition were optimized as described in Section 3.2. (Figure 3).

Calibration Curves, LOD and LOQ

A methanol stock solution containing all the reference standards was prepared and diluted to an appropriate concentration range for the construction of calibration curves. At least six concentrations of the solution were analyzed in triplicate, and then the calibration curves were constructed from peak areas of the reference standards *vs*. their concentrations. The high correlation coefficient values ($r^2 > 0.9916$) indicated good linearity between their peak areas (y) and investigated compound concentration (x, μg/mL) in relatively wide concentration ranges (Table 3).

The limits of detection (LOD) and the limits of quantification (LOQ) under the chromatographic conditions were determined by injecting a series of standard solutions until the signal-to-noise ratios (S/N) for each compound was 3 for LOD and 10 for LOQ (Table 3).

Table 3. Calibration curves, LOD and LDQ of the investigated compounds

Analytes	Linear Regres-sion Data Regression Equation	Test Range (μg/mL)	r^2	LOD [a](ng/mL)	LOQ [b](ng/mL)
Gallic acid	y = 3598.5x + 205.82	0.03–4.00	0.9997	0.03	0.10
Emodin	y = 25639x + 2283.1	0.01–1.40	0.9916	0.01	0.02
trans-2,3,5,4'-Tetrahydroxy-stilbene-2-*O*-β-d-gluco-pyranoside	y = 652.2x − 5013.2	1.15–105.00	0.9997	0.01	0.02
Emodin-8-*O*-β-d-gluco-pyranoside	y = 7690.7x + 14652	0.29–114.00	0.9998	0.01	0.06

[a] LOD refers to the limit of detection, *s/n* = 3; [b] LOQ refers to the limit of quantification, *s/n* = 10.

Precision

The precision of the developed assay was evaluated by analyzing the mixed standard solution at three different concentration levels (high, middle and low) (Table 4). The experiment was repeated six times within one day to determine intra-day precision, while for inter-day variability test, the solution was examined in triplicate for consecutive 3 day. Relative standard deviation (RSD) for each of the marker compounds was calculated respectively, and it was no more than 3.27%.

Table 4. Intra- and inter-day variability for the assay of four components

Compound	Intra-Day ($n = 6$)		Inter-Day ($n = 3$)	
	Mean (μg/mL)	RSD [a] (%)	Mean (μg/mL)	RSD (%)
Gallic acid	0.01	1.27	0.07	3.27
	0.25	1.15	0.25	1.62
	1.00	0.92	0.97	2.74
Emodin	0.00	1.90	0.04	2.14
	0.34	1.67	0.34	1.87
	2.56	1.48	2.50	2.39
trans-2,3,5,4'-Tetrahydroxystilbene-2-*O*-β-d-glucopyranoside	2.17	1.45	2.08	1.68
	10.10	1.23	10.12	2.71
	107.67	0.77	106.33	1.65
Emodin-8-*O*-β-d-glucopyranoside	0.37	1.14	0.37	2.45
	1.52	1.49	1.50	1.49
	67.88	1.30	68.13	2.57

[a] RSD (%) = (SD/mean) × 100%.

Repeatability, Stability, Specificity and Selectivity

The repeatability of this method was determined by analyzing six samples from the same batch (Sample A6) using the same preparation procedure described in Section 3.3 of the paper. The RSD values of the four component contents were all less than 3.81% (Table 5). The stability was tested by analysis of the same sample solution (Sample A6) which was at room temperature at different times for 24 h. The RSD values of the four analytes were less than 2.01%, which suggested that it was feasible to analyze samples within 24 h (Table 5).

Table 5. Repeatability and stability of the four analytes, expressed as RSD (%)

Compound	Repeatability ($n = 6$)		Stability ($n = 6$)	
	Mean (μg/mL)	RSD [a] (%)	Mean (μg/mL)	RSD (%)
Gallic acid	0.33	3.81	0.34	2.01
Emodin	1.04	2.53	1.02	1.17
trans-2,3,5,4'-Tetrahydroxystilbene-2-*O*-β-d-glucopyranoside	27.34	1.85	27.83	1.49

Emodin-8-O-β-d-glucopyranoside	0.83	2.98	0.80	1.37

[a] RSD (%) = (SD/mean) × 100%.

Accuracy

Recovery test was used to evaluate the accuracy of this method. A known amount (low, medium and high) of each standard solution was spiked into known amounts of CPM samples (simple A6), and then extracted according to the section of sample pretreatment. The recoveries were counted by the formula: recovery (%) = concentration found/original concentration × 100%. The recoveries of analytes varied from 97.34% to 100.75% (Table 6). The result indicated the reliability and accuracy for the quantitative determination of the constituents.

Table 6. Recoveries of the four determined constituents

Compound	Amount (μg/mL)		Recovery [a] (%)	RSD [b] (%)
	Spiked	Found		
Gallic acid	0.26	0.26 ± 0.00	98.39	1.19
	0.33	0.32 ± 0.01	98.87	1.81
	0.39	0.39 ± 0.00	98.54	0.04
Emodin	0.90	0.91 ± 0.00	100.75	0.06
	1.12	1.13 ± 0.02	100.53	1.43
	1.35	1.33 ± 0.01	98.73	1.41
trans-2,3,5,4'-Tetrahydroxystilbene-2-O-β-d-glucopyranoside	23.46	22.83 ± 0.14	97.34	0.63
	29.32	29.43 ± 0.37	100.37	1.26
	35.19	34.54 ± 0.40	98.17	1.18
Emodin-8-O-β-d-glucopyranoside	7.35	7.17 ± 0.06	97.56	0.77
	9.19	9.16 ± 0.13	99.75	1.46
	11.03	11.07 ± 0.05	100.40	0.43

[a] Recovery (%) = (detected amount − original amount)/spiked amount × 100%. [b] RSD (%) = (SD/mean) × 100%.

Sample Analysis

Twenty batches of *P. multiflorum* acquired from different regions of China were determined using the described method. This work disclosed that the variation between different locations and processing methods of this traditional Chinese medicine was obvious. Such variations may presumably be attributed to differences in the cultivation conditions (Table 7).

Table 7. Contents of four components in 20 samples

No.	Content (mg/g) ($n = 3$)			
	Gallic Acid (GA)	Emodin (EM)	2,3,5,4′-Tetrahydroxystilbene-2-O-β-d-glucoside (THSG)	Emodin-8-O-β-d-glucoside (EMG)
A1	0.63 ± 0.01	1.01 ± 0.02	34.43 ± 0.59	21.11 ± 0.34
A2	0.44 ± 0.05	0.23 ± 0.00	4.53 ± 0.15	1.32 ± 0.05
A3	0.32 ± 0.01	0.18 ± 0.00	24.54 ± 0.93	7.29 ± 0.20
A4	0.30 ± 0.04	0.52 ± 0.06	16.27 ± 0.57	7.98 ± 0.17
A5	0.50 ± 0.01	0.82 ± 0.01	24.01 ± 0.76	14.44 ± 0.65
A6	0.31 ± 0.00	2.46 ± 0.04	26.57 ± 0.46	15.77 ± 0.25
A7	0.58 ± 0.01	0.32 ± 0.00	20.55 ± 0.67	3.85 ± 0.08
A8	0.43 ± 0.01	0.10 ± 0.00	24.97 ± 0.58	7.83 ± 0.25
A9	0.60 ± 0.01	2.38 ± 0.05	19.71 ± 0.23	15.87 ± 0.15
A10	0.23 ± 0.00	3.27 ± 0.04	23.51 ± 0.55	14.59 ± 0.31
B1	0.02 ± 0.00	0.49 ± 0.01	2.54 ± 0.04	2.12 ± 0.03
B2	0.09 ± 0.00	0.36 ± 0.00	1.85 ± 0.03	0.35 ± 0.00
B3	0.28 ± 0.00	0.45 ± 0.01	2.63 ± 0.04	3.11 ± 0.04
B4	0.15 ± 0.01	0.50 ± 0.01	1.97 ± 0.05	2.59 ± 0.04
B5	0.90 ± 0.01	0.85 ± 0.03	10.42 ± 0.37	2.51 ± 0.04
C1	0.31 ± 0.01	0.56 ± 0.02	2.05 ± 0.03	3.49 ± 0.04
C2	0.02 ± 0.01	1.13 ± 0.02	5.94 ± 0.10	3.92 ± 0.04
C3	1.31 ± 0.04	0.32 ± 0.01	2.55 ± 0.06	0.47 ± 0.01
C4	0.47 ± 0.01	1.31 ± 0.04	11.71 ± 0.41	4.92 ± 0.09
C5	0.39 ± 0.01	0.67 ± 0.02	8.21 ± 0.10	2.62 ± 0.09

EXPERIMENTAL

Reagents and Chemicals

HPLC-grade acetonitrile, methanol and formic acid were purchased from Sigma Aldrich (St. Louis, MO, USA). Ultrapure Water (18.2 MΩ) was produced by Milli-Q water system (Millipore, Bedford, MA, USA). Twenty samples of roots of CPM and PPM were collected from various habitats in China, and authenticated by Professor Zhihai Huang in our lab. The related information is summarized in the Supplementary Data (Table S1). The standards of gallic acid (GA), emodin (EM) and *trans*-2,3,5,4′-tetrahydroxystilbene-2-O-β-d-glucopyranoside (THSG) were obtained from the National Institutes for Food and Drug Control (Beijing, China). Emodin-8-O-β-d-glucopyranoside (EMG; over 98% purity by HPLC) was isolated in our laboratory.

Chromatography and MS Conditions

LC analyses were performed on a Thermo Accela UHPLC system (Thermo Fisher Scientific, San Jose, CA, USA) equipped with a quaternary pump, a diode-array detector (DAD), an auto-sampler, and a thermostatically column compartment.

After also running optimization analyses on both Hypersil C_{18} (2.1 mm × 100 mm, 5 μm) and Kinetex XB C_{18} (2.1 mm × 100 mm, 1.7 μm) columns, we found the best overall resolution on an AcQuity UPLC™ BEH C_{18} column (2.1 mm × 50 mm, 1.7 μm) at room temperature (see Section 2.1) and used this column for all subsequent runs. The mobile phase was also optimized by comparing an acetonitrile-H_2O mobile to a methanol-H_2O system and with the addition of various modifiers to the acetonitrile-H_2O mobile, including formic acid, acetic acid, ammonium acetate, and ammonium formate, all at 0.1% (see Section 2.1). The final mobile phase was composed of acetonitrile (A) and water containing 0.1% formic acid (B) using the following gradient program: 13% A (0 min), 35% A (3.5 min), 90% A (7.5 min), 95% A (8.5 min) and 95% A (10 min). A pre-equilibration period of 4 min was used between individual runs. The mobile phase flow rate was 400 μL/min, and the injection volume was 2 μL. The online UV spectra were recorded in the range of 200–400 nm.

Mass spectra were acquired using a Thermo-Fisher LTQ-Orbitrap XL hybrid mass spectrometer, which was connected to LC system via an electrospray ionization (ESI) source as interface. The basic conditions of MS analysis were as follows: the mass spectrometer parameters were negative ion mode, ion spray voltage at 3500 V, capillary voltage at 37 V, capillary temperature at 300 °C, sheath gas flow rate at 40 psi and auxiliary gas flow rate at 4 psi. The scan spectra were from m/z 150 to 1200.

For qualitative analysis of CPM and PPM, the Orbitrap resolution of survey scan was set as 30000 and MS^n scan was set 15000. The data-dependent MS^n scanning was performed to trigger fragmentation spectra of target ions and to prevent repetition by dynamic exclusion settings. For quantitative determinations of four compounds, the MS detection was operated in linear ion trap (LTQ) with selected reaction monitoring (SRM). The ion trap collision induced dissociation (CID) mode was used for SRM fragmentation and the ions transitions are as follows: GA (m/z 169→125), EM (m/z 269→225), THSG (m/z 405→243) and EMG (m/z 431→269). The selected ion width was m/z ±1 and the normalized collision energy was set 35%.

Sample Preparation

The dried roots were powdered to a homogeneous size by a mill, sieved through a No.60 mesh (250 μm), and further dried at 50 °C in the oven for 6 h to constant weight. The powdered sample accurately weighed 0.2 g was extracted with 25 mL of methanol in a round-bottomed flask and the mixture was heated under reflux for 0.5 h at 70–75 °C, and cooled at room temperature. Methanol was added to compensate for the lost weight. The extracted solution was centrifuged at 12,000 rpm for 10 min, after filtration through a 0.45 μm membrane, an aliquot of 10 μL of the filtrate was injected into the UHPLC-MS system for LC-MS analysis.

Preparation of Standard Solutions

The reference standards were accurately weighed and dissolved in methanol to prepare stock solutions. All standards were completely dissolved in the mixed standard working solution of 8 μg/mL for gallic acid, 2.75 μg/mL for EM, 210 μg/mL for THSG, 228 μg/mL for EMG, respectively. For the construction of calibration plots, the standard stock solution was further diluted with methanol to make seven different concentrations at 1/2, 1/4, 1/8, 1/16, 1/32, 1/64, and 1/128 of the working solutions. All solutions were stored in a refrigerator at 4 °C for analysis.

Statistical Analysis

The mass data acquired were imported into the Xcalibur software (version 2.1) (Thermo Fisher Scientific, CA, USA) for peak detection and alignment. Prism 5.0 was used to run two-way ANOVAs with Bonferroni corrections on ion peak areas of twelve investigated components (sucrose, gallic acid, procyanidin B, catechin, 6-methoxyl-2-acetyl-3-methyl-juglone-8-O-glu, tetrahydroxy stilbene-O-di-glu, THSG, citreorosein-O-glu, torachrysone-O-glu, emodin-8-O-glu, emodin-O-(malonyl)-glu and emodin) to test for differences between PM and its processed products. When the peak areas of investigated compounds were difficult to integrate or not detected in the samples, the values of such data was considered to be zero.

CONCLUSIONS

In this study, a simple, rapid and accurate UHPLC-MS method was established to qualitatively and quantitatively determine the major components of PM and PPM. The method was used to successfully quantify four components in twenty batches of PM and PPM samples. This novel approach is a highly useful technique to identify constituents and control the quality of CPM and PPM, it also offers incredible advantages, including speed, simplicity, and a reduction in solvent consumption. The method would become an important quality control technique for Chinese medicine and could be adopted widely.

ACKNOWLEDGMENTS

This research was supported by the National Natural Science Foundations of China (81073052, 81373967), the Funds for Distinguished Young Scholars from Guangzhou University of Chinese Medicine (No.10) and supported by Science and Technology Planning Project of Guangdong Province, China (2013B021800236).

AUTHOR CONTRIBUTIONS

Wen Xu and You-Heng Gao conceived and designed the experiments; Teng-Hua Wang performed the experiments; Wen Xu and Teng-Hua Wang analyzed the data; Jing Zhang, Xiao-Hui Qiu, Jun-Oi Bai and You-Heng Gao contributed reagents/materials/analysis tools; Wen Xu and Teng-Hua Wang wrote the paper.

REFERENCES

1. Wang, H.Y.; Song, L.X.; Feng, S.B.; Liu, Y.C.; Zuo, G.; Lai, F.L.; He, G.Y.; Chen, M.J.; Huang, D. Characterization of proanthocyanidins in stems of *Polygonum multiflorum* Thunb. as strong starch hydrolase inhibitors. *Molecules* **2014**, *18*, 2255–2265.

2. Sun, Y.N.; Li, W.; Kim, J.H. Chemical constituents from the root of *Polygonum multiflorum* and their soluble epoxide hydrolase inhibitory activity. *Arch. Pharm. Res.* **2015**, *38*, 998–1004.

3. Li, L.F.; Ni, B.R.; Lin, H.M.; Zhang, M. Traditional usages, botany, phytochemistry and toxicology of *Polygonum multiflorum* Thunb.: A review. *J. Ethnopharmacol.* **2015**, *159*, 158–183.

4. Choi, S.G.; Kim, J.; Sung, N.D.; Son, K.H.; Cheon, H.G.; Kim, K.R.; Kwon, B.M. Anthraquinones, Cdc25B phosphatase inhibitors, isolated from the roots of *Polygonum multiflorum* Thunb. *Nat. Prod. Res.* **2007**, *6*, 487–493.

5. Yan, S.Y.; Su, Y.F.; Chen, L. Polygonumosides A−D, Stilbene Derivatives from Processed Roots of *Polygonum multiflorum. J. Nat. Prod.* **2014**, *77*, 397–401.

6. Han, J.W.; Shim, D.W.; Shin, W.Y. Anti-inflammatory effect of emodin via attenuation of NLRP3 inflammasome activation. *Int. J. Mol. Sci.* **2015**, *4*, 8102–8109.

7. Srinivas, G.; Anto, R.J.; Srinivas, P.; Vidhyalakshmi, S. Emodin induces apoptosis of human cervical cancer cells through poly (ADP-ribose) polymerase cleavage and activation of caspase-9. *Eur. J.*

Pharmacol. **2003**, *473*, 117–125.

8. Thiruvengadam, M.; Praveen, N.; Kim, E.H.; Kim, S.H.; Chung, I.M. Production of anthraquinones, phenolic compounds and biological activities from hairy root cultures of *Polygonum multiflorum* Thunb. *Protoplasma* **2014**, *251*, 555–566.

9. Chan, Y.C.; Wang, M.F.; Chen, Y.C.; Yang, D.Y.; Lee, M.S.; Cheng, F.C. Long-term administration of *Polygonum multiflorum* Thunb. reduces cerebral ischemia-induced infarct volume in gerbils. *Am. J. Chin. Med.* **2003**, *31*, 71–77.

10. Zhang, Y.Z.; Shen, J.F.; Xu, J.Y.; Xiao, J.H.; Wang, J.L. Inhibitory effects of 2,3,5,4′-tetrahydroxystilbene-2-*O*-β-d-glucoside on experimental inflammation and cyclo-oxygenase 2 activity. *J. Asian Nat. Prod. Res.* **2007**, *9*, 355–363.

11. Wang, X.M.; Zhao, L.B.; Han, T.Z.; Wang, J.L. Protective effects of 2,3,5,4′-tetrahydroxystilbene-2-*O*-β-d-glucoside, an active component of *Polygonum multiflorum* Thunb. on experimental colitis in mice. *Eur. J. Pharmacol.* **2008**, *578*, 339–348.

12. Zeng, C.; Xiao, J.H.; Chang, M.J.; Wang, J.L. Beneficial effects of THSG on acetic acid-induced experimental Colitis: Involvement of upregulation of PPAR-γ and inhibition of the NF-κB Inflammatory Pathway. *Molecules* **2011**, *16*, 8552–8568.

13. Lv, L.S.; Gu, X.H.; Tang, J.; Ho, C.T.; Tang, J. Stilbene glycoside from the roots of *Polygonum multiflorum* Thunb. and their *in vitro* antioxidant activities. *J. Food Lipids* **2006**, *13*, 131–144.

14. Lv, L.S.; Gu, X.H.; Tang, J.; Ho, C.T. Antioxidant activity of stilbene glycoside from *Polygonum multiflorum* Thunb *in vivo*. *Food Chem.* **2007**, *104*, 1678–1681.

15. Liu, Q.L.; Xiao, J.H.; Ma, R.; Ban, Y.; Wang, J.L. Effect of 2,3,5,4′-tetrahydroxystilbene -2-*O*-β-d-glucoside on lipoprotein oxidation and proliferation of coronary arterial smooth cells. *J. Asian Nat. Prod. Res.* **2007**, *9*, 689–697.

16. Chiang, Y.C.; Huang, G.H.; Ho, Y.L.; Hsieh, P.C.; Chung, H.P.; Chou, F.I.; Chang, Y.S. Influence of gamma irradiation on microbial load and antioxidative characteristics of *Polygoni multiflori*. Radix. *Process. Biochem.* **2011**, *46*, 777–782.

17. Xie, W.D.; Zhao, Y.N.; Du, L.J. Emerging approaches of traditional Chinese medicine formulas for the treatment of hyperlipidemia. *J. Ethnopharmacol.* **2012**, *140*, 345–367.

18. Cheung, F.W.; Leung, A.W.; Liu, W.K.; Che, C.T. Tyrosinase inhibitory activity of a glucosylated hydroxystilbene in mouse melan—A melanocytes. *J. Nat. Prod.* **2014**, *77*, 1270–1274.

19. Um, M.Y.; Choi, W.H.; Aan, J.Y.; Kim, S.R.; Ha, T.Y. Protective effect of *Polygonum multiflorum* Thunb. on amyloid β-peptide 25–35 induced cognitive deficits in mice. *J. Ethnopharmacol.* **2006**, *104*, 144–148.

20. Wang, R.; Tang, Y.; Feng, B.; Ye, C.; Fang, L.; Zhang, L.; Li, L. Changes in hippocampal synapses and learning-memory abilities in age-increasing rats and effects of tetrahydroxystilbene glucoside in aged rats. *Neuroscience* **2007**, *149*, 739–746.

21. Luo, A.X.; Fan, Y.J.; Luo, A.S. *In vitro* free radicals scavenging activities of polysaccharide from *Polygonum multiflorum* Thunb. *J. Med. Plants Res.* **2011**, *5*, 966–972.

22. Guan, S.Y.; Su, W.W.; Wang, N.; Li, P.B.; Wang, Y.G. A potent tyrosinase activator from *Radix Polygoni multiflori* and its melanogenesis stimulatory effect in B16 melanoma cells. *Phytother. Res.* **2008**, *22*, 660–663.

23. Jiang, Z.Q.; Xu, J.M.; Long, M.H.; Tu, Z.M.; Yang, G.X.; He, G.Y. 2,3,5,4′-Tetrahydroxystilbene-2-*O*-β-d-glucoside (THSG) induces melanogenesis in B16 cells by MAP kinase activation and tyrosinase upregulation. *Life Sci.* **2009**, *85*, 345–350.

24. Park, H.J.; Zhang, N.N.; Park, D.K. Topical application of *Polygonum multiflorum* extract induces hair growth of resting hair follicles through upregulating Shh and β-catenin expression in C57BL/6 mice. *J. Ethnopharmacol.* **2011**, *135*, 369–375.

25. Li, S.; Zhao, S.J.; Cui, T.B.; Liu, Z.Y.; Zhao, W. 2,3,5,4′-Tetrahydroxystilbene- 2-*O*-β-d-glycoside biosynthesis by suspension cells cultures of *Polygonum multiflorum* Thunb. and production enhancement by methyl jasmonate and salicylic acid. *Molecules* **2012**, *17*, 2240–2247.

26. Han, L.F.; Wu, B.; Pan, G.X.; Wang, Y.F.; Song, X.B.; Gao, X.M. UPLC-PDA analysis for simultaneous quantification of four active compounds in crude and processed rhizome of *Polygonum multiflorum* Thunb. *Chromatographia* **2009**, *70*, 657–659.

27. Yan, H.J.; Fang, Z.J.; Fu, J.; Yu, S.X. The correlation between bioactive components of *Fallopia multiflora* root and environmental factors. *Am. J. Chin. Med.* **2010**, *38*, 473–483.

28. Zhang, F.; Chen, W.S.; Sun, L.N. LC-VWD-MS determination of three anthraquinones and one stilbene in the quality control of crude and prepared

roots of *Polygonum multiflorum* Thunb. *Chromatographia* **2008**, *67*, 869–874.

29. Lin, Z.L.; Liu, Y.Y.; Wang, C.; Guo, N.; Song, Z.Q.; Wang, C.; Xin, L.; Lu, A.P. Comparative analyses of chromatographic fingerprints of the roots of *Polygonum multiflorum* Thunb. and their processed products using RRLC/DAD/ESI-MS[n]. *Planta. Med.* **2011**, *77*, 1855–1860.

30. Yi, T.; Leung, K.S.Y.; Lu, G.H.; Zhang, H.; Chan, K. Identification and determination of the major constituents in traditional Chinese medicinal plant *Polygonum multiflorum* Thunb. by HPLC coupled with PAD and ESI/MS.*Phytochem. Anal.* **2007**, *18*, 181–187.

31. Xu, W.; Zhang, J.; Huang, Z.H.; Qiu, X.H. Identification of new dianthrone glycosides from *Polygonum multiflorum*Thunb. using high-performance liquid chromatography coupled with LTQ-Orbitrap mass spectrometry detection: A strategy for the rapid detection of new low abundant metabolites from traditional Chinese medicines. *Anal. Methods***2012**, *4*, 1806–1812.

32. Qiu, X.H.; Zhang, J.; Huang, Z.H.; Zhu, D.Y.; Xu, W. Profiling of phenolic constituents in *Polygonum multiflorum*Thunb. by combination of ultra-high-pressure liquid chromatography with linear ion trap-Orbitrap mass spectrometry. *J. Chromatogr. A* **2013**, *1292*, 121–131.

33. Liu, Z.L.; Song, Z.Q.; Wang, C. Content variances of catechin and gallic acid in *Polygonum multiflorum* after steaming.*Chin. Tradit. Pat. Med.* **2009**, *31*, 1392–1394.

34. Liang, Z.T.; Chen, H.B.; Yu, Z.L. Comparison of raw and processed *Radix Polygoni multiflori* (Heshouwu) by HPLC-MS. *Chin. Med.* **2010**, *5*, 29–32.

Chapter 4

QUALITATIVE AND QUANTITATIVE ANALYSIS OF RHIZOMA SMILACIS GLABRAE BY ULTRA HIGH PERFORMANCE LIQUID CHROMATOGRAPHY COUPLED WITH LTQ ORBITRAPXLHYBRID MASS SPECTROMETRY

Shao-Dan Chen [1,2], Chuan-Jian Lu [1,2] and Rui-Zhi Zhao [1,2]

[1]The Second College of Clinic Medicine, Guangzhou University of Chinese Medicine, Guangzhou 510000, China

[2]Guangdong Provincial Hospital of Chinese Medicine, Guangzhou 510000, China

ABSTRACT

Rhizoma Smilacis glabrae, a traditional Chinese medicine (TCM) as well as a functional food, has been commonly used for detoxification treatments, relieving dampness and as a diuretic. In order to quickly define the chemical profiles and control the quality of *Smilacis glabrae*, ultra high performance liquid chromatography coupled with electrospray ionization hybrid linear trap quadrupole orbitrap mass spectrometry (UHPLC-ESI/LTQ-Orbitrap-MS) was applied for simultaneous identification and quantification of its bioactive constituents. A total of 56 compounds, including six new compounds, were identified or tentatively deduced on the basis of their retention behaviors, mass spectra, or by comparison with reference substances and literature data. The identified compounds belonged to flavonoids, phenolic acids and phenylpropanoid glycosides. In addition, an optimized UHPLC-ESI/LTQ-Orbitrap-MS method was established for quantitative determination of six marker compounds from five batches. The validation of the method, including linearity, sensitivity (LOQ), precision, repeatability and spike recoveries, was carried out and demonstrated to be satisfied the requirements of quantitative analysis. The results suggested that the established method would be a powerful and reliable analytical tool for the characterization of multi-constituent in complex chemical system and quality control of TCM.

INTRODUCTION

The rhizome of *Smilacis glabrae* Roxb (family Smilacaceae) is a well-known traditional Chinese medicine (TCM) with great medicinal values. It is officially listed in the Chinese Pharmacopoeia and has been widely used for detoxification treatments, relieving dampness and as a diuretic [1]. It was also consumed as a functional food. People in China like to use it to boil soup or tea for clearing damp. Besides, it is one of the main ingredients of turtle jelly (Gui-ling-gao), a traditional functional food popular in Southern China and Hong Kong. Phytochemical studies have shown the presence of abundant compounds in *S. glabrae*, such as flavonoids, phenolic acids and phenylpropanoid glycosides [2,3], among which flavonoids were considered to be the primary bioactive constituents of the herbal medicine. Astilbin, neoastilbin, isoastilbin, neoisoastilbin, engeletin and isoengeletin were considered as marker constituents included in *S. glabrae*. These six flavonoids were reported to possess various biological activities, involving anti-inflammatory, antioxidative, antibacterial and antitumor properties [4,5,6,7,8,9,10]. Some analytical methods have been used for qualitative or quantitative analysis of some of these bioactive constituents in *S. glabrae*. Li *et al.* identified the main constituents in *Rhizoma Smilacis glabrae* by means of UHPLC-DAD-MS [3]. Chen *et al.* established an HPLC method for determination of five compounds in *Rhizoma Smilacis glabrae* [11]. Although these methods have made significant contributions to the studies of the quality control of *Smilacis glabrae*, they have limitations, such as taking a long time to perform or being either qualitative or quantitative. Less effort has been dedicated to further characterize minor new components or the rapid determination of active components, so a new method is required to address the limitations of the previous techniques.

The present work aimed at developing a rapid and simple UHPLC-ESI-MS method for analyzing and discovering minor new constituents, and quantifying the active components in *Smilacis glabrae*. The advantages of this method comprised high-speed detection, excellent peak shapes, and less solvent usage. With the new method it took less than 10 min to detect 56 compounds of *Smilacis glabrae*, including six new compounds. Further, six marker flavonoids were quantitatively determined in negative ionization mode and five batches of *Smilacis glabrae* were analyzed for assessment of quality consistence. This is the first time for determination of multiple components in *Smilacis glabrae* using UHPLC-ESI/LTQ-Orbitrap-MS.

RESULTS AND DISCUSSION

Optimization of Chromatographic Conditions

To improve the resolution and sensitivity of the analysis but reduce the analytical time, the mobile phase system was optimized. To inhibit ionization of the acidic ingredients in *Smilacis glabrae* extract, formic acid was added to the mobile phase. Two mobile phase systems, methanol-aqueous solution and acetonitrile-aqueous solution were compared. Both negative and positive modes were examined. Generally, in positive mode, low abundance of $[M+H]^+$, $[M+NH_4]^+$ ions and few product ions were observed, while, in negative ion mode, a series of $[M-H]^-$ ions and/or adduct ions ($[M+HCOOH-H]^-$) appeared with sufficient abundance. Thus the negative ion mode was chosen and the $[M-H]^-/([M+HCOOH-H]^-)$ ions were further subjected to LC-MS^n analysis.

Identification of Chemical Constituents in Smilacis glabrae Extract

The reference standards and *Smilacis glabrae* sample were analyzed by using the optimized UHPLC-ESI-MS^n method. The TIC chromatograms of the six reference standards and the extract of *Smilacis glabrae* in negative ESI mode were shown inFigure 1. Fifty six peaks were observed. The MS data showed high precision with all the mass accuracies within 5 ppm. For most of the constituents, a $[M-H]^-$ peak was observed. Due to the use of formic acid in mobile phase, there were additional ions of $[M+46-H]^-$ corresponding to $[M+HCOOH-H]^-$ in negative ion mode. These results provided valuable information for confirming accurate molecular weights and composition of the constituents. The 56 compounds including six new ones were tentatively identified on the basis of their retention behaviors, accurate molecular weight and MS^n fragment data, or by comparison with reference standards or literature data (chemical structures of the compounds corresponding to the peaks shown in Figure 1 below can be found in Figure 1 in the Supplementary). The corresponding quasimolecular ions and their fragment ions in the MS^n spectra are listed in Table 1.

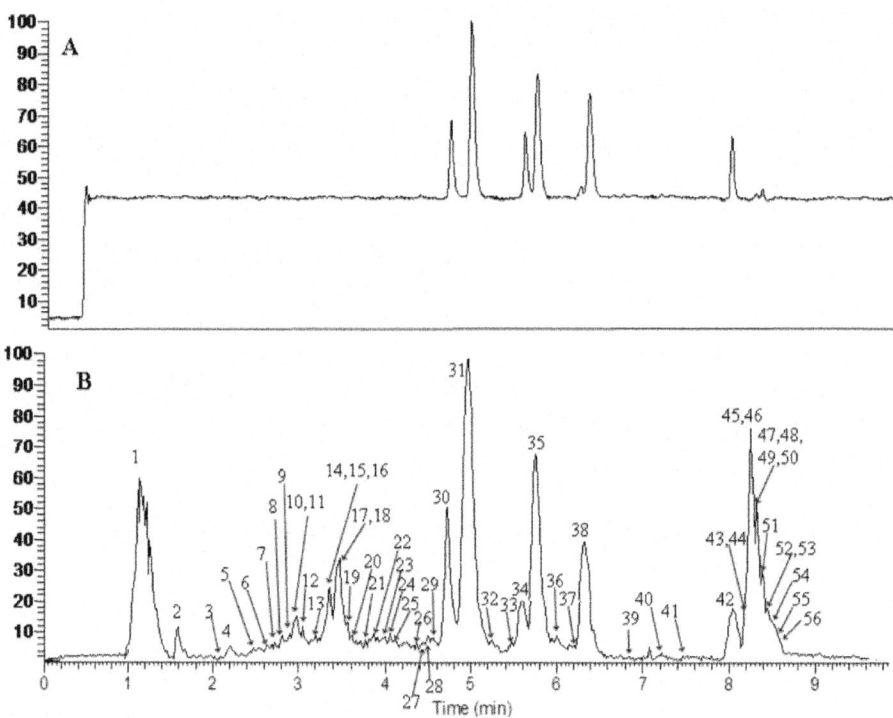

Figure 1. UHPLC-(-) ESI-MS total ion chromatograms of a mixture of six standards (**A**) and the extract of *Smilacis glabrae* (**B**).

The identified compounds can be classified into three classes, namely flavonoids, phenolic acids and phenylpropanoid glycosides. Four flavanonol isomers (compounds **30**, **31**, **34** and **35**) were unambiguously identified by the same deprotonated ions at m/z 449 ($C_{21}H_{21}O_{11}$) and the same product ions at m/z 303 and m/z 285, and they could be distinguished through their UV absorption and elution order when compared to reference standards. Neoastilbin (**30**) with 2S,3S configuration and astilbin (**31**) with 2R,3R configuration had the same UV$_{max}$ absorption at 290 nm, while neoisoastilbin (**34**) with 2S,3R configuration and isoastilbin (**35**) with 2R,3S configuration had the same UV absorption at 295–296 nm (see Figure 2 in the Supplementary), the latter caused a red shift of 5–6 nm, and the elution order of the four flavanonol isomers were 2S,3S > 2R,3R > 2S,3R > 2R,3S. The four flavanonols were the main constituents of *S. glabrae*. To our surprise, compounds **19**, **21**, **25** and **29** had the same deprotonated ions at m/z 629 ($C_{30}H_{29}O_{15}$) and the same fragment ions (Figure 2), which demonstrated they were also diastereomers. In the MS^2 spectra, the product ions at m/z 449 [M-H-$C_9H_8O_4$] and m/z 303 [M-H-

$C_9H_8O_4$-rhamnose] suggested the four diastereomers were the derivatives of the four configurationally different astilbins. In addition, two prominent MS^2 product ions were observed at m/z 475 and m/z 483, respectively, for the neutral loss of $CO_2 + C_6H_6O_2$ and for the loss of a rhamnose, which indicated they had the same substituent group and substituent site. The four isomers could also be distinguished through their UV absorption. Compounds **19** and **21** had the same UV absorption at 289 nm, while compounds **25** and **29** had the same UV absorption at 295 nm (see Figure 2 in the Supplementary), which indicated that compounds **19** and **21** had the 2*S*,3*S* or 2*R*,3*R* configuration, while compounds **25** and **29** had the 2*S*,3*R* or 2*R*,3*S* configuration. As the elution order was 2*S*,3*S* > 2*R*,3*R* > 2*S*,3*R* > 2*R*,3*S*, thus compounds **19**, **21**, **25**and **29** were tentatively identified as 8-[β-(3,4-dihydroxyphenyl)-α-carboxyl-3-oxopropyl]-substituted neoastilbin, 8-[β-(3,4-dihydroxyphenyl)-α-carboxyl-3-oxopropyl]-substituted astilbin, 8-[β-(3,4-dihydroxyphenyl)-α-carboxyl-3-oxopropyl]-substituted neoisoastilbin and 8-[β-(3,4-dihydroxyphenyl)-α-carboxyl-3-oxopropyl]-substituted isoastilbin, respectively. Similarly, compounds **38** and **42** were unambiguously identified as engeletin (**38**) and isoengeletin (**42**) based on reference standards, and compounds**24**and**28**weretentativelyidentifiedas8-[β-(3,4-dihydroxyphenyl)-α-carboxyl-3-oxopropyl]-substituted engeletin and 8-[β-(3,4-dihydroxyphenyl)-α-carboxyl-3-oxopropyl]-substituted isoengeletin, respectively (Figure 3 in the Supplementary). Compounds **19**, **21**, **24**, **25**, **28** and **29** were identified as new compounds, but their absolute configurations could not be determined.

Method Validation of the Quantitative Analysis

The calibration curves, linear ranges, limit of quantification (LOQ) and repeatability of six analytes were performed using the above-developed UHPLC-ESI-MS method (Table 2). Reasonable correlation coefficient values ($r^2 \geq 0.9981$) indicated good correlations between investigated standards concentrations and their peak areas within the ranges tested. The ranges of LOQ for all the analytes were from 0.011 to 0.067 µg/mL, respectively. The repeatability present as RSD ($n = 6$) was between 1.77% and 2.37% of the 6 analytes. The overall intra- and inter-day precisions (RSD) of the six analytes were in the range from 1.03% to 3.19%, and 0.76% to 3.91% (Table 2), respectively. The developed method had good accuracy with the RSD of the recoveries were between 1.49% and 4.73% (Table 2). Therefore, the results demonstrated that the UHPLC-ESI-MS method was sensitive, precise, and accurate enough for quantitative evaluation of *Smilacis glabrae*.

Table 1. Identification of the chemical constituents of *Smilacis glabrae* by UHPLC-ESI-MS^n analysis

Peak No.	t_R(min)	Selected Ion	Observed Mass (*m/z*)	Calculated Mass (*m/z*)	Formula	MS/MS Fragmentation Patterns	Identifieation
1 [a]	1.15	[M-H]⁻	173.0457	173.0450	$C_7H_9O_5$	173→155, 129, 111	shikimic acid
2	1.62	[M-H]⁻	117.0195	117.0188	$C_4H_5O_4$	117→99, 73	succinic acid
3	2.32	[M-H]⁻	359.0984	359.0978	$C_{15}H_{19}O_{10}$	359→197, 182	syringic acid-4-*O*-β-d-glucopyranoside
4	2.34	[M+COOH]⁻	255.0512	255.0505	$C_{11}H_{11}O_7$	255→209, 193, 179, 165	3,4-dihydroxy-5-methoxycinnamic acid
5	2.47	[M+COOH]⁻	345.1191	345.1186	$C_{15}H_{21}O_9$	345→299	rhodioloside
6 [a]	2.70	[M-H]⁻	153.0194	153.0188	$C_7H_5O_4$	153→109	protocatechuic acid
7	2.91	[M+COOH]⁻	197.0458	197.0450	$C_9H_9O_5$	197→153	syringic acid
8	2.97	[M-H]⁻	387.1296	387.1291	$C_{17}H_{23}O_{10}$	387→207, 177	3-(β-d-glucopyranosyloxy)-1-(4-hydroxy-3,5-dimethoxyphenyl)-1-propanone
9	3.08	[M-H]⁻	577.1346	577.1346	$C_{30}H_{25}O_{12}$	577→559, 451, 425, 407, 289	procyanidin B
10 [a]	3.13	[M-H]⁻	289.0720	289.0712	$C_{15}H_{13}O_6$	289→271, 245, 205,179,151	catechin
11	3.19	[M-H]⁻	239.0564	239.0556	$C_{11}H_{11}O_6$	239→221, 195, 179, 177, 149	syringic acid acetate
12	3.36	[M-H]⁻	315.1074	315.1080	$C_{14}H_{19}O_8$	315→153	3,4-dihydroxy-phenethyl glucoside
13 [a]	3.45	[M-H]⁻	469.1141	469.1135	$C_{24}H_{21}O_{10}$	469→315, 289	(2 *R*,3*S*)-8-[β-(3,4-dihydroxyphenyl)-α-carboxyl-3-oxopropyl]-substituted catechin
14 [a]	3.55	[M-H]⁻	335.0777	335.0767	$C_{16}H_{15}O_8$	335→291, 179, 135	3-*O*-caffeoylshikimic acid
15 [a]	3.58	[M-H]⁻	561.1397	561.1397	$C_{30}H_{25}O_{11}$	561→543, 435, 289	3',4',5,7-tetra-hydroxyflavan (4°8)-3,3',4',5,7-pentahydroxyflavan
16 [a]	3.61	[M-H]⁻	289.0722	289.0712	$C_{15}H_{13}O_6$	289→271, 245, 205, 179, 151	epicatechin
17 [a]	3.74	[M-H]⁻	335.0777	335.0767	$C_{16}H_{15}O_8$	335→291, 179, 135	4-*O*-caffeoylshikimic acid
18 [a]	3.76	[M-H]⁻	179.0350	179.0344	$C_9H_7O_4$	179→161, 135	caffeic acid

19 [b]	3.93	[M-H]⁻	629.1514	629.1506	$C_{30}H_{29}O_{15}$	629→483, 475, 449, 303, 285	8-[β-(3,4-dihydroxyphenyl)-α-carboxyl-3-oxopropyl]-substituted neoastilbin
20	4.04	[M-H]⁻	465.1041	465.1033	$C_{21}H_{21}O_{12}$	465→421, 297	4-O-β-d-(6-O-gentisoylglucopyranosyl)-vanillic acid
21 [b]	4.20	[M-H]⁻	629.1514	629.1506	$C_{30}H_{29}O_{15}$	629→483, 475, 449, 303, 285	8-[β-(3,4-dihydroxyphenyl)-α-carboxyl-3-oxopropyl]-substituted astilbin
22	4.23	[M-H]⁻	339.0721	339.0716	$C_{15}H_{15}O_{9}$	339→193	smiglanin
23 [a]	4.39	[M-H]⁻	335.0777	335.0767	$C_{16}H_{15}O_{8}$	335→291, 179, 135	5-O-caffeoylshikimic acid
24 [b]	4.44	[M-H]⁻	613.1565	613.1557	$C_{30}H_{29}O_{14}$	613→467, 459, 433, 287	8-[β-(3,4-dihydroxyphenyl)-α-carboxyl-3-oxopropyl]-substituted engeletin
25 [b]	4.56	[M-H]⁻	629.1514	629.1506	$C_{30}H_{29}O_{15}$	629→483, 475, 449, 303, 285	8-[β-(3,4-dihydroxyphenyl)-α-carboxyl-3-oxopropyl]-substituted neoisoastilbin
26	4.84	[M-H]⁻	301.0354	301.0348	$C_{15}H_{9}O_{7}$	301→283, 255, 215, 175, 151	quercetin
27	4.97	[M+COOH]⁻	435.1297	435.1291	$C_{21}H_{23}O_{10}$	435→389, 227,195	polydatin
28 [b]	5.06	[M-H]⁻	613.1565	613.1557	$C_{30}H_{29}O_{14}$	613→467, 459, 433, 287	8-[β-(3,4-dihydroxyphenyl)-α-carboxyl-3-oxopropyl]-substituted isoengeletin
29 [b]	5.12	[M-H]⁻	629.1514	629.1506	$C_{30}H_{29}O_{15}$	629→483, 475, 449, 303	8-[β-(3,4-dihydroxyphenyl)-α-carboxyl-3-oxopropyl]-substituted isoastilbin
30 [c]	5.29	[M-H]⁻	449.1099	449.1084	$C_{21}H_{21}O_{11}$	449→303, 285	neoastilbin
31 [c]	5.63	[M-H]⁻	449.1099	449.1084	$C_{21}H_{21}O_{11}$	449→303, 285	astilbin
32 [a]	5.72	[M-H]⁻	193.0511	193.0501	$C_{10}H_{9}O_{4}$	193→178, 161, 134	ferulic acid
33 [a]	6.10	[M-H]⁻	303.0513	303.0505	$C_{15}H_{11}O_{7}$	303→285, 177, 125	taxifolin
34 [c]	6.55	[M-H]⁻	449.1099	449.1084	$C_{21}H_{21}O_{11}$	449→303, 285	neoisoastilbin

35 [c]	6.81	[M-H]⁻	449.1099	449.1084	$C_{21}H_{21}O_{11}$	449→303, 285	isoastilbin
36	6.86	[M-H]⁻	243.0665	243.0657	$C_{14}H_{11}O_4$	243→225, 201, 199, 175	piceatannol
37	7.27	[M-H]⁻	433.1149	433.1135	$C_{21}H_{21}O_{10}$	433→287, 269	neoengeletin
38 [c]	7.43	[M-H]⁻	433.1149	433.1135	$C_{21}H_{21}O_{10}$	433→287, 269	engeletin
39 [a]	7.49	[M-H]⁻	359.0771	359.0767	$C_{18}H_{15}O_8$	359→341, 291, 239, 197	rosmarinic acid
40	7.53	[M-H]⁻	433.1149	433.1135	$C_{21}H_{21}O_{10}$	433→287,269	neoisoengeletin
41	8.16	[M-H]⁻	693.2029	693.2031	$C_{32}H_{37}O_{17}$	693→517, 337	helonioside A
42 [c]	8.20	[M-H]⁻	433.1149	433.1135	$C_{21}H_{21}O_{10}$	433→287,269	isoengeletin
43 [a]	8.23	[M-H]⁻	451.1038	451.1029	$C_{24}H_{19}O_9$	451→341	cinchonain Ia
44	8.25	[M-H]⁻	693.2029	693.2031	$C_{32}H_{37}O_{17}$	693→357	securoside A
37	7.27	[M-H]⁻	433.1149	433.1135	$C_{21}H_{21}O_{10}$	433→287, 269	neoengeletin
45 [a]	8.30	[M-H]⁻	451.1035	451.1029	$C_{24}H_{19}O_9$	451→341	cinchonain Ib
46	8.32	[M-H]⁻	227.0717	227.0708	$C_{14}H_{11}O_3$	227→209,185, 183, 159, 157, 143	resveratrol
47 [a]	8.35	[M-H]⁻	809.2293	809.2293	$C_{40}H_{41}O_{18}$	809→767, 663, 633	smilaside G
48 [a]	8.36	[M-H]⁻	839.2408	839.2398	$C_{41}H_{43}O_{19}$	839→797, 693, 663, 517	smilaside J
49 [a]	8.38	[M-H]⁻	869.2502	869.2504	$C_{42}H_{45}O_{20}$	869→827, 693, 675	smilaside L
50	8.40	[M-H]⁻	777.2248	777.2242	$C_{36}H_{41}O_{19}$	777→735, 717, 601, 559	(3,6-di-O-feruloyl)-β-d-fructofuranosyl-(3,6-di-O-acetyl)-α-d-glucopyranoside
51	8.42	[M-H]⁻	819.2354	819.2348	$C_{38}H_{43}O_{20}$	819→777, 643, 601, 513	smilaside C
52	8.44	[M-H]⁻	923.2604	923.2610	$C_{45}H_{47}O_{21}$	923→881, 863, 747, 601, 483	smilaside E
53	8.45	[M-H]⁻	953.2712	953.2715	$C_{46}H_{49}O_{22}$	953→911, 777, 735, 717, 289	smilaside B
54 [a]	8.48	[M-H]⁻	271.0614	271.0606	$C_{15}H_{11}O_5$	271→177, 151	naringenin
55	8.52	[M-H]⁻	965.2719	965.2715	$C_{47}H_{49}O_{22}$	965→923, 905, 789, 747, 483	smilaside D
56	8.55	[M-H]⁻	995.2829	995.2821	$C_{48}H_{51}O_{23}$	995→953, 819, 777, 513	smilaside A

[a] Compared with reference [3]; [b] Identified as new compound; [c] Compared with reference standards.

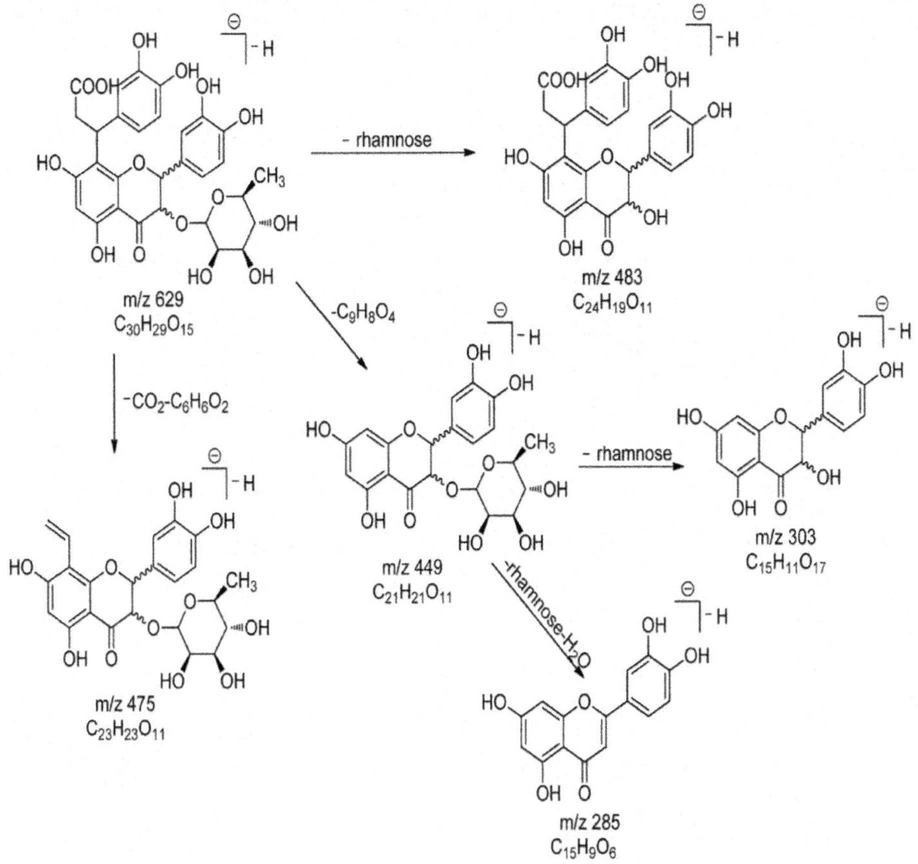

Figure 2. Proposed fragmentation pathways for compounds **19, 21, 25** and **29**.

Quantitative Analysis

The newly established analytical method was subsequently applied to determine the six compounds of *Smilacis glabrae*. The target compounds were identified based on comparison of retention time and mass information obtained from UHPLC-ESI-MS analysis of the reference standards. Table 3 showed the content determined for each compound. The results indicated that the amount of most components determined was similar in the five different batches.

Table 2. Summary of calibration curves, linear range, LOQ, repeatability, intra-day and inter-day precisions and recoveries for six analytes analyzed with the LC-MS system

Analyte	Linear Range (μg/mL)	Calibration Curve (n = 7)	r²	LOQ (μg/mL)	Repeatability RSD (%)	Intra-day (RSD, %) (n = 6)	Inter-day (RSD, %) (n = 3)	Recoveries (n = 3)				
								Initial (μg)	Spiked (μg)	Detected (μg)	Recoveries (%)	RSD (%)
Neoastilbin	0.82–32.8	y = 8593.3 x + 281942	0.9993	0.016	2.37	3.19	3.05	3.470	2.628	5.872	96.27	3.51
								3.284	6.890		101.99	3.16
								3.940	7.085		95.63	1.65
Astilbin	3.10–124.1	y = 8921.6 x + 16423	0.9991	0.062	1.86	1.03	0.76	13.677	9.932	22.963	97.27	2.57
								12.416	25.468		97.60	3.86
								14.900	27.916		97.69	2.66
Neoisoastilbin	0.33–13.3	y = 8299.7 x + 165713	0.9988	0.067	1.91	2.43	2.49	1.517	1.064	2.342	90.77	4.73
								1.340	2.426		91.94	3.39
								1.606	3.205		102.65	3.83
Isoastilbin	1.78–71.2	y = 8479.3 x + 161354	0.9981	0.018	2.15	1.07	0.79	7.188	5.702	12.485	96.86	1.80
								7.128	14.153		98.86	2.83
								8.555	15.011		95.35	2.30
Engeletin	0.86–4.4	y = 4620.5 x – 107846	0.9992	0.017	1.77	2.00	3.91	4.110	2.756	7.038	102.51	2.14
								3.444	7.300		96.59	1.49
								4.132	8.271		100.39	2.37
Isoengeletin	0.28–11.1	y = 4472.8 x – 12397	0.9991	0.011	1.94	2.83	2.86	1.237	0.896	2.152	100.95	2.34
								1.120	2.368		100.37	3.09
								1.134	2.506		97.18	4.50

Table 3. Contents of the six compounds in different batches of *Smilacis glabrae*

Analyte	Content (µg/g)				
	Batch 1	Batch 2	Batch 3	Batch 4	Batch 5
Neoastilbin	2173.1	2735.9	2356.9	2537.4	2253.7
Astilbin	8548.2	8996.1	9262.1	10,962.2	9988.6
Neoisoastilbin	948.3	1046.4	971.2	1188.7	1097.3
Isoastilbin	4493.2	4189.5	4257.9	2800.9	3461.3
Engeletin	2587.2	2494.3	2682.1	1821.6	2047.6
Isoengeletin	771.6	727.6	834.9	594.3	488.5

EXPERIMENTAL SECTION

Chemicals and Materials

HPLC grade acetonitrile and methanol were purchased from Fisher Chemicals (Fairlawn, NJ, USA). Formic acid of HPLC grade was purchased from Sigma Aldrich (St. Louis, MO, USA). Water (18.2 MΩ) was from a Milli-Q water system (Millipore, Bedford, MA, USA). Neoastilbin (**30**), astilbin (**31**), neoisoastilbin (**34**), isoastilbin (**35**), engeletin (**38**) and isoengeletin (**42**) were provided by Dr. Lixiong from the Guangdong Provincial Hospital of Chinese Medicine. Three batches of *Smilacis glabrae* originating from Guangdong Province, China were supplied by Kangmei Pharmaceutical Co. Ltd. (Puning, China). Two batches of *Smilacis glabrae* from the Hunan and Guangxi provinces of China were purchased from Er-tian-tang Pharmacy (Guangzhou, China). Voucher samples were deposited in the Laboratory of Chinese Materia Medica Preparation, Second Affiliated Hospital, Guangzhou University of Traditional Chinese Medicine.

Standard Solutions and Sample Preparation

The standard solution mixture of the six flavonoids was prepared by dissolving the reference substances in methanol to final concentration of 32.8 µg/mL for neoastilbin, 124.1 µg/mL for astilbin, 13.3 µg/mL for neoisoastilbin, 71.2 µg/mL for isoastilbin, 34.4 µg/mL for engeletin and 11.1 µg/mL for isoengeletin, respectively. Then, the standard solution mixture was diluted to 80%, 60%, 40%, 20%, 10%, 5% and 2.5% of the concentration of the original solution. All the standard solutions were stored at 4 °C.

The dried rhizome (0.2 g, 60 mesh) was accurately weighed and ultrasonically extracted by infusion with 25 mL water for 30 min. The extracted solution was centrifuged at 10,000 rpm for 10 min, and then filtered through a 0.22 m nylon membrane filter prior to injection for UHPLC-MS analysis.

Analytical System

Chromatographic separation was performed on an Accela™ ultra high pressure liquid chromatography (UHPLC) system (Thermo Fisher Scientific, San Jose, CA, USA) comprising a UHPLC pump, a PDA detector, scanning from 200 to 400 nm, and an autosampler settled to 30 °C. The LC conditions were as follows: column: Agilent Eclipse Plus C18 (100 mm × 3.0 mm, 1.7 μm); mobile phase: acetonitrile (A) and water (B) both containing 0.1% (v/v) formic acid; gradient: 0 min, 10: 90; 1 min, 20: 80; 3–6.5 min, 23: 77; 7 min, 80: 20; 9–10 min, 100: 0 (A: B, v/v); flow rate: 0.3 mL/min; injection volume: 10 μL.

Qualitative Characteristic of Chemical Constituents

Identification of chemical constituents in *Smilacis glabrae* extract was performed by UHPLC-ESI-MS^n analysis. MS analysis was performed using an LTQ Orbitrap^{XL} hybrid mass spectrometer (Thermo Fisher Scientific), fitted with an ESI source, and operated in negative ion mode, with a mass range of 100–1500 with resolution set at 30000 using the normal scan rate.

The data-dependent MS/MS events were always performed on the most intense ions detected in full scan MS. The MS/MS isolation width was 1 amu, and the normalized collision energy was 35% for all compounds. Nitrogen was used as sheath gas and helium served as the collision gas. The key optimized ESI parameters were as follows: source voltage: 3.8 kV; sheath gas (nitrogen): 50 L/min; auxiliary gas flow: 10 L/min; capillary voltage: −35.0 V; capillary temperature: 300.0 °C; tube lens: −110.0 V. The ion injection time used was 50.0 ms. MS scan functions and HPLC solvent gradients were controlled by the Xcalibur data system (Thermo Fisher Scientific). Data was collected and analyzed with Xcalibur 2.0.7 software (Thermo Fisher Scientific). The Orbitrap mass analyzer was calibrated according to the manufacturer's directions using a mixture of caffeine, methionine-arginine-phenylalanine-alanine-acetate (MRFA), sodium dodecyl sulfate, sodium taurocholate and Ultramark 1621 in an acetonitrile-methanol-water solution containing 1% acetic acid by direct injection at a flow rate of 5 μL/min in negative mode before analysis.

Validation of the Quantitative Analysis

A calibration curve was used to determine the calculated concentration of the samples. The calibration curve of each compound was performed with at least six appropriate concentrations. The limit of quantification (LOQ) under the present chromatographic conditions was determined at signal-to-noise ratios (S/N) of 10. Intra- and inter-day variations were chosen to determine the precision of the developed method. The precision was examined by five repetitive injections in the same day and in three consecutive days, respectively. The relative standard deviation (R.S.D.) was considered as the measure of precision. The accuracy was evaluated by calculating the mean recoveries of six reference standards from the spiked standard solutions. A known amount of *Smilacis glabrae* sample was spiked with the standard solution at three different concentration levels. The high spiked amount was 1.2 times of the known amount sample, the middle spiked amount was 1.0 times of the known amount sample and the low spiked amount was 0.8 times of the known amount sample. The recovery percentages were calculated using to the following equation: (total detected amount − original amount)/added amount ×100%.

CONCLUSIONS

In this study, a total of 56 compounds, including six minor new ones, were simultaneously detected and identified by UHPLC-LTQ-Orbitrap-MS. Based on the qualitative analysis, a rapid method was established for quantitative analysis of six marker components in *Smilacis glabrae* extract. This is the first report on the comprehensive determination of chemical constituents in *S. glabrae* by UHPLC-LTQ-Orbitrap-MS. The results would provide the chemical support for the further pharmacokinetic studies and for the improvement of quality control of *Smilacis glabrae* and its preparations. The study also suggested that UHPLC-ESI/LTQ-Orbitrap mass spectrometry would be a powerful and reliable analytical tool for the characterization of chemical profile in complex chemical system, such as TCM preparations.

SUPPLEMENTARY MATERIALS

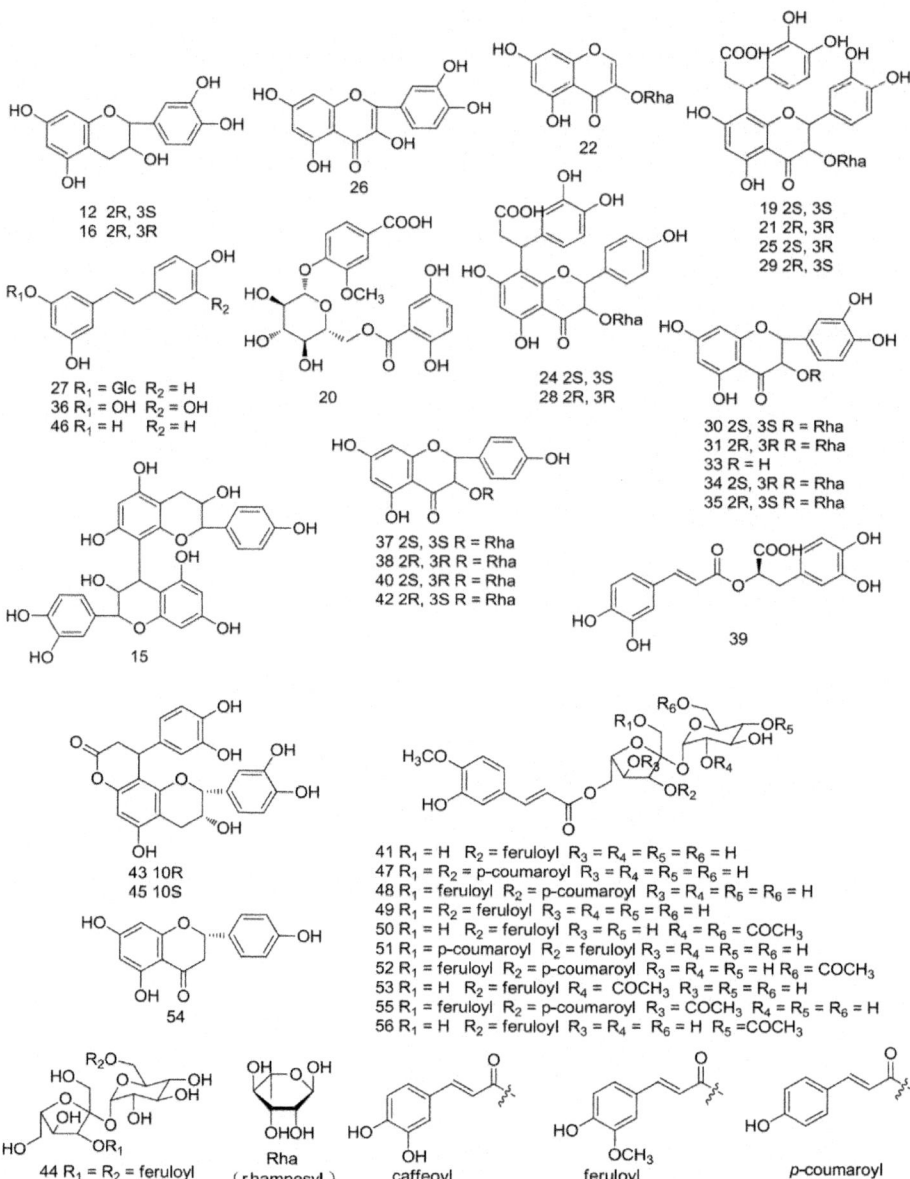

Figure S1. The chemical structures of identified compounds of Smilacis glabrae extract.

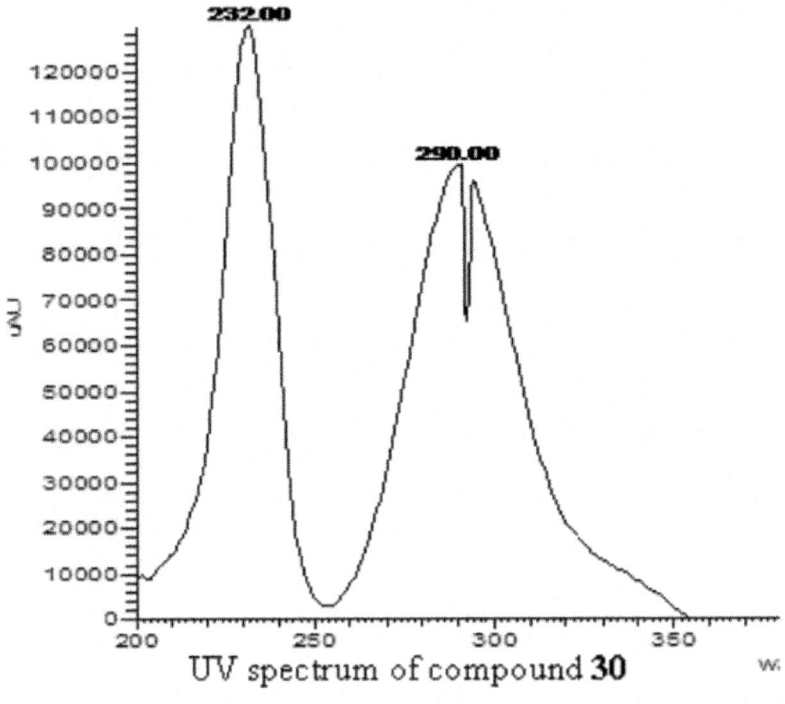

UV spectrum of compound **30**

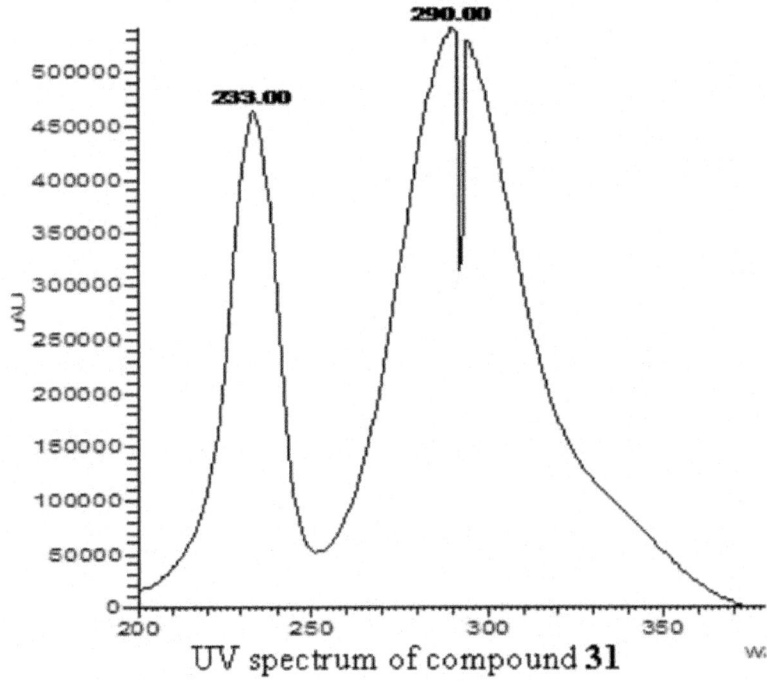

UV spectrum of compound **31**

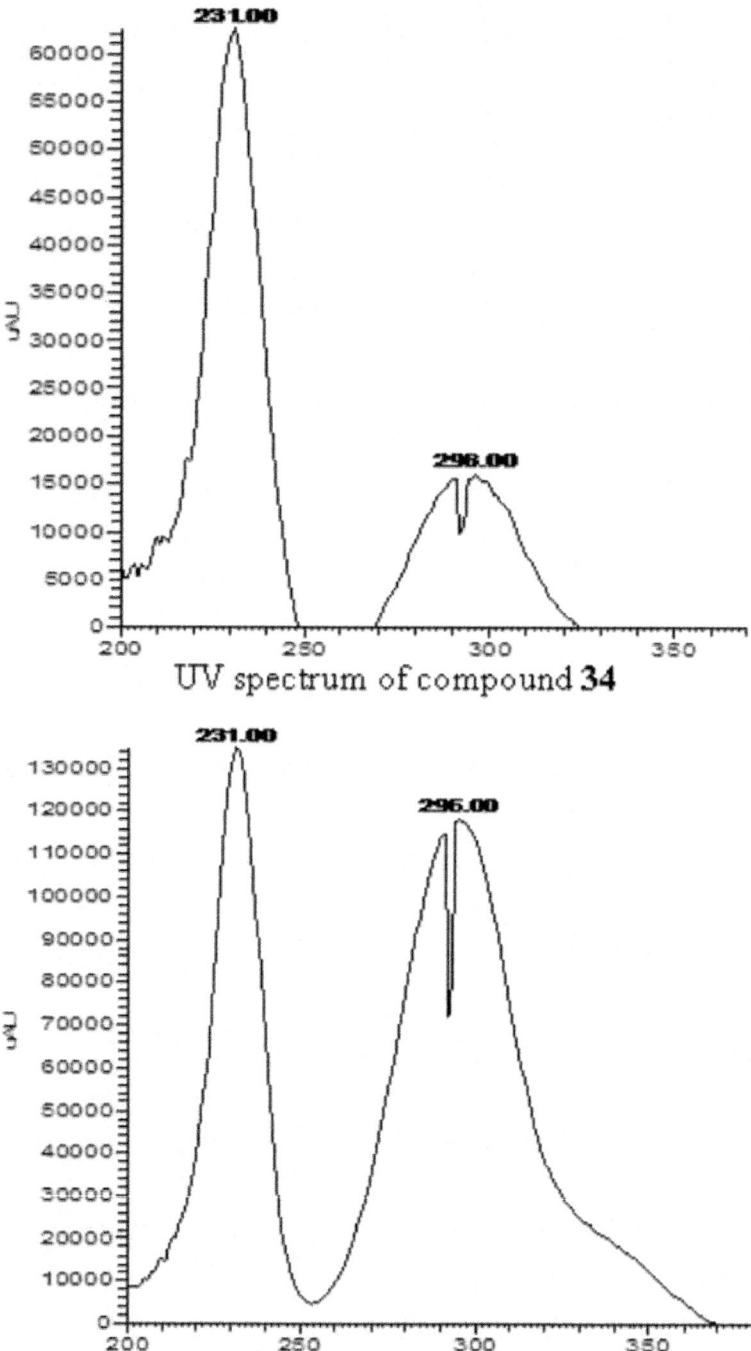

UV spectrum of compound **34**

UV spectrum of compound **35**

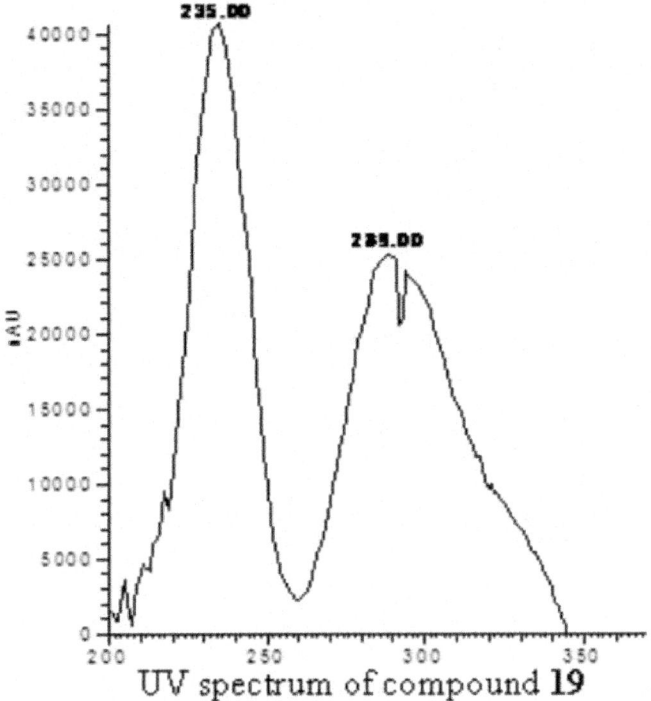

UV spectrum of compound **19**

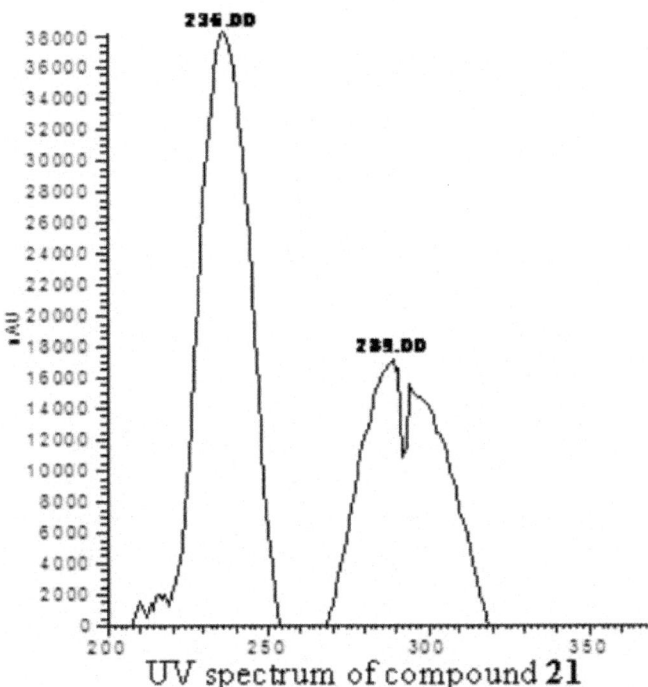

UV spectrum of compound **21**

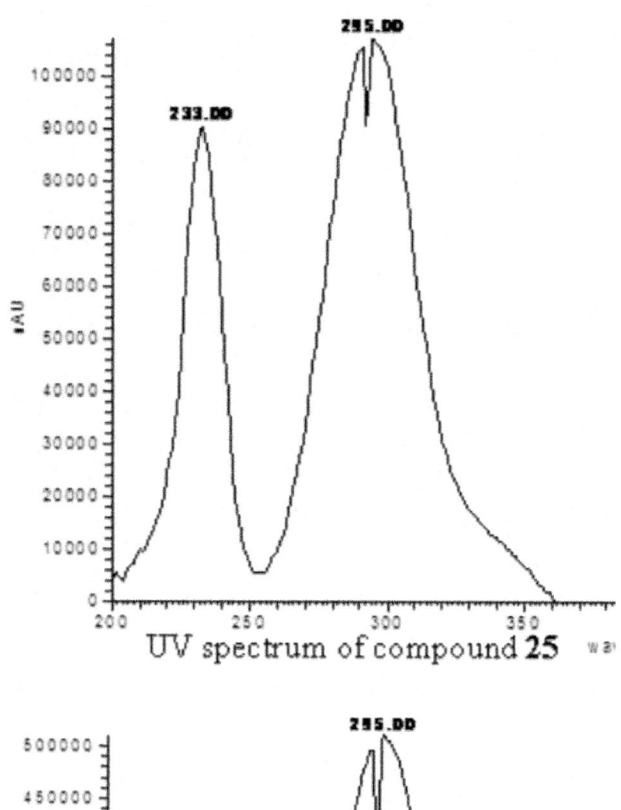

UV spectrum of compound 25

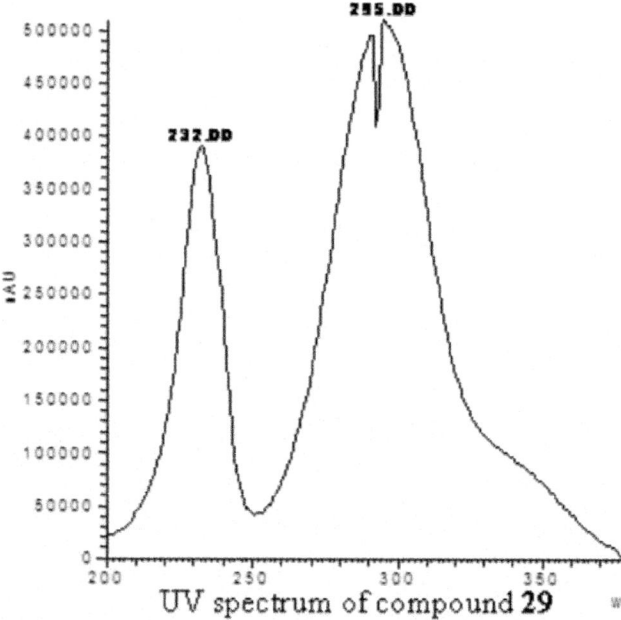

UV spectrum of compound 29

Figure S2. UV spectra of compounds 30, 31, 34, 35 and 19, 21, 25, 29.

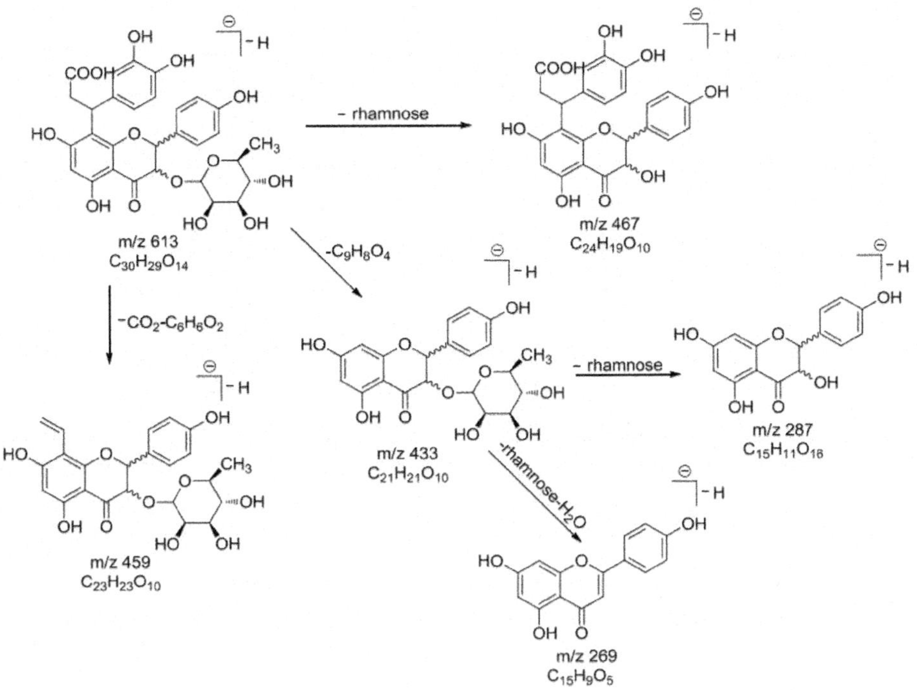

Figure S3. Proposed fragmentation pathway for compounds 24 and 28.

ACKNOWLEDGMENTS

This research was financially supported by Guangdong Natural Science Fund (S2013030011515), Guangdong Financial Industry Technology Research Development Fund [2011(285)05], Guangdong Science and Technology Department-Guangdong Provincial Hospital of Chinese Medicine Joint Special Fund (2011B032200009) and Guangdong Provincial Hospital of Chinese Medicine Special Fund (YK2013B1N11).

AUTHOR CONTRIBUTIONS

C.-J. Lu and R.-Z. Zhao designed the experiments and provided critical advice on operation of the analytical equipment. S.-D. Chen was responsible for performing most of the experiment and analysis, and preparing the draft of the manuscript.

REFERENCES

1. Chinese Pharmacopoeia Commission. *Pharmacopoeia of People's Republic of China*; China Medical Pharmaceutical Science and Technology Publishing Press: Beijing, China, 2010; Volume 1, p. 15.

2. Xu, S.; Shang, M.Y.; Liu, G.X.; Xu, F.; Wang, X.; Shou, C.C.; Cai, S.Q. Chemical constituents from the rhizomes of *Smilax glabra* and their antimicrobial activity. *Molecules* **2013**, *18*, 5265–5287.

3. Li, X.; Zhang, Y.F.; Yang, L.; Feng, Y.; Deng, Y.H.; Liu, Y.M.; Zeng, X. Chemical profiling of constituents of *Smilacis glabrae* using ultrahigh pressure liquid chromatography coupled with LTQ Orbitrap mass spectrometry. *Nat. Prod. Commun.* **2012**, *7*, 181–184.

4. Huang, H.Q.; Cheng, Z.H.; Shi, H.M.; Xin, W.B.; Wang, T.T.Y.; Yu, L.L. Isolation and characterization of two flavonoids, engeletin and astilbin, from the leaves of *Engelhardia roxburghiana* and their potential anti-inflammatory properties. *J. Agric. Food. Chem.* **2011**, *59*, 4562–4569.

5. Haraguchi, H.; Mochida, Y.; Sakai, S.; Masuda, H.; Tamura, Y.; Mizutani, K. Protection against oxidative damage by dihydroflavonols in Engelhardtia chrysolepis. *Biosci. Biotechnol. Biochem.* **1996**, *60*, 945–948.

6. Igarashi, K.; Uchida, Y.; Murakami, N.; Mizutani, K.; Masuda, H. Effect of astilbin in tea processed from leaves of Engelhardtia chrysolepis on the serum and liver lipid concentrations and on the erythrocyte and liver antioxidative enzyme activities of rats. *Biosci. Biotechnol. Biochem.* **1996**, *60*, 513–515.

7. Nia, R.; Adesanya, S.A.; Okeke, I.N.; Illoh, H.C.; Adesina, S.K. Antibacterial constituents of *Calliandra haematocephala.Niger. J. Natur. Prod. Med.* **1999**, *3*, 58–60.

8. Mizutani, K.; Kambara, T.; Masuda, H.; Tamura, Y.; Tanaka, O.; Tokuda, H.; Nishino, H.; Kozuka, M. *Food Factors for Cancer Prevention*; Springer: Tokyo, Japan, 1995; pp. 607–612.

9. Wirasathien, L.; Pengsuparp, T.; Suttisri, R.; Ueda, H.; Moriyasu, M.; Kawanishi, K. Inhibitors of aldose reductase and advanced glycation end-products formation from the leaves of *Stelechocarpus cauliflorus* R.E. Fr. *Phytomedicine* **2007**, *14*, 546–550.

10. Ruangnoo, S.; Jaiaree, N.; Makchuchit, S.; Panthong, S.; Thongdeeying, P.; Itharat, A. An *in vitro* inhibitory effect on RAW 264.7 cells by anti-inflammatory compounds from *Smilax corbularia* Kunth. *Asian Pac. J. Allergy Immunol.* **2012**, *30*, 268–274.

11.   Chen, L.; Yin, Y.; Yi, H.W.; Xu, Q.; Chen, T. Simultaneous quantification of five major bioactive flavonoids in*Rhizoma**Smilacis Glabrae* by high–performance liquid chromatography. *J. Pharm. Biomed. Anal.* **2007**, *43*, 1715–1720.

Chapter 5

PHOTODIODE ARRAY DETECTION IN CLINICAL APPLICATIONS; QUANTITATIVE ANALYTE ASSAY ADVANTAGES, LIMITATIONS AND DISADVANTAGES

Zarrin Es'haghi[1]

[1]Department of Chemistry, Payame Noor University, 19395-4697 Tehran, I.R. of IRAN

INTRODUCTION

Optical Spectroscopy

Study of the electromagnetic radiation by matter, as related to the dependence of these processes on the wavelength of the radiation. More recently, the definition has been expanded to include the study of the interactions between particles such as electrons, protons, and ions, as well as their interaction with other particles as a function of their collision energy. Spectroscopic analysis has been crucial in the development of the most fundamental theories in physics, including quantum mechanics, the special and general theories of relativity, and quantum electrodynamics.

Spectroscopic techniques have been applied in virtually all technical fields of science and technology. One of the most famous kinds of spectroscopy, optical spectroscopy is used routinely to identify the chemical composition of matter and to determine its physical structure. Spectroscopic techniques are extremely sensitive. Single atoms and even different isotopes of the same atom can be detected among 10^{20} or more atoms of a different species. Isotopes are all atoms of an element that have unequal mass but the same atomic number. Isotopes of the same element are virtually identical chemically. Trace amounts of pollutants or contaminants are often detected most effectively by spectroscopic techniques. Because of this sensitivity, the most accurate physical measurements have been frequency measurements.

Spectroscopy now covers a sizable fraction of the electromagnetic spectrum. The table (1) summarizes the electromagnetic spectrum over a frequency range of 16 orders of magnitude. Spectroscopic techniques are not confined to electromagnetic radiation, however. Because the energy E of a photon (a quantum of light) is related to its frequency v by the relation $E = hv$, where h is Planck's constant, spectroscopy is actually the measure of the interaction of photons with matter as a function of the photon energy. In instances where the probe particle is not a photon, spectroscopy refers to the measurement of how the particle interacts with the test particle or material as a function of the energy of the probe particle.

Electromagnetic radiation is composed of oscillating electric and magnetic fields that have the ability to transfer energy through space. The energy propagates as a wave, such that the crests and troughs of the wave move in vacuum at the speed of 299,792,458 metres per second.

Table 1. Frequency and wavelength domain of electromagnetic radiations

Electromagnetic phenomena		
Gamma rays(γ rays)	$<5 \times 1^{0-1}2$	"/$6 \times 1^{0l}9$
X-rays	5×10^{-12}–$1 \times 1^{0-}8$	3×10^{16}–$6 \times 1^{0l}9$
Ultraviolet	1×10^{-8}–$4 \times 1^{0-}7$	7×10^{14}–$3 \times 1^{0l}6$
Visible light	4×10^{-7}–$7 \times 1^{0-}7$	4×10^{14}–$7 \times 1^{0l}4$
Infrared	8×10^{-7}–$1 \times 1^{0-}3$	3×10^{11}–$4 \times 1^{0l}4$
Microwaves, Radar	1×10^{-3}–1	3×10^{8}–$3 \times 1^{0l}1$
Television waves	1–10	3×10^{7}–$3 \times 1^{0}8$
Radio waves	10–$1,000$	3×10^{5}–$3 \times 1^{0}7$

The decomposition of electromagnetic radiation into its component wavelengths is fundamental to spectroscopy. Evolving from the first crude prism spectrographs that separated white light into its constituent colours, modern spectrometers have provided ever-increasing wavelength resolution. Large-grating spectrometers are capable of resolving wavelengths as close as 10^{-3} nanometre, while modern laser techniques can resolve optical wavelengths separated by less than 10^{-10} nanometre. The frequency with which the electromagnetic wave oscillates is also used to characterize the radiation. The product of the frequency (v) and the wavelength (λ) is equal to the speed of light (c); i.e., $v\lambda = c$. The frequency is often expressed as the number of oscillations per second, and the unit of frequency is hertz (Hz), where one hertz is one cycle per second.

Spectroscopy is used as a tool for studying the structures of atoms and molecules. The large number of wavelengths emitted by these systems makes it possible to investigate their structures in detail, including the electron configurations of ground and various excited states. Spectroscopy also provides a precise analytical method for finding the constituents in material having unknown chemical composition. In a typical spectroscopic analysis, a concentration of a few parts per million of a trace element in a material can be detected through its emission spectrum.

Production and analysis of a spectrum usually require the following:

- a source of electromagnetic radiation,
- a disperser to separate the light into its component wavelengths, and
- a detector to sense the presence of light after dispersion (See Figure 1).

The apparatus used to accept light, separate it into its component wavelengths, and detect the spectrum is called a spectrometer. Spectra can be obtained either in the form of emission spectra, which show one or more bright lines or bands on a dark background, or absorption spectra, which have a continuously bright background except for one or more dark lines.

Optical Detectors

The principal detection methods used in optical spectroscopy are photographic (*e.g.,* film), photoemissive (photomultipliers), and photoconductive (semiconductor). Prior to about 1940, most spectra were recorded with photographic plates or film, in which the film is placed at the image point of a grating or prism spectrometer. An advantage of this technique is that the entire spectrum of interest can be obtained simultaneously, and low-intensity spectra can be easily taken with sensitive film.

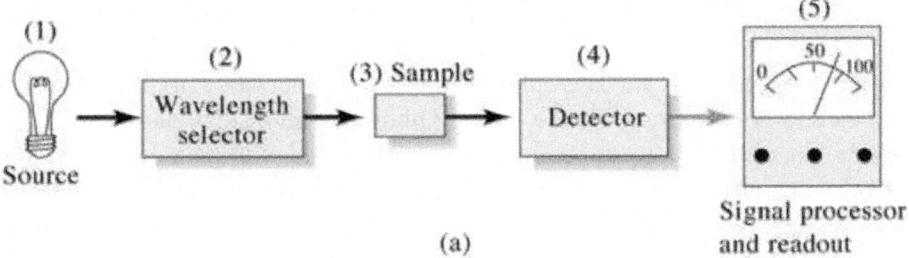

(a)

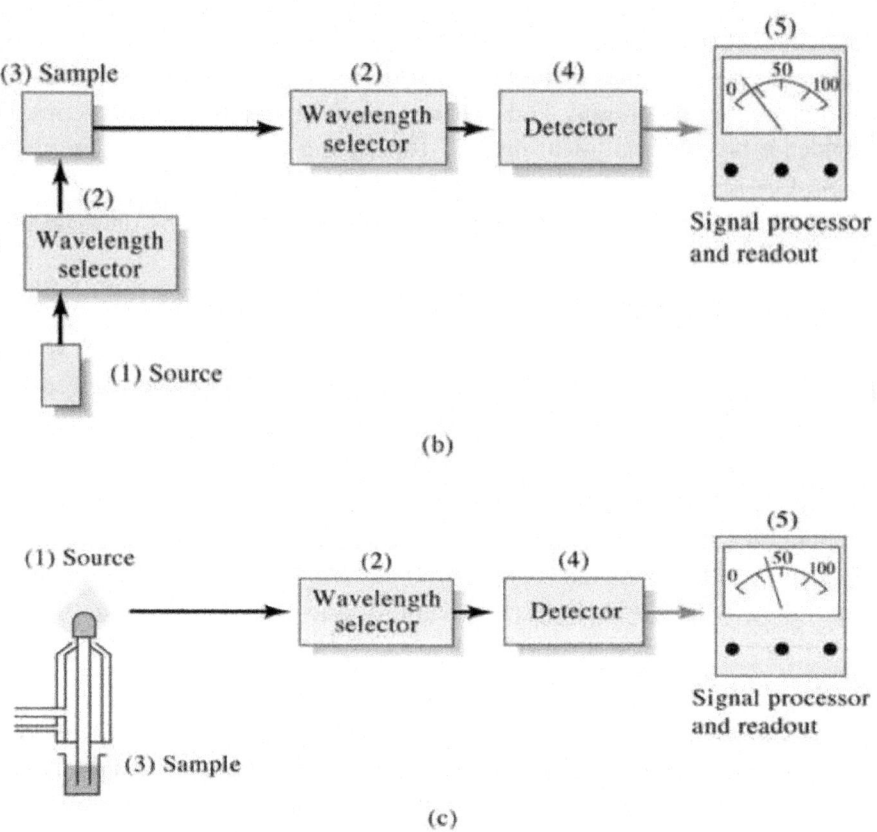

Figure 1. Components of Optical Instruments. The generic spectrometer, (a) Molecular absorption, (b) Molecular emission and (c) Atomic absorption.

Photoemissive detectors have replaced photographic plates in most applications. When a photon with sufficient energy strikes a surface, it can cause the ejection of an electron from the surface into a vacuum. A photoemissive diode consists of a surface (photocathode) appropriately treated to permit the ejection of electrons by low-energy photons and a separate electrode (the anode) on which electrons are collected, both sealed within an evacuated glass envelope. A photomultiplier tube has a cathode, a series of electrodes (dynodes), and an anode sealed within a common evacuated envelope. Appropriate voltages applied to the cathode, dynodes, and anode cause electrons ejected from the cathode to collide with the dynodes in succession. Each electron collision produces several more electrons; after a dozen or more dynodes, a single electron ejected by one photon can be converted into a fast pulse (with a duration of less than 10^{-8} second) of as many as 10^7 electrons at the anode. In this way, individual photons can be counted with good time resolution.

Other photodetectors include imaging tubes (*e.g.,* television cameras), which can measure a spatial variation of the light across the surface of the photocathode, and microchannel plates, which combine the spatial resolution of an imaging tube with the light sensitivity of a photomultiplier. A night vision device consists of a microchannel plate multiplier in which the electrons at the output are directed onto a phosphor screen and can then be read out with an imaging tube.

Solid-state detectors such as semiconductor photodiodes detect light by causing photons to excite electrons from immobile, bound states of the semiconductor (the valence band) to a state where the electrons are mobile (the conduction band). The mobile electrons in the conduction band and the vacancies, or "holes," in the valence band can be moved through the solid with externally applied electric fields, collected onto a metal electrode, and sensed as a photoinduced current. Microfabrication techniques developed for the integrated-circuit semiconductor industry are used to construct large arrays of individual photodiodes closely spaced together. The device, called a charge-coupled device (CCD), permits the charges that are collected by the individual diodes to be read out separately and displayed as an image.

Multichannel Detectors

Multichannel detectors can be used to sense optical and ionizing radiation or convert to an electrical signal an incoming chemical, physical, mechanical, or thermal stimulus. In other words; multichannel detector, can measure all wavelengths dispersed by a dispersing element simultaneously.

The multichannel detector employs a light source that emits light over a wide range of wavelengths. Employing an appropriate optical system (a prism or diffraction grating), light of a specific wavelength can be selected for detection purposes. The specific wavelength might be chosen where a solute has an absorption maximum to provide maximum sensitivity. Alternatively, the absorption spectra of an eluted substances could be obtained for identification purposes by scanning over a range of wavelengths. The latter procedure, however, differs with the type of multichannel detector being used.

There are two basic types of multi–wavelength detector, the *dispersion* detector and the *diode array detector*, the latter being the more popular. In fact, very few dispersion instruments are sold today but many are still used in the field and so their characteristics will be discussed. All multichannel detectors require a broad emission light source such as deuterium or the xenon lamp, the deuterium lamp being the most popular.

The two types of multichannel detectors have important differences. In the dispersive instrument, the light is dispersed before it enters the sensor cell and thus virtually monochromatic light passes through the cell. However, if the incident light is of a wavelength that can excite the solute and cause fluorescence at another wavelength, then the light falling on the photo cell will contain the incident light together with any fluorescent light that may have been generated. It follows, that the light monitored by the photocell may not be monochromatic and light of another wavelength, if present, would impair the linear nature of the response. This effect would be negligible in most cases but with certain fluorescent materials the effect could be significant. The diode array detector operates quite a differently. Light of *all wavelengths* generated by the deuterium lamp is passed through the cell and then dispersed over an array of diodes. Thus, the absorption at discrete groups of wavelengths is continuously monitored at each diode. However, light falling on a discrete diode may not be derived solely from the incident light but may contain light generated by fluorescence excited by light of a shorter wavelength.

The ideal multichannel detector would be a combination of both the dispersion system and the diode array detector. This arrangement would allow a true monochromatic light beam to pass through the detector and then the transmitted beam would itself be dispersed again onto a diode array. Only that diode sensing the wavelength of the incident light would be used for monitoring the transmission. Under some circumstances, measurement of transmitted light may involve fluorescent light and the absorption spectrum obtained for a substance may be a degraded form of the true absorption curve. In this way any fluorescent light would strike other diodes, the true absorption would be measured and accurate monochromatic sensing could be obtained.

In a multichannel dispersive detector light from the deuterium lamp is collimated by two curved mirrors onto a holographic diffraction grating. The dispersed light is then focused by means of a curved mirror, onto a plane mirror and light of a specific wavelength is selected by appropriately positioning the angle of the plane mirror. Light of the selected wavelength is then focused by means of a lens through the flow cell. The exit beam from the cell is then focused by another lens onto a photocell, which gives a response that is some function of the intensity of the transmitted light. The detector is usually fitted with a scanning facility that allows the spectrum of the solute contained in the cell to be obtained. There is an inherent similarity between UV spectra of widely different types of compounds, and so UV spectra are not very reliable for the identification of most solutes.

A usual use of multichannel choice is to enhance the sensitivity of the detector by selecting a wavelength that is characteristically absorbed by the substance of interest. Conversely, a wavelength can be chosen that substances of little interest in the mixture do not adsorb and, thus, make the detector more specific to those substances that do.

Multichannel dispersive detectors provides adequate sensitivity, versatility and a linear response. But, it has mechanically operated wavelength selection and requires a stop/flow procedure to obtain spectra. In contrast, the *diode array detector* has the same advantages but none of these disadvantages.

Find some important multichannel detector on the list below.

- Photodiode Array (PDA)

  o   Semiconductors (Silicon and Germanium) (see Figure 3)

  ▪   Group IV elements

  ▪   Formation of holes (via thermal agitation/excitation)

  ▪   Doping

  n-type: Si (or Ge) doped with group V element (As, Sb) to add electrons.

  $$As: [Ar]4S^2\,3d^{10}\,4p^3$$

- p-type: Doped with group III element (In, Ga) to added holes

  In: $[Kr]5S^2\,4d^{10}\,5p^1$ (see Figure 4)

- coupled device (CCD)

- vidicon

Photodiode Array Detectors

A photodiode array is a linear array of several hundred light sensing diodes light ranging from 128 to 1024 – and even up to 4096 having a thousand phototubes, for every different wavelength. The design of this kind of machine is somewhat different and simpler. (Figures 2-4) Light passes through the sample first. Then it hits the monochromator, and then it is dispersed onto the photodiode array.

This multichannel detector makes an ideal sensor for an entire spectrum in a UV-VIS dispersive spectrophotometer. With that application, newer arrays have been made with adjacent diodes 25.6 mm long and spaced 25 mm on centers.

A polychromatic beam from the source is irradiated onto the inlet slit of the polychromator after passing through the sample compartment. The polychromator disperses the narrow band of the spectrum onto the diode array. The photodiode converts light into electrical signals and temporarily stores them. These signals are then read out as time-series signals via the output line by sequentially turning on the switch array connected to each photodiode with address pulses generated from the shift register.

A silicon photodiode consists of a reversed biased *pn* junction formed on a silicon chip. A photon promotes an electron from the valence bond (filled orbitals) to the conduction bond (unfilled orbitals) creating an electron(-) - hole(+) pair. The concentration of these electron-hole pairs is dependent on the amount of light striking the semiconductor. Spectral resolution limited by size of diode.

PDA detectors are useful in both research and quality assurance laboratories. In the research laboratory, the PDA provides the analyst with a variety of approaches to the analysis. In the quality assurance laboratory, the PDA provides several results from a single run, thereby increasing the throughput of the HPLC.

PDA detection offers the following advantages:

Peak Measurement at All Wavelengths

In methods development, detailed information about the detector conditions required for the analysis may not be known. When a variable wavelength detector is used, a sample must often be injected several times, with varying wavelengths, to ensure that all peaks are detected. When a PDA detector is used, a wavelength range can be programmed and all compounds that absorb within this range can be detected in a single run.

Determination of the Correct Wavelengths in One Run

After all peaks have been detected, the maximum absorbance wavelength for each peak can be determined. A PDA detector can collect spectra of each peak and calculate the absorbance maximum.

Detection of Multiple Wavelengths

A PDA detector can monitor a sample at more than one wavelength. This is especially useful when the wavelength maxima of the analytes are different. Wavelengths can be selected to analyze each compound at its highest sensitivity.

Peak Purity Analysis

It is difficult to determine component purity from a chromatogram. However, a PDA detector can analyze peak purity by comparing spectra within a peak. A pure peak has matching spectra throughout the peak (at all wavelengths).

Positive Peak Identification

In liquid chromatography, peak identification is usually based on relative retention times. When a PDA detector is used, spectra are automatically collected as each peak elutes. The PDA software compares the spectra with those stored in a library to determine the best fit matches; this method increases the likelihood of correctly identifying peaks.

Scan Spectrum Very Quickly

entire spectrum in <1 second

- Provides single beam.
- Powerful tool for studies of transient intermediates in moderately fast reactions.
- Useful for kinetic studies.
- Useful for qualitative and quantitative determination of the components exiting from a liquid chromatographic column.

In addition to above points, there are many major advantages of diode array detection. In the first, it allows for the best wavelength(s) to be selected for actual analysis. This is particularly important when no information is available on molar absorptivities at different wavelengths.

The second major advantage is related to the problem of peak purity. Often, the peak shape in itself does not reveal that it actually corresponds to two or even more components. In such a case, absorbance rationing at several wavelengths is particularly helpful in deciding whether the peak represents a single compound or, is in fact, a composite peak.

As already mentioned, a special feature of some variable wavelength UV detectors is the ability to perform spectroscopic scanning and precise absorbance readings at a variety of wavelengths while the peak is passing though the flow cell. Diode array adds a new dimension of analytical capability to liquid chromatography because it permits qualitative information to be obtained beyond simple identification by retention time.

In absorbance rationing, the absorbance is measured at two or more wavelengths and ratios are calculated for two selected wavelengths.

Simultaneous measurement at several wavelengths allows one to calculate the absorbance ratio. Evaluation can be carried out in two ways:

In the first case, the ratios at chosen wavelength are continuously monitored during the analysis: if the compound under the peak is pure, the response will be a square wave function (rectangle). If the response is not rectangle, the peak is not pure.

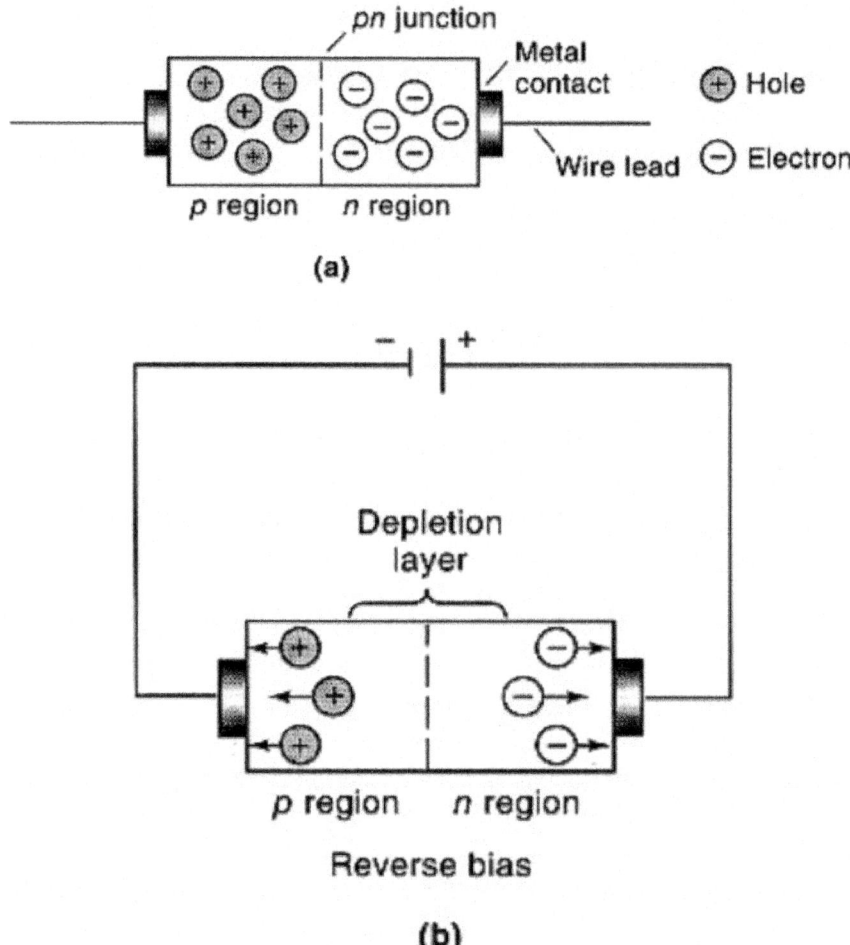

Figure 2. a) Schematic of a silicon diode, (b) Formation of depletion layer which prevents of flow of electricity under reverse bias [Skoog & Leary,1992].

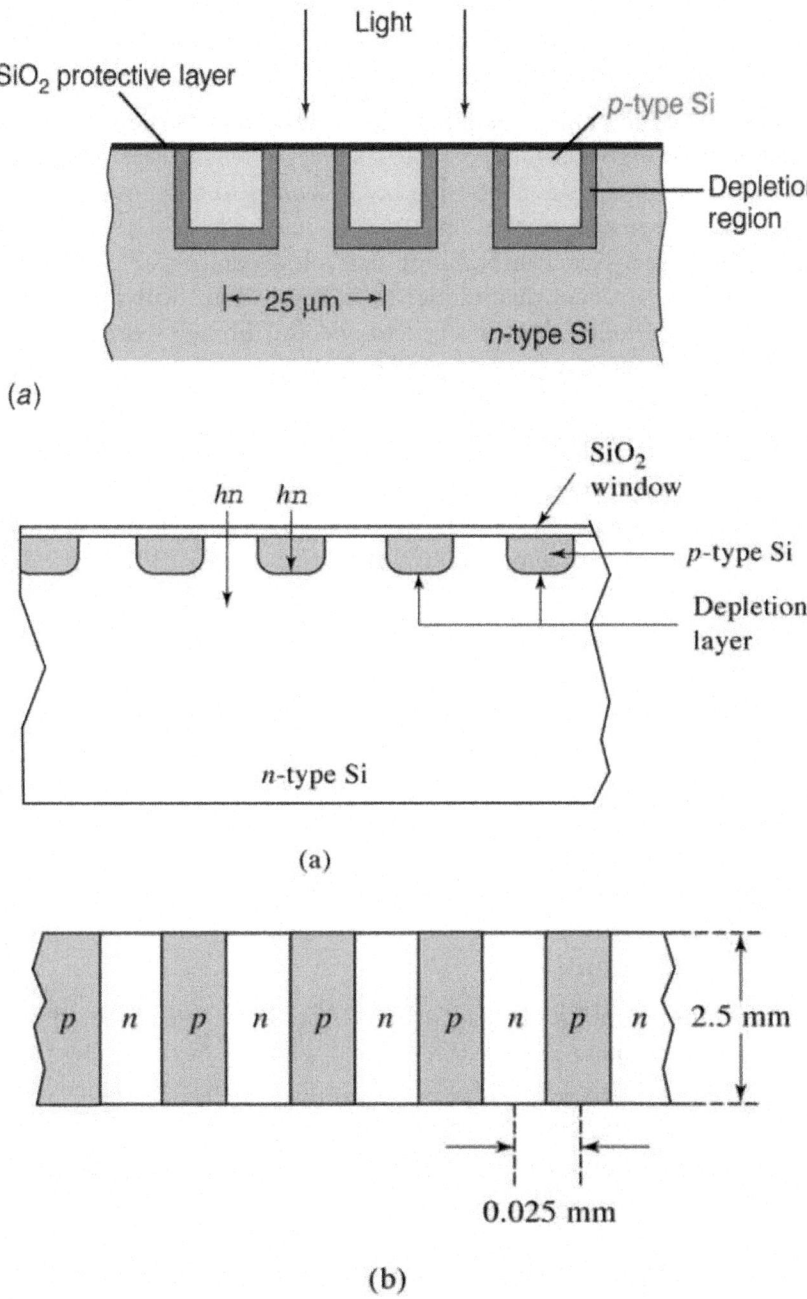

Figure 3. a) n-type and (b) p-type photodiode array.

Photodiode array (PDA) detectors scan a range of wavelengths every few milliseconds and continually generate spectral information. Wavelength, time, and absorbance can all be plotted.

In methods development, detailed information about the detector conditions required for the analysis may not be known. When a variable wavelength detector is used, a sample must often be injected several times, with varying wavelengths, to ensure that all peaks are detected. When PDA detectors provide three-dimensional information that allows an accurate assessment of peak identity, purity, and quantitation in a single run. Software support for PDA detectors includes peak purity and spectral library search functions to help determine peak homogeneity and identity.

Photodiode Array Applications

Spectrometers have developed in many ways since the introduction of simple spectrophotometers which were commercially available from the mid 1950's. Such improvements have enabled us to use PDA type UV-Vis. spectrophotometers.

The scope and performance of conventional single channel detector type UV-Vis spectrophotometers were found to be somewhat limited. This encouraged a search for novel techniques which could be applied to the development of UV-Vis. spectrophotometers.

Dispersed light is focused directly onto the detector array, saving considerable time and greatly reducing instrument complexity. The combination of dispersing element and detector array is employed in most spectrophotometers today.

UV-VIS Spectroscopy

The introduction of multichannel detectors such as the linear photodiode array (PDA), charge coupled device (CCD) and vidicon enabled new detection systems to be developed for UV-Vis spectrophotometers and encouraged the rapid development of polychromators from the 1970s [Talmi, 1975,1982].

As was expressed earlier a polychromator is an enhanced monochromator which it is accomplished by electronic scanning of the multichannel detector. Multichannel detectors such as the photodiode array, charge coupled device or vidicon are usually flat and are best used with a dispersing arrangement which yields a flat focal plane. Under optimum conditions, they can detect as many wavelengths simultaneously as their number of individual diodes, resolution elements or pixels. Stray light and background per element are negligible because they are arrays and they have very low dark currents.

PDA, on the other hand, is more suited for applications where the light level is relatively high. Because in the PDA the photon saturation charge is greater than CCD so the detection range of PDA is larger than CCD. Furthermore, PDA delivers lower noise than CCD. So it PDA was recommend in applications where higher output accuracy is needed.

This multichannel detector having numbers of elements ranging from 128 to 1024 and even up to 4096. It makes an ideal sensor for an entire spectrum in a UV-VIS dispersive spectrophotometer.

A polychromatic beam from the source is irradiated onto the inlet slit of the polychromator after passing through the sample compartment. The polychromator disperses the narrow band of the spectrum onto the diode array. The photodiode converts light into electrical signals and temporarily stores them. These signals are then read out as time-series signals (see Figure.4).

A spectrum for the whole wavelength range should be acquired for best results. The correlation between wavelengths and particular detector channels in a polychromator facilitates nearly simultaneous measurement of the intensities of the various wavelengths.

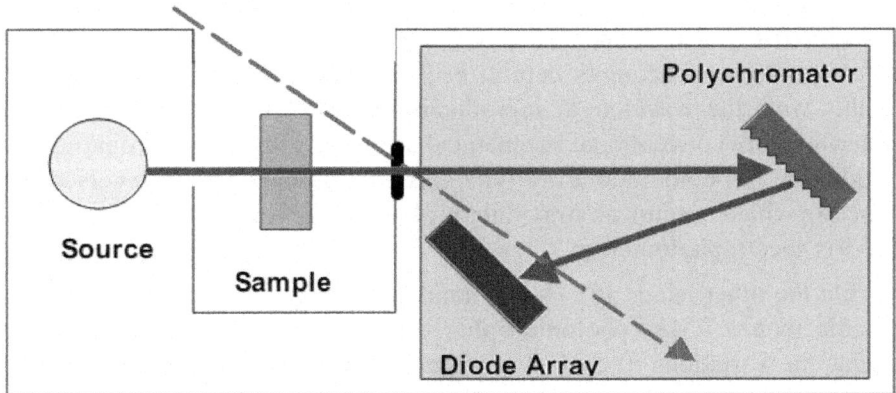

Figure 4. Schematic of a photodiode array spectrophotometer.

The conventional UV-Vis. spectrophotometer only has one detector. But data for many wavelengths can be acquired with the photodiode array spectrophotometer simultaneously since there are several hundred or a thousand detectors present. Fast spectral acquisition makes diode array spectrophometers the first choice for measurement of fast chemical reactions and kinetics study of materials.

The duration and intensity of illumination determine both the final S/N ratio and the exposure interval needed to acquire a spectrum. This interval is also the integration time for the signal. A longer integration time allows a higher S/N since the signal will be larger and noise averaged more completely towards zero.

There is no Integration function in the conventional UV-Vis spectrophotometer which accumulates the signal. For example, the total required time will be 1000 sec. for 1000 data points and it takes 1 sec. to measure one datum. In this case, all 1000 data have the same signal to noise ratio (S/N). But in a PDA instrument which has a 1000 photodiode array, 1000 data points can be measured in 1 sec. and it would take 1/1000 sec. to achieve the same result obtainable in 1 sec. in a conventional instrument. Therefore, when the same sample is measured for 1000 sec in a PDA instrument, the signal is accumulated and is 1000 times greater than when measuring for 1 sec. The noise will be 1000. This means that the S/N ratio is improved by 1000.

This resulting benefit of fast data acquisition is termed Felgett's Advantage or Multichannel Advantage.

In a conventional UV-Vis spectrophotometer mechanical movement is required to select a specific wavelength. But a photodiode array UV-Vis spectrophotometer acquires data at each wavelength by electrical scanning. In this way, the wavelength reproducibility of a PDA instrument is much better than the conventional mechanical scanning UV-Vis spectrophotometer. In addition, a photodiode array type spectrophotometer has a reversed optic structure which minimizes stray light problems, a serious issue in conventional UV-Vis spectrophotometers.

On the other hand, a PDA is a solid-state device and is more secure and reliable than a PMT (photomultiplier tube). Furthermore, a polychromator avoids the variations in optical performance with wavelength and time that are introduced in a scanning monochromator by moving the grating. Indeed, in a polychromator no mechanical movement is required except perhaps the opening of a shutter at the entrance slit.

The Spectroscopy methods which are used of PDA can be divided into 3 sections: mass spectrometry, atomic spectroscopy and molecular spectroscopy. The applications of PDA for all 3 sections have been growing steadily. UV-Vis, FT-IR, Fluorescence, Raman and NIR spectroscopy instruments are in the molecular group. UV-Vis, is the largest category in this section. UV-Vis spectroscopy finds applications not only in traditional chemistry but also in newer fields such as pharmaceuticals & life science, environment, agriculture, energy and the petrochemical Industry.

Photodiode Array and HPLC

The great importance of diode-array detection in HPLC can be characterized by the fact that this is solely the subject of an excellent book edited by Huber and George [Huber & George, 2003].

The most important advantage of the diode-array UV detector over conventional multiwavelength UV detectors is the speed of scanning the spectra. Using the reversed optics of the diode-array spectrophotometer enables all points in the spectrum to be measured simultaneously on the array of fixed photodiodes. The speed of scanning the spectrum is thus determined by the speed of data acquisition. In modern diode-array UV detectors equipped with powerful computers the time necessary to take the full spectrum from 190 to 600 nm can be reduced to as short as about 10 msec. This speed is more than sufficient in the overwhelming majority of cases in pharmaceutical analysis when the half-band width of peaks separated by HPLC is usually in the order of 1 min and it is only very rarely in the order of 1-10 sec in fast HPLC systems and especially in capillary electrophoresis where the peaks are in general narrower.

The quality of the UV spectrum of the separated impurities obtained by the diode-array detector is influenced by several of photodiodes. For example, the number of diodes in a DAD of the HPLC instrument is only 205 while in the other it is 1024. If the spectrum has fine structure, better quality spectra are obtainable with the latter. In addition to this the quality of the spectra of especially the low level impurities greatly depends on the baseline noise. This can be reduced by using a light source with high intensity, by selecting a suitable reference wavelength (which is as close to the cut-off wavelength of the separated analyte as possible and a suitable slit width. Generally speaking the sensitivity of the new generation of diode-array detectors is much higher than that of the older ones.

There are three main areas within drug impurity profiling where the advantages of diode-array detectors can contribute to the success of the HPLC (CE) analysis (see Figures 5-7).

Peak purity determination. The determination of peak homogeneity is an integral part of the protocol in the validation of any kind of HPLC (and CE) analysis of pharmaceuticals. In the course of impurity profiling studies it is especially important to check the peak of the main component for its homogeneity from the simple and most widely used absorbance ratio method [Drouen et al.,1984;Wilson et al.,1989] to more sophisticated deconvolution, spectral suppression, spectrum subtraction and other chemometric methods[Huber & George, 2003]. If any kind of peak in-homogeneity is found (impurity on the leading or tailing edges of the main peak or fused impurity peaks, conveniently demonstrated in the three-dimensional mode) the diode-array spectra themselves furnish further information for the identification of the unresolved impurities.

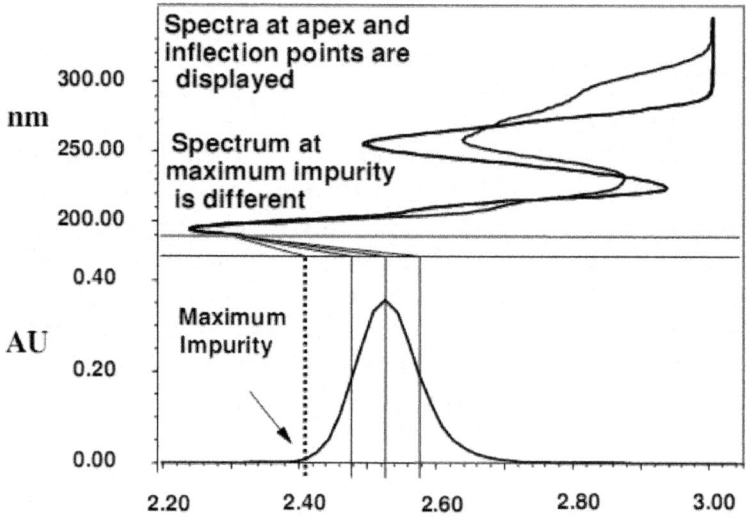

Figure 5. Peak purity measurement.

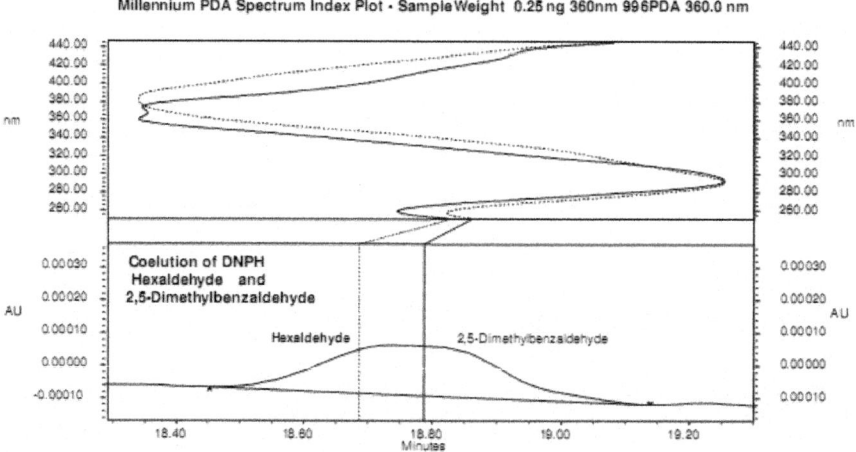

Figure 6. Maximum impurity detection.

Spectral matching. Matching the diode-array spectra of components separated by HPLC with those taken by computer search from spectral libraries is a widely used method [Huber & George, 2003] especially in toxicological analysis. This approach is of limited value in drug impurity profiling since it is unlikely that impurities of especially new drugs are included in spectrum libraries. However, matching the diode-array spectra of the separated impurities with standard materials can greatly support the identification of the impurities on the basis of retention matching.

Structure elucidation of the separated impurities. It is reasonable to begin the search for the structure unknown impurity separated by HPLC or CE with drawing as many conclusions from its diode-array UV spectrum as possible.

Peak Purity

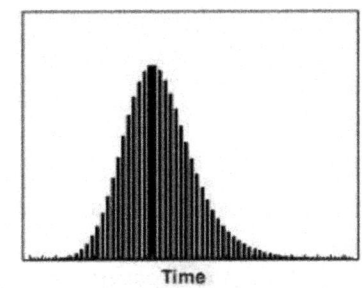

Peak Purity analyzes all spectra (minimum 15) within a peak against the apex spectrum of the peak itself

Spectral Matching

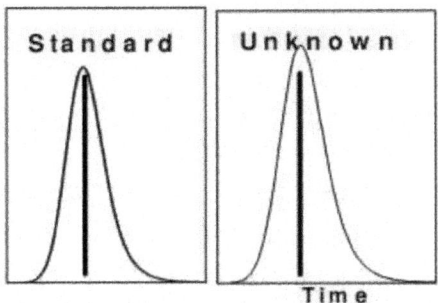

Spectral match of apex spectrum of the unknown against the apex spectrum of a standard, stored in a user's library.

Figure 7. Determination of peak purity.

The short-wavelength parts of the (diode-array) UV spectra can be subject of several distorting effects, moreover even false maxima can occur. In addition to this, short-wavelength UV bands can originate from different chromophoric functional groups and for this reason they are of limited value in the structure elucidation of organic compounds. As a consequence of these factors it is a prerequisite of drawing useful conclusions from the UV spectrum of an impurity that it should have at least one maximum above 210-220 nm.

Another limitation is that the difference between the structures of the drug material and the impurity should be at or near the chromophoric part of the molecule in order that the difference between their spectra can be of diagnostic value in the structure elucidation of the impurity. For example, the chromophoric group of various steroids is the 4-ene-3-oxo group with an absorption maximum around 240 nm. As it will be shown later, the position of this band is influenced by substituents in the B and C ring of the steroid nucleus but by no means by substituents at C-17. For this reason various esters of 17-hydroxy-4-ene-3-oxo steroids (testosterone, 19-nortestosterone, 17-hydroxyprogesterone, etc.) cannot be differentiated on the basis of their UV spectra.

HPLC with photodiode array detection (HPLC-PDA or HPLC-DAD) is regularly employed for substance identification in the context of Systematic Toxicological Analysis [Koves,1995; Gaillard & Pépin,1997; Herre & Pragst,1997]. With HPLC-PDA the most important parameters in identifying a compound are its retention time and its UV spectrum. Critics of the method often question the specificity of UV detection because of poorly structured spectra and broad absorption bands. Therefore a systematic investigation into the selectivity of PDA detection was carried out by analyzing large numbers of UV spectra with respect to their correlation with chemical structure.

For data analysis the following tools are needed:

- A spectra library ; the library is embedded into the chromatography software in a way that spectral similarity is compared nm by nm and a "hit list" is returned to the operator.
- A database of retention times and specific peak areas.
- A database of all molecular structures with an ability for substructure searches.
- A structural database of all registered chromophores.

As an alternative to Mass Spectrometers, absorbance detectors (including PDA) are much less expensive and relatively simple to use. LC-DAD is a fast and robust method for screening biological samples in conjunction with a library search algorithm to quickly identify those samples that require confirmatory testing. Numerous methods for using LC-PDA as a screening method have been published and were recently reviewed by Pragst et al. [Pragst et al.,2004]. Because a PDA detector can collect an entire spectrum at each time point in a chromatogram, the data are information rich and more selective than single wavelength chromatograms. Herzler et al. [Herzler et al.,2003] showed that PDA data could be used to selectively identify abused substances in spectrochromatograms based on comparison to a library of over 2500 "toxicologically relevant" substances. Their method relied on the calculation of a 'similarity index' (related to the correlation coefficient) to determine the similarity between a spectrum in an unknown chromatogram and a library spectrum. In addition to spectral matching, a relative retention time was also used to identify the substances of interest.

Medical Chemistry Applications of HPLC-PDA

High performance liquid chromatography (HPLC) with photodiode array detection has been proved to be the demanded method of systematic analysis for unknown drugs in biological sample because of separation efficiency, sensitivity, flexibility and identification potential. HPLC can be an easy way of quantitation as well. Ultraviolet spectra acquired with photodiode array detector together with retention data are used to identify unknown or suspected drugs and metabolites in various biological material. These analytical systems are suitable for toxicological examinations of forensic cases, acute poisonings, drug abuse. They are convenient to subsequent monitoring of serum drug levels during treatment of intoxication as well.

High-performance liquid chromatography coupled with diode array detection (HPLC-DAD) has been widely used as a powerful means for the analysis of multi-component medicines, which can provide a UV chromatogram and comprehensive data about the compounds in complex mixtures [Han et al.,2007; Su etal.,2010; Wei et al.,2010; Zhang et al.,2010]. This technology facilitates identification of unknown components in the matrices system remarkably with high sensitivity and accuracy.

Photodiode array (PDA) detectors record light absorption at different wavelengths and can provide spectra of the analytes. This is useful in identifying unknowns. Mass spectrometry (MS) is a better detector for unknowns. It gives an unambiguous molecular weight of an analyte and provides structural information. When coupled with CE or HPLC, MS can separate co-eluting analytes with different mass to charge ratios. But the Mass spectrometer is an expensive instrument and the possibility of using it is not available in all laboratories. Of course, if possible HPLC/ESI-MS/UV-DAD analysis gives the best sensitivity [Cuyckens& Claeys,2002; Beretta et al.,2009; Christiansen et al.,2011].

The potentials and limitations of high-performance liquid chromatography-photodiode array detection are highlighted in respect to its use in the analysis of different biological matrices followed by the identification of unknowns. The logical analytical approach used in clinical and forensic toxicology, vital for the identification of one or more toxic substances as a cause of intoxication, is largely based on both simple and fast "general unknown screening" methods which cover most relevant drugs and potentially hazardous chemicals. In this field of systematic toxicological analysis, a literature overview shows that HPLC can play a substantial role. Both column packing material and eluent composition have their impact on intra- and inter laboratory reproducibility. In view of the sometimes different retention characteristics of various HPLC columns, several possibilities are addressed to enhance the discriminating power of primary retention parameters. The advantages of photodiode array detection as compared to UV detection have been of paramount importance to the success of HPLC in toxicological analysis. Dedicated libraries with spectral information and searching software are powerful tools in the process of identification of an unknown substance. In the present section, these aspects are also verified in a number of real cases.

HPLC-DAD used as a general unknown screening tool should cover as many drugs and toxicants as possible, but should be also very selective, sensitive and reliable. Liquid chromatography is used in forensic laboratories for numerous applications including examination of drugs. LC with photodiode array detection (PDA) is a hybrid technique which can provide complete UV-visible spectral information on a given peak in a chromatogram, enabling determinations of peak purity to be made, and identification of unknown peaks to be assigned by library searches of spectral information in combination with retention behavior. These are valuable features normally associated with gas chromatography-mass spectrometry. The additional information available on each peak makes LC-PDA a particularly attractive technique for the forensic laboratory where higher levels of certainty are often demanded in test results. This paper reviews some of those applications for LC-PDA in the forensic sciences, including drug screening, drug and pharmaceutical analysis, idenfication of pesticides, fungi, quality control testing and profiling of cosmetics, street drugs and profiling of other complex mixtures. The practical and technical limitations of the technique are explored and its place in the hierarchy of methods available in forensic laboratories is evaluated [Proença et al.,2003; Madej et al.,2003; Proenc et al.,2004; Nieddu et al., 2007; Es'haghi et al.,2010; ; Vosough et al.,2010].

HPLC-DAD offers many advantages in terms of specificity, sensitivity, speed and ruggedness. The data produced, comprising both retention behavior and absorption spectra of eluting chemical entities, result in an identification power at low cost and with widened availability through many laboratories. In addition, the examples showed a great versatility in application fields and excellent quantitative potential. The fast progress in DAD detector technology, computer and software power and HPLC packing material quality have led to an exponential rise of the number of reports on the use of HPLC-DAD. The advent of routine use of HPLC-MS will probably promote HPLC as a viable if not better alternative to GC-MS.

We examined that combined with a sample preparation method; HPLC-PDA can be easy achieved to very low detection limits [Es'haghi et al., 2009, 2010]. In a research, we used of direct suspended droplet microextraction (DSDME) method, based on a three-phase extraction system which is compatible with HPLC-PDA for determination of ecstasy; MDMA (3,4methylendioxy-N-methylamphetamine) in human hair samples. After the extraction, pre-concentrated analyte was directly introduced into HPLC for further analysis. In concentration range between 1.0 and 15,000 ng mL^{-1} calibration curve is drowned. Linearity was observed with r = 0.9921 for analyte. Limit of detection (LOD) were calculated as the minimum concentration providing chromatographic signals three times higher than background noise. Limit of quantification (LOQ) was estimated as the minimum concentration preparing chromatographic signals ten times higher than background noise. Thus, LOD obtained was 0.1 and LOQ was 1.0 ng mL^{-1} too [Es'haghi et al., 2010].

In the other work we successfully used of DSDME method combined with HPLC-PDA for determination of low-residue benzodiazepine, diazepam and lorazepam, in the environmental water samples [Es'haghi et al., 2009, 2009]. After the optimized extraction conditions, the suspended micro-droplet is withdrawn by a HPLC microsyringe, injected to and analyzed by HPLC-DAD. Method was evaluated and enrichment factor 839.8, linearity range from 25 to 5000 ng mL^{-1} with an average of relative standard deviation (n=5) 5.62% for diazepam using a photodiode array detector were determined. HPLC-PDA has good matches with complex matrices such as hair.

A method combining liquid–liquid–liquid microextraction and automated movement of the acceptor and donor phases (LLLME/AMADP) with ion-pair HPLC/DAD has been developed to detect trace levels of chlorophenols in water [Lin etal.,2008]. The extracted chlorophenols, present in anionic form, were then separated, identified, and quantitated by ion-pair high-performance liquid chromatography with photodiode array detection (HPLC/DAD). For trace chlorophenol determination using HPLC/DAD, the chlorophenolate anion provides a better ultraviolet spectrum for quantitative and qualitative analyses than does uncharged chlorophenol. The proposed method was capable of identifying and quantitating each analyte to 0.5 ng mL^{-1}, confirming the HPLC/DAD technique to be quite robust for monitoring trace levels of chlorophenols in water samples.

HPLC/DAD could simultaneously detect UV absorptions at multiple wavelengths and extract the UV spectra of separated analytes in a chromatogram. Absorbance measurements at the band maxima of UV spectra obey the linear Beer's law more accurately than measurements off the band maxima, and UV spectra of the separated analytes can be utilized to identify target analytes in HPLC/DAD. Accordingly, each extracted chlorophenolate anion after ion-pair liquid chromatography separation was quantitated by the maximum adsorption of its own red shift characteristic band, and each target chlorophenlate anion was identified by its own red shift characteristic band as well as its enhanced B band. The chlorophenols were determined under selected experimental conditions to assess repeatability, linearity, coefficient of determination, and detection limit.

A HPLC-DAD method for drug screening in plasma were developed by M. A. Alabdalla [Alabdalla,2005]. This analytical method extracted and tested a number of drugs of different classes. The method included; an acidic and basic Solid Phase Extraction (SPE) of plasma with C18 cartridges, a gradient elution of a modified cyano column with acidic buffer/acetonitrile eluent and a photodiode array ultraviolet (UV) detection. The drug screening procedure applied used retention index and UV spectral data for the identification of compounds, may be appropriate in particular laboratory settings.

Continuous administration of polyphenols from aqueous rooibos (Aspalathus linearis) extract ameliorates dietary-induced metabolic disturbances in hyperlipidemic mice was studied by HPLC-DAD and introduced by R. Beltrán-Debón et al. [Beltrán-Debón et al., 2011]. In this biological matrices and they could find good results.

In a recent study neurons from the olfactory system of the fish crucian carp, Carassius carassius L. were used as components in an in-line neurophysiologic detector (NPD) to measure physiological activities following the separation of substances by high-performance liquid chromatography (HPLC). The skin of crucian carp, C. carassius L. contains pheromones that induce an alarm reaction in conspecifics. Extra-cellular recordings were made from neurons situated in the posterior part of the medial region of the olfactory bulb known to mediate this alarm reaction. The nervous activity of these specific neurons in the olfactory bulb of crucian carp was used as an in-line neurophysiologic detector. HPLC was performed with a diode array detector (DAD) [Brondz et al.,2004]. UV spectral detection was performed at 214, 254 and 345 nm, and scans (190–400 nm) were collected continuously. This system enabled the selection of peaks in the chromatogram with fish alarm pheromone activity. Neurophysiologic detectors (NPDs) in-line with diode array detectors (DADs) are able to provide the physiologically active substances and their spectral characteristics.

Li-wei Yang et al. were developed a method using high-performance liquid chromatography–photodiode array detection (HPLC–DAD) for the quality control of Hypericum japonicum thunb (Tianjihuang), a Chinese herbal medicine. For the first time, the feasibility and advantages of employing chromatographic fingerprint were investigated for the evaluation of Tianjihuang by systematically comparing chromatograms with a professional analytical. The results revealed that the chromatographic fingerprint combining similarity evaluation could efficiently identify and distinguish raw herbs of Tianjihuang from different sources. The effects resulted from collecting locations; harvesting time and storage time on herbal chromatographic fingerprints were also examined [Yang et al.,2005].

Photo Diode Array Detector in Kinetic Study

In kinetic experiments, transient optical absorption is recorded versus time to evaluate rate constants related to the species under investigation. In addition, the recording of a spectrum sometimes becomes necessary in order to identify the species. In most cases, the spectrum is constructed from point-to-point recordings of kinetic curves at selected wavelengths. This procedure is time consuming, and becomes boring especially at long recording times in the second and minute time domain. The use of a device, which enables the recording of a complete spectrum, can be very helpful as it reduces experiment time remarkably. Unwanted side effects, such as photolysis during long recording times, can also be prevented. The application of optical multichannel analyzers which use either a linear charge coupled device (CCD) or a linear photodiode array (PDA) in kinetic experiments was reported by some laboratories [Hunter et al.,1985; Sedlmair et al.,1986; Johnson et al.,1994]. The advantage of using such a detector is the ability to immediately record a complete spectrum from UV to IR with one measurement. The PDA detector has the ability to record a spectrum over a large range of wavelengths. The uniformity of the analyzing light intensity over the whole range is important because the dynamics and the sensitivity of the measurements depend largely on the intensity. The spectral distribution of the analyzing light, as recorded by the multichannel detector is shown in Figure.8.

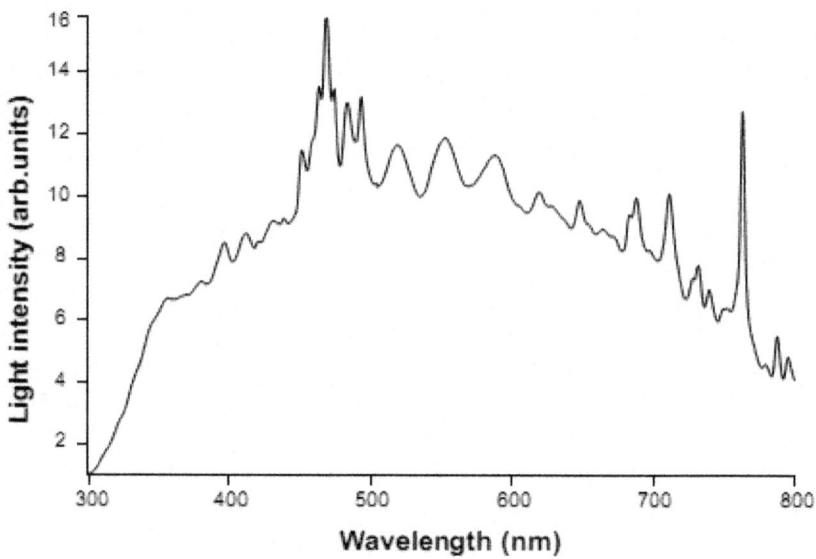

Figure 8. Light intensity vs. wavelength of an xenon lamp, recorded by the multichannel detector.

The source of the analyzing light is an xenon lamp. The light intensity is attenuated tenfold as compared to kinetic experiments. Although, the recorded intensity of the analyzing light decreases drastically below 350 nm, a spectral range from 300 to 800 nm can be covered. Below 300 nm, recording should be accomplished in small segments and with the help of band-pass filters in order to adjust for the reduced level of analyzing light and for the decreased sensitivity of the detector, and, in addition, to avoid scattered light effects. The measurement depends largely on proper focusing of the light path, i.e., how well the lamp arc is imaged onto the diode array.

Each spectrum is the average of some (for example five) individual measurements; each irradiation consists of a train of ten pulses. The interval between the recordings of the individual spectra or between the pulses in each pulse train was set to zero. The recording at time zero, i.e. before irradiation, shows a straight line. The change in absorption increases with increasing irradiation. In general, kinetic trace scan be constructed from the recorded spectra at selected wavelengths. Similar to the construction of spectra from kinetic traces [Janata,1994].

At measurements in the UV region, Cerenkov emission is a common problem at short measuring times. The intensity of the Cerenkov emission increases with decreasing wavelength and can be much larger than the kinetic signal itself, but probably will not exceed the intensity of the analyzing light. Although this apparatus makes data at longer time scale available, overdriving of the photodiodes and long recovery times are conceivable.

The use of an optical multichannel detector consisting of a linear diode array embedded in the instrumentation for kinetic spectroscopy, as well as the highlights of the computer program used for controlling the gathering and the evaluation of data are described. Complete spectra can be recorded and irradiation can be triggered according to a preset timetable. Due to the read-out time of the photodiode array and the time required by the computer to control the experiment, this apparatus is suitable for application starting in the millisecond time domain and extending up to very long time periods.

Chemometrics Investigations Using Photo Diode Array Detection

Chemometrics is a statistical approach to the interpretation of patterns in multivariate data. When used to analyze instrument data, chemometrics often results in a faster and more precise Assessment of composition of a product or even physical or sensory properties. For example, composition of drugs can be quickly measured using LC and chemometrics. Food properties can also be monitored on a continuous basis. In all cases, the data patterns are used to develop a model with the goal of predicting quality parameters for future data. The two general applications of chemometrics technology to predict a property of interest; and to classify the sample into one of several categories (e.g., good versus bad, Type A versus Type B versus Type C etc.). Chemometrics can be used to speed methods development and make routine the use of statistical models for data analysis. Keeping in view of the complexity of the chromatographic fingerprint and the irreproducibility of chromatographic and spectral instruments and experimental conditions, several chemometric approaches such as variance analysis, peak alignment, correlation analysis and pattern recognition were employed to deal with the chromatographic fingerprint. Many mathematical algorithms are used for data processing in chemometric approaches. The basic principles for this approach are variation determination of common peaks/regions and similarity comparison with similarity index and linear correlation coefficient. Similarity index and linear correlation coefficient can be used to compare common pattern of the chromatographic fingerprints obtained. In general, the mean or median of the chromatographic fingerprints under study is taken as the target and both are considered to be reliable [Brereton,1987].

The rapid scanning detectors, as diode array detection, present an alternative technology for rapid, multi-wavelength detection in HPLC. If hyphenated chromatography is further combined with chemometric approaches, clear pictures might be developed for chromatographic fingerprints obtained. A chemical fingerprint obtained by hyphenated chromatography, out of question, will become the primary tool for quality control of medicines.

The full UV-Vis spectrum became accessible as a three-dimensional (3D) data matrix (A, A, t). Data are available in the time, concentration and wavelength domains. This allows the simultaneous use of more than two wavelengths for detection or for the full application of detector information to the analytical problem by means of available chemometric techniques to data from second-order bilinear instruments, as chromatographic and excitation-emission data.

As an alternative to MS, absorbance detectors (including PDA) are much less expensive and relatively simple to use. LC-DAD is a fast and robust method for screening biological samples in conjunction with a library search algorithm to quickly identify those samples that require confirmatory testing. Numerous methods for using LC-DAD as a screening method have been published and were recently reviewed by Pragst et al. [Pragst et al., 2004]. Because a DAD can collect an entire spectrum at each time point in a chromatogram, the resultant data are information rich and more selective than single wavelength chromatograms.

For the above reasons could be adopted PDA detectors with the various chemometric methods to match spectra contained within a spectrochromatogram to a library.

In a research, triply coupled diode array detection high performance liquid chromatography mass spectroscopy was applied to a complex mixture of at least eight chlorophyll degradation products. Derivatives were employed to determine parts of the chromatogram of composition one. Mass selection was performed on the mass spectroscopic data. Principal components analysis was performed on both the raw and simultaneously normalised/standardised data; three dimensional projections of the data were obtained and compared to conventional two dimensional graphs. Angular plots between diode array loadings characteristic of individual compounds and scores of the diode array data were described. In mass spectra, angular plots between loadings characteristic of individual compounds and the remaining diagnostic masses revealed further mass spectral structure [Zissis et al.,1999].

Liquid chromatography–chemometric methods [LC-Partial least squares (LC-PLS), LC-principle component regression (LC-PCR) and LC-artificial neural network (LC-ANN)] were developed for the determination of anomalin (ANO) and deltoin (DEL) in the root by Alev Tosun et al.[Tosun et al.,2007].

Firstly, chemometric conditions were optimized by testing different mobile phases at various proportions of solvents with various flow rates in different wavelengths by using a normal phase column to obtain the best separation and recovery results. As a result, a mobile phase consisting of n-hexane and ethyl acetate (75:25 v/v) at a constant flow rate of 0.8 mL min $^{-1}$ on the at ambient temperature were found to be the optimal chromatographic conditions for good separation and determination of ANO and DEL in samples. Multi-chromatograms for the concentration set containing ANO and DEL compounds in the concentration range of 50–400 ng mL^{-1} were obtained by using a diode array detector (DAD) system at selected wavelength sets, 300 (A), 310 (B), 320 (C), 330 (D) and 340 (E). Three LC-chemometric approaches were applied to the multichromatographic data to construct chemometric calibrations. As an alternative method, traditional LC at single wavelength was used for the analysis of the related compounds in the plant extracts. All of the methods were validated by analyzing various synthetic ANO–DEL mixtures. After the above step, traditional and chemometric LC methods were applied to the real samples consisting of extracts from roots and aerial parts of analytes.

In a recent research, metabolism disorders in Kunming mice induced by two tumor cells were characterized. Metabolic fingerprint based on high performance liquid chromatography-diode array detector (HPLC-DAD) was developed to map the disturbed metabolic responses. Based on 27 common peaks, principal component analysis (PCA) and partial least squares-discriminant analysis (PLS-DA) were used to distinguish the abnormal from control and to find significant endogenous compounds which have significant contributions to classification. The tumor growth inhibition ratios of Taxol groups were used to validate the predictive accuracies of the PLS-DA models. The predictive accuracies of PLS-DA models for tumors model groups were 97.6 and 100%, respectively. Nine and seven of two models tumors were discovered, including uric acid and cytidine. In addition, the correlations between relative tumor weights and chromatographic data were significant ($p<0.05$). Investigations on the stability and precision of the established metabolic fingerprints demonstrate that the experiment is well controlled and reliable. This work was shown that the platform of HPLC-DAD coupled with chemometric methods provides a promising method for the study of metabolism disorders [Sun et al., 2011].

REFERENCES

1.   M. A. Alabdalla, 2005HPLC-DAD for analysis of different classes of drugs in plasma,J. Clinical Forensic Med., 12 6December), 310 315 1353-1131

2. R. Beltrán-Debón, A. Rull, F. Rodríguez-Sanabria, I. Iswaldi, M. Herranz-López, G. Aragonès, J. Camps, C. Alonso-Villaverde, J. A. Menéndez, V. Micol, A. Segura-Carretero, G. Aragonès, J. Joven, 2011Continuous administration of polyphenols from aqueous rooibos (Aspalathus linearis) extract ameliorates dietary-induced metabolic disturbances in hyperlipidemic mice, Phytomedicine,Article in press,0944-7113

3. G. Beretta, R. Artali, E. Caneva, S. Orlandini, M. Centini, R. M. Facino, 2009Quinoline alkaloids in honey: Further analytical (HPLC-DAD-ESI-MS, multidimensional diffusion-ordered NMR spectroscopy), theoretical and chemometric studies, J. Pharm. Biomed. Anal., 3October), 432 439 0731-7085

4. E. Bothe, E. Janata, 1994Instrumentation of kinetic spectroscopy-12. Software for data acquisition in kinetic experiments, Radiat. Phys. Chem. 44 4November), 449 454 0096-9806X

5. R. G. Brereton, 1987Chemometrics in analytical chemistry :a review, Analyst, 112 12December), 1635 1657 0003-2654

6. I. Brondz, E. H. Hamdani, K. Døving, 2004Neurophysiologic detector- A selective and sensitive tool in high-performance liquid chromatography, J.Chromatogr. B, 800 1-2February),41 47 1570-0232

7. A. Christiansen, T. Backensfeld, S. Kühn, W. Weitschies, 2011Investigating the stability of the nonionic surfactants tocopheryl polyethylene glycol succinate and sucrose laurate by HPLC-MS, DAD, and CAD,J.Pharm. Sci, 100 5May),1773 1782 0022-3549

8. F Cuyckens, M Claeys, (2002), Optimization of a liquid chromatography method based on simultaneous electrospray ionization mass spectrometric and ultraviolet photodiode array detection for analysis of flavonoid glycosides, Rapid. Comm. Mass. Spectrom., 0951-4198 24 16 2341 2348

9. A. C. J. H. Drouen, H. A. H. Billiet, L. De Galan, 1984Dual-wavelength absorbance ratio for solute recognition in liquid chromatography, Anal. Chem., 56 6May), 971 978 0003-2700

10. Z. Es'Haghi, S. Bandegi, L. Daneshvar, P. Salari, 2009Analysis of diazepam residue from water samples by triple phase-suspended droplet microextraction coupled to high performance liquid chromatography and diode array detection. Asian J. Chem., 21 8October), 6392 6402 0970-7077

11. Z. Es'Haghi, L. Daneshvar, P. Salari, S. Bandegi, 2009Determination of low-residue benzodiazepine, lorazepam, in the environmental water samples by suspended droplet icroextraction and high performance liquid hromatography-diod array detector. Chemija, 20 3nd), 181 186

0235-7216

12. Z. Es'haghi, M. Mohtaji, M. Hasanzade-Meidani, M. Masrournia, 2010The measurement of ecstasy in human hair by triple phase directly suspended droplet microextraction prior to HPLC-DAD analysis, J. Chromatogr. B, 878 1April),903 908 1570-0232

13. Y. Gaillard, G. Pépin, 1997Use of high-performance liquid chromatography with photodio de-array UV detection for the creation of a 600-compound library. Application to forensic toxicology, J. Chromatogr. A, 1-2February), 149 163 0021-9673

14. J. Han, M. Ye, H. Guo, M. Yang, B.-r. Wang, D.-a. Guo, 2007Analysis of multiple constituents in a Chinese herbal preparation Shuang-Huang-Lian oral liquid by HPLC-DAD-ESI-MS, J. Pharm. Biomed. Anal. 44 2June), 430 438 0731-7085

15. S. Herre, F. Pragst, 1997Shift of the high performance liquid chromatographic retention ti mes of metabolites in relation to the original drug at an RP-8 column with acidic mobile phase, J. Chromatogr. B, 692 1April), 111 126 1572-6495

16. M. Herzler, S. Herre, F. Pragst, 2003Selectivity of substance identification by HPLC-DAD in toxicological analysis using a UV spectra library of 2682 compounds, J. Anal. Toxicol. 27 4May),233 242 0146-4760

17. L. . Huber, S. A. George, 2003 Diode Array Detection in HPLC (Second edition), Marcel Dekker,, 082474York

18. E.P.L. Hunter, M.G. Simic, B.D. Michael, (1985), Use of an optical multichannel analyzer for recording absorption spectra of short-lived transients, Rev. Sci. Instrum.0034-6748 12 56 2199 2204

19. J. B. Johnson, G. Edwards, M. Mendenhall, 1994A low-cost, high-performance array detector for spectroscopy based on a charged-coupled photodiode, Rev. Sci. Instrum., 65 5 1782 1783 0034-6748

20. E. M. Koves, 1995Use of high-performance liquid chromatography-diode array detection in forensic toxicology, J. Chromatogr. A, 692 1-2nd), 103 119 0021-9673

21. W. E. Lambert, J. F. Van -Bocxlaer, A. P. . De Leenheer, 1997Potential of high-performance liquid chromatography with photodiode array detection in forensic toxicology, J. Chromatogr. B, 1February), 45 53 1572-6495

22. C. Lin, Y. , S.D Huang, 2008Application of liquid-liquid-liquid microextraction and ion-pair liquid chromatography coupled with photodiode array detection for the determination of chlorophenols in

water, J. Chromatogr. A, 1193 1-2June), 79 84 0021-9673

23. K. Madej, A. Parczewski, M. Kała, 2003HPLC/DAD Screening Method for Selected Psychotropic Drugs in Blood, Toxicol. Mech. Meth., 13 2April),121 127 1537-6524

24. M. Nieddu, G. Boatto, D. Serra, A. Soro, S. Lorenzoni, F. Lubinu, 2007HPLC-DAD determination of mepivacaine in cerebrospinal fluid from a fatal case. J. Forensic Sci., 52 5September), 1223 1224 0022-1198

25. F. Pragst, M. Herzler, B. Erxleben, T. , 2004Systematic toxicological analysis by high-performance liquid chromatography with diode array detection (HPLC-DAD), Clin. Chem. Lab. Med. 42 11nd), 1325 1340 1434-6621

26. P. Proença, Marques. E. Pinho, H. Teixeira, F. Castanheira, M. Barroso, S. Ávila, D. N. Vieira, 2003A fatal forensic intoxication with fenarimol: Analysis by HPLC/DAD/MSD, Forensic Sci. Int., 133 1-2April),95 100 0379-0738

27. P. Proença, H. Teixeira, J. Pinheiro, E. P. Marques, D. N. Vieira, 2004Forensic intoxication with clobazam: HPLC/DAD/MSD analysis, Forensic Sci. Int., 143 2-3July), 205 209 0379-0738

28. J. Sedlmair, S. G. Ballard, D. C. Mauzerall, 1986Diode-array spectrometer (DAPS) for visible and near-IR absorption measurements with 10-ns time resolution, Rev. Sci. Instrum, 57 12nd), 2995 3003 0034-6748

29. D. A. . Skoog, J. J. Leary, 1992 Principles of Instrumental Analysis (third edition), Saunders College Publishing, 4-83370-282-7Worth

30. S. Su, J. Guo, J.-a. Duan, T. Wang, D. Qian, E. Shang, Y. Tang, 2010Ultra-performance liquid chromatography-tandem mass spectrometry analysis of the bioactive components and their metabolites of Shaofu Zhuyu decoction active extract in rat plasma, J. Chromatogr. B, 3-4February), 355 362 1570-0232

31. X. Sun, Y. Liu, D. Di , G. Wu, H. Guo, 2011Chemometric analysis of metabolism disorders in blood plasma of S180 and H22 tumor-bearing mice by high performance liquid chromatography-diode array detection, J.Chemometrics, Article in press, DOI:cem.1387, SSN:1099 128X

32. Y. Talmi, 1975Applicability of TV-type multichannel detectors to spectroscopy, Anal.Chem., 47 7nd), 658A 670A 0003-2700

33. Y. Talmi, 1982Spectrophotometry and spectrofluorometry with the self-scanned photodiode array, App. Spectrosc., 36 1January), 1 19 0003-7028

34. A. Tosun, Ö. Bahadir, E. Dinç, 2007Determination of anomalin and deltoin in Seseli resinosum by LC combined with chemometric methods,

Chromatographia, 66 9-10November),677 683 0009-5893

35. M. Vosough, M. Bayat, A. Salemi, 2010Matrix-free analysis of aflatoxins in pistachio nuts using parallel factor modeling of liquid chromatography diode-array detection data, Anal.Chim.Acta, 663 1March),11 18 0003-2670

36. H. Wei, L. Sun, Z. Tai, S. Gao, W. Xu, W. Chen, 2010A simple and sensitive HPLC method for the simultaneous determination of eight bioactive components and fingerprint analysis of Schisandra sphenanthera, Anal. Chim. Acta, 662 1March),97 104 0003-2670

37. T. D. Wilson, W. F. Trompeter, H. F. Gartelman, 1989Analysis of barbiturate mixtures using HPLC with diode array detection, J. Liq. Chromatogr. 12 7nd), 1231 1251 0148-3919

38. L. Yang, W. , D. Wu, H. , X. Tang, W. Peng, X. Wang, R. , Y. , W. Su, W. , 2005Fingerprint quality control of Tianjihuang by high-performance liquid chromatography-photodiode array detection,, J.Chromatogr. A, 1070 1-2April),35 42 0021-9673

39. W. Zhang, M. W. Saif, G. E. Dutschman, X. Li, W. Lam, S. Bussom, Z. Jiang, Y. Cheng, C. , 2010Identification of chemicals and their metabolites from PHY906, a Chinese medicine formulation, in the plasma of a patient treated with irinotecan and PHY906 using liquid chromatography/tandem mass spectrometry (LC/MS/MS, J. Chromatogr. A, 1217 37September), 5785 5793 0021-9673

40. K. D. Zissis, S. Dunkerley, R. G. Brereton, 1999Chemometric techniques for exploring complex chromatograms: application of diode array detection highperformance liquid chromatography electrospray ionization mass spectrometry to chlorophyll a allomers, Analyst, 7nd), 971 979 0003-2654

Chapter 6

QUANTITATIVE CHEMICAL DEFENSE TRAITS, LITTER DECOMPOSITION AND FOREST ECOSYSTEM FUNCTIONING

Mohammed Mahabubur Rahman[1,2] and Rahman Md. Motiur[3]

[1]United Graduate School of Agricultural Science, Ehime University, Matsuyama, Ehime, Japan

[2]Education and Research Center for Subtropical Field Science, Faculty of Agriculture, Kochi University, Kochi, Japan

[3]Silvacom Ltd., Edmonton, Canada

INTRODUCTION

In forest ecosystems, litter decomposition, which plays a critical role in nutrient cycling, is influenced by a number of biotic and abiotic factors, including quantitative chemical defense (Ross et al., 2002), environmental conditions such as soil properties and climate (Badre et al., 1998, Vanderbilt et al., 2008), and decomposer community and its complex nature. However, Lavelle et al., (1993) proposed a hierarchical model for the factors controlling litter decomposition. The levels of the hierarchal model are: climate (temperature and moisture)> physical properties of soil (clay and nutrients)>litter quality> macro and microorganisms (Lavelle et al., 1993). Leaf litter quality, which is inherited from living leaves, has repeatedly been emphasized as one of the most important factors controlling the decomposition process (Swift et al., 1979, Melillo et al., 1982, Osono & Takeda 2005). Decomposition rate may be decreased with latitude and lignin content of litter but increased with temperature, precipitation and nutrient concentrations at the large spatial scale (Zhang et al., 2008).

Bernhard-Reversa & Loumeto (2002) mentioned that litter fall serves three main functions in the ecosystem such as energy input for soil microflora and fauna, nutrient input for plant nutrition, and material input for soil organic matter building up. The first two functions are completed through decomposition and mineralization, and the third one through decomposition and humification.

Litter decomposition is a primary source of soil nutrients such as nitrogen and phosphorus, which are often limiting to plant growth in terrestrial ecosystems. The litter is broken down by insects, worms, fungi and microorganisms, organically-bound nutrients are released as free ions to the soil solution which are then available for uptake by plants. The variation in soil carbon and nutrient cycling has been clearly linked to variation in particular aspects of litter chemistry. For example, net N mineralization rates in monocultures of different grass species were correlated with root lignin content, suggesting that substrate chemistry is an important control over mineralization and/or immobilization processes (Hobby & Gough 2004).

De Santo et al., (2009) revealed that litter decomposition rates at the boreal forest were significantly lower than at the temperate one and did not differ between needle litter and leaf litter. In the boreal forest mass-loss was positively correlated with the nutrient release. In this site, Mn concentration at the start of the late stage was positively correlated with lignin decay and Ca concentration was negatively correlated to litter mass loss and lignin decay. In the temperate forest neither lignin, N, Mn, and Ca concentration at the start of the late stage, nor their dynamics were related to litter decomposition rate and lignin decay (Santo et al., 2009). On the other hand, litter quality had stronger effects on decomposition than the temperature in temperate forest (Rouifed et al., 2010). In arid and semi-arid sites, photodegradation could be an influential factor for litter decomposition (Austin & Vivanco 2006). Powers et al., (2009) used a short-term litter bag experiment to quantify the effects of litter quality, placement and mesofaunal exclusion on decomposition in 23 tropical forests in 14 countries. They concluded that decomposition in tropical forest is controlled by soil fauna and litter chemistry, which would vary with the precipitation regime.

Zhang et al., (2008) stressed that the combination of total nutrient elements and C:N accounted for 70.2% of the variation in the litter decomposition rates. On the other hand, the combination of latitude, mean annual temperature, C: N and total nutrient accounted for 87.54% of the variation in the litter decomposition rates. They also indicated that litter quality is the most important direct regulator of litter decomposition at the global scale.

The management of scheme options can influence vegetation and wildlife value for delivering ecosystem services by modifying the composition of floral and faunal communities (Smith et al., 2009). Most European temperate forests have been managed according to "classical" sustainable yield principles for a very long time. In France, vast areas of deciduous forests have been cultivated on short rotations for the production of fuelwood under so-called "low forest" regimes. Many of these forests are now being converted back to high forest management (FAO, 2011). Clear felling is common in harvesting operations in mature softwood stands in Northern America. Thinning is a very common and recommended management practice to manage the plantation and forest stands in general (Blanco et al., 2011). Intensive forest management, which may include operations such as: site preparation; tree planting; tending; thinning; and fertilizer application is often influence the litter decomposition process. Litter decomposition in unmanaged systems is affected mainly by climatic variables (Aerts 1997, Blanco et al., 2011).

However, for better understanding of litter decomposition process and forest ecosystem functioning, we need to know more detail about quantitative chemical defense and their effects on the litter decomposition process which influence the ecosystem functioning. This paper will discuss the quantitative defense traits of forest litter, their effects on litter decomposition and ecosystem functioning.

QUANTITATIVE CHEMICAL DEFENSE TRAITS OF LITTER

Quantitative chemicals are those that are present in high concentration in plants (5 – 40% dry weight). The most quantitative metabolites are digestibility reducers that make plant cell walls indigestible to animals. The effects of quantitative metabolites are dosage dependent and the higher these chemicals' proportion in the herbivore's diet, the less nutrition the herbivore can gain from ingesting plant tissues. Because they are typically large molecules, these defenses are energetically expensive to produce and maintain, and often take longer to synthesize and transport (Nina & Lerdau 2003). These secondary compounds may be secreted within the cells, for example, in vacuoles, or excreted extracellularly. They include poisonous compounds whose concentration in the cell tends to be relatively low, e.g. alkaloids, cyanogenic glycosides and cardenolides. Some of these secondary metabolites accumulate to levels high enough to reduce the plant's digestibility and palatability for herbivores (McKey 1979; Lindroth & Batzli 1984; Lambers 1993).

However, quantitative chemical defensive traits of litter may divide into three main categories:
- Lignin

- Total phenolics
- Tannins (Hydrolyzable tannins and Proanthocyanidins or Condensed tannins).

Lignin

Lignin is a polymer of aromatic subunits usually derived from phenylalanine. It is an important constituent of plant secondary cell walls and comprises the largest fraction of plant litter. Lignified leaves are rigid in structure, and highly recalcitrant to decay. Its chemical assay is difficult and different methods may lead to different results. Because it constitutes a barrier preventing decomposition of cellulose, lignin content of litter has been reported to control litter decomposition rate (Sterjiades & Erikson 1993).

Chemistry and Occurrence of Lignin

Lignins are complex polymers formed by the dehydrogenative polymerization of three main monolignols, p-coumaryl, coniferyl, and sinapyl alcohols (Fig. 1).

Gymnosperm lignins are mainly formed from coniferyl alcohol, together with small proportions of p-coumaryl alcohol. Angiosperm lignins are mainly formed from coniferyl and sinapyl alcohols with small amounts of p-coumaryl alcohol (Lewis 1999). Table 1 shows the % lignin content of different leaf litter.

Effects of Lignin on Litter Decomposition

Lignin is an essential component of plant litter, and is among the most recalcitrant compounds, and consequently is of major importance in soil humus building. Lignin concentration in leaves (or lignin to mineral ratios) has been widely used as an index of organic-matter quality. For instance, lignin concentrations alone, or lignin to N ratios in leaves could explain the rate of decomposition; negative correlations have been reported between lignin concentrations (or lignin to mineral ratios) and decomposition rates (Meentemeyer 1978; Melillo et al., 1982; Vitousek et al., 1994; Hobbie 1996; Kitayama et al., 2004). The ratio of lignin and N as a factor that is more related to decomposition than lignin content (Fig. 2).

On the other hand, hemicellulose and lignin concentrations were reported to be negatively correlated with decomposition (Vivanco & Austin 2008). The initial lignin content of leaf litter influenced the rate of decomposition. The species exhibiting higher initial lignin contents showed lower rates of decomposition of leaf litter. For example, the decomposition of Quercus dealbata litter is slower than that of Quercus fenestrata (Laishram & Yadava, 1988). However, the concentrations of the lignin fraction increased as decomposition proceeded, reaching relatively steady levels in the range of 45–51% (Berg 2000; Devi & Yadava, 2007). These increases showed partially linear relationships with accumulated mass loss (Berg et al., 1984).

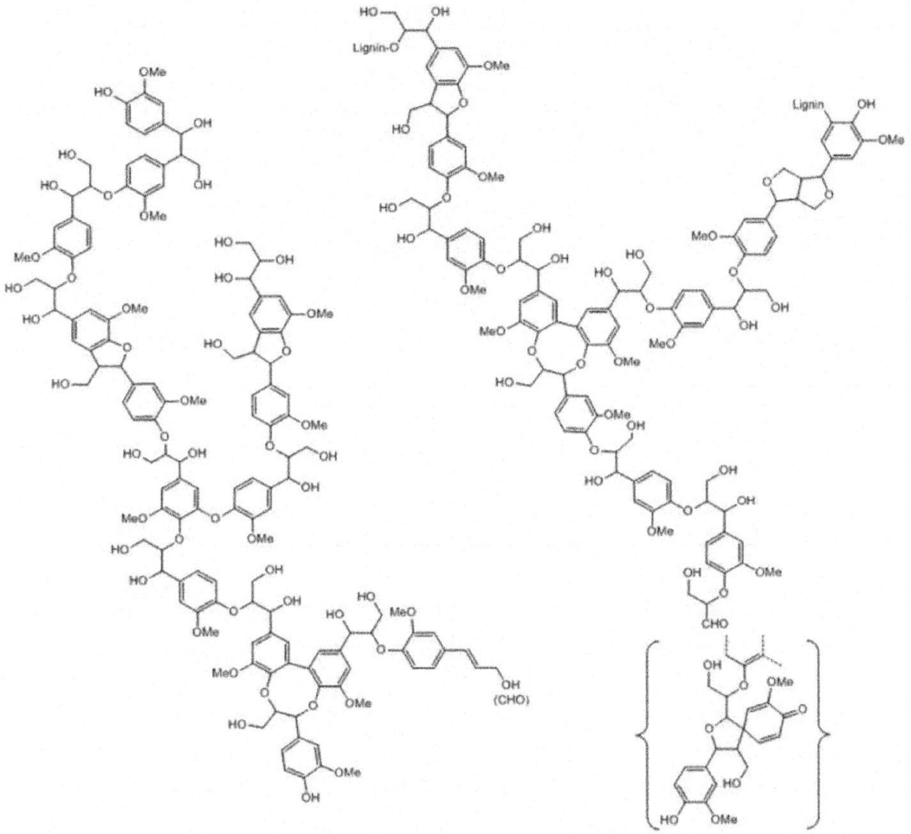

Figure. 1. Common structure of lignin.

Table 1. The lignin content of different plant species litter (percentage of dry weight) (adapted from Rahman et al., 2011)

Name of the species	Category of Plant	Lignin content of litter (%)
Acacia auriculiformis	Broadleaf tree	54.4
Acacia mangium	Broadleaf tree	43.1
Acer saccharum	Broadleaf tree	10.8
Alphitonia petriei	Broadleaf tree	40.4
Betula pubescens Ehrh	Broadleaf tree	14.0
Castanea sativa	Broadleaf tree	21.1
Dipterocarpus tuberculatus	Broadleaf tree	7.45
Eucalyptus grandis	Broadleaf tree	21.1
Gaultheria griffithiana	Shrub	5.0
Nothofagus dombeyi	Broadleaf tree	19.3
Nothofagus nervosa	Broadleaf tree	29.2
Nothofagus obliqua	Broadleaf tree	27.6
Picea orientalis	Coniferous tree	21.5
Pinus contorta Dougl	Coniferous tree	37
Pinus sylvestris L.	Coniferous tree	29.3
Populus nigra	Broadleaf tree	21.5
Quercus rubra	Broadleaf tree	23.1
Quercus dealbata	Broadleaf tree	6.0
Quercus fenestrata	Broadleaf tree	4.0
Quercus griffithii	Broadleaf tree	3.8
Rhododendron arboreum	Shrub	8.0
Tilia americana	Broadleaf tree	20.0
Broadleaves		21.57±14.23
Coniferous tree		29.26±7.75
Shrub		6.50±2.12

As decomposition proceeds the litter becomes enriched with lignin and N along with other components (Devi & Yadava, 2007). Earlier works have shown that as the lignin concentration increases during litter decomposition, the decomposition rates get suppressed (Fogel & Cromack, 1977; Devi & Yadava, 2007). The suppressing effect of lignin on litter mass-loss rates can be described as a linear relationship in the later stages of decomposition, which, for pine litter, may start at ca. 20–30% mass loss (Fig. 3a). For these later stages, the slope and intercept of this negative relationship varies among sites under different climates (Berg et al., 1993). The lowest effect of lignin concentration on mass-loss rates was found near the Arctic Circle (where long-term average actual evapotranspiration was about 385–390 mm).

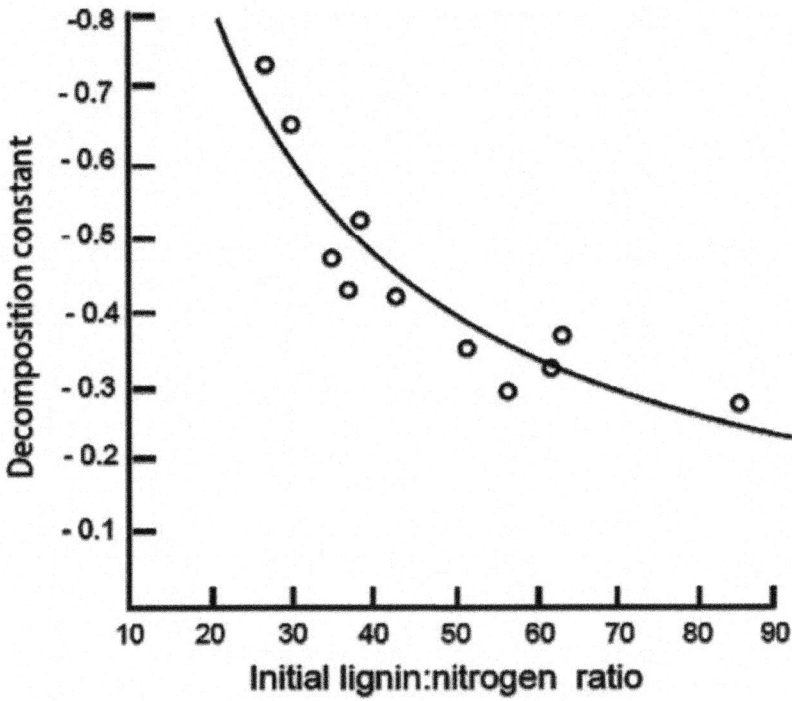

Figure. 2. Relationship between the decomposition constant (k = lose of dry mass, to initial lignin concentration of litter) and the lignin: nitrogen ratio of litter (adapted from Melillo et al.,1982).

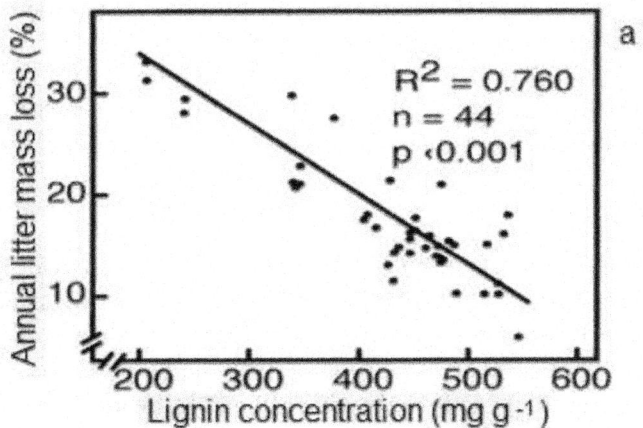

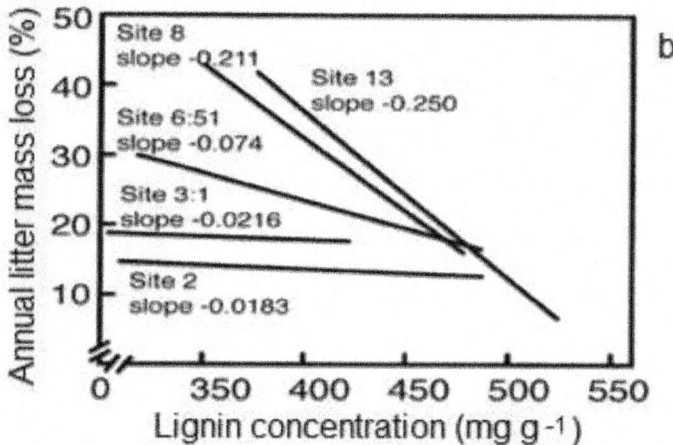

Figure. 3. Annual litter mass loss (%) as a function of initial lignin concentration at the start of each one year period. (a) The linear relationship indicated here, namely a decreasing rate with increasing lignin concentrations for one type of liter. (b) The same relationships under different climatic conditions (five climatically different sites with the AET values 385, 387, 472, 509 and 560 mm for sites 2, 3, 6:51, 8 and 13, respectively) indicate that the rate retarding effect of lignin is stronger in warmer wetter climates (adapted from Berg 2000).

In contrast, in Northern Germany and on the European continent the rate-regulating effect of lignin was found to be higher (Fig. 3 b). In a research on decomposition study from India, it is reported that lignin and fibre contents have showed a negative relation with weight loss of litter (Devi & Yadava, 2007). Many studies have reported a decline in the rate of weight loss of litter due to high initial lignin content (Singh & Gupta, 1977; Devi & Yadava, 2007). More recently, the scientists have found a highly significant, positive correlation between lignin contents and litter decay rates (Raich et al., 2007).

Total Phenolics

Phenolic compounds are one of the most abundant and widely distributed groups of substances in the plant kingdom with more than 8000 phenolic structures currently known (Harbone, 1980). They are products of the secondary metabolism of plants and arise biogenetically from two main primary synthetic pathways: the shikimate pathway and the acetate pathway (Paixao et al., 2007).

Chemistry and Occurrence of Total Phenolics

Phenolics are a heterogeneous group of natural substances characterized by an aromatic ring with one or more hydroxyl groups (Fig. 4). These substances are chemically diverse carbon-based secondary plant compounds occurring in plant tissues (Harborne 1997).

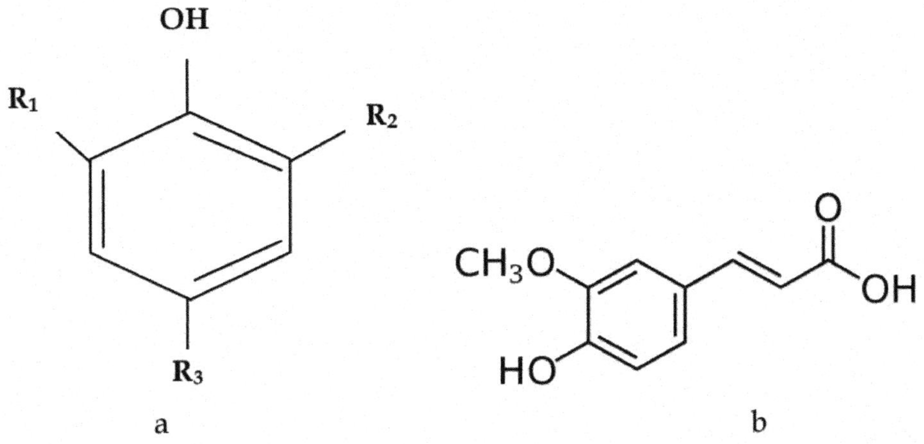

Figure. 4. Simple phenols (C_6), a. phenolic acids b. ferulic acid.

Phenols can be roughly divided into two groups: (1) low molecular weight compounds; and (2) oligomers and polymers of relatively high molecular weight (Hättenschwiler & Vitousek, 2000). Low molecular weight phenolics occur universally in higher plants, some of them are common in a variety of plant species and others are species specific. Because of the large variety of analytical methods and problems with choosing the appropriate standards, polyphenol concentrations reported in the literature vary immensely and might not be comparable with each other. Nevertheless, the two most frequently used polyphenol measurements (i.e. 'total phenolics' and Proanthocyanidins) are accepted reasonably well, and they commonly yield results in the range of about 1% to 25% of total green leaf dry mass (Hättenschwiler & Vitousek, 2000). The amount of phenolics in plant tissues varies with leaf species, age and degree of decomposition (Table 2) (Barlocher & Graca, 2005).

Table 2. The phenolics concentrations (%) in selected plant tissues, including senescent leaves (s), live (l) and yellow-green to brown-dead grass leaves (y) (after Barlocher and Graca, 2005)

Name of the Species	Plant type	Phenolics (% leaf dry mass)
Acer saccharum (s)	Deciduous tree	15
		2.7
Alnus glutinosa (s)	Tree	6.6
		6.8-7.6
Carya glabra (s)	Deciduous tree	9.1
Eucalyptus globulus (s)	Evergreen tree	6.4
		9.8
Fagus sylvatica (s)	Deciduous tree	8.0
Quercus alba (s)	Tree	16.2
Sapium sebiferum (l, s)	Deciduous tree	3.0
Spartina alterniflora (y)	Perennial deciduous grass	0.4-1.5

Effects of Total Phenolics on Litter Decomposition

Total phenolics are considered to be biologically active, e.g. by protecting plants against biotic (e.g. microbial pests, herbivores) or abiotic stresses (e.g. air pollution, heavy metal ions, UV-B radiation) (Hutzler et al., 1998), by contributing to allelopathic reactions (Waterman & Mole 1994) and by retarding decomposition rates of organic matter (Hattenschwieler & Vitousek 2000). In particular, phenolics play a major role in the defense against herbivores and pathogens (Lill & Marquis 2001).

In addition, some phenolics may prevent leaf damage resulting from exposure to excessive light (Lee & Gould 2002). The bulk of phenolics remain present during leaf senescence and after death, these compounds may also affect microbial decomposers (Harrison 1971) and therefore delay microbial decomposition of plant litter (Salusso 2000). Canhoto & Graca (1996) observed a strong negative correlation between the phenol content of different native litter types and litter decomposition rates in a stream, whereas Canhoto & Graca (1999) showed that phenolics from Eucalyptus leaves decrease feeding by detritivores. Thus, effects of phenolics on detritivores may be one reason for the low decomposability of Eucalyptus litter. The initial concentration of total phenolics in litter is positively correlated with dry organic carbon loss (Madritch & Hunter 2004). High amount of phenolics compounds in plants tissue decrease N concentration, which impedes the litter decomposition (Xuefeng et al., 2007). Barta et al., (2010) confirmed that a low amount of phenolics and low phenolics/N ratio in plant litter is closely related to higher differences in microbial respiration rates and mineral N release during the four months of litter decomposition in spruce forest.

Lin et al., (2006) observed a negative correlation between total phenolics and N contents for Kandelia candel and Bruguiera gymnorrhiza leaf litter at various stages of decomposition (Fig. 5). The perception of phenols as inhibitors, however, is far too simple, and the variety of phenolic compounds can have many different functions within the litter layer and the underlying soil (Hattenschwiler & Vitousek 2000). Even intraspecific variation in litter polyphenol concentrations can strongly influence soil processes and ecosystem functioning (Schweitzer et al., 2004). Phenols may influence rates of decomposition as they bind to N in the leaves forming compounds resistant to decomposition (Palm & Sanchez 1990). Gorbacheva & Kikuch (2006) found that dynamics of easily oxidized phenolics may influence the litter decomposition rate in the monitored subarctic field. Some scientist mentioned that phenolics stimulate microbial activity and subsequently reduce plant available N (Madritch & Hunter 2004; Lin et al.,, 2006). These results contribute important information to the growing body of evidence, indicating that the quality of C moving from plants to soils is a critical component of plant-mediated effects on soil biogeochemistry and, possibly, competitive interactions among species. Gorbacheva & Kikuch (2006) found that the essential part of phenolics that participates in the formation of mobile forms of organic matter, leaches from the organic horizon and migrates through the soil profile.

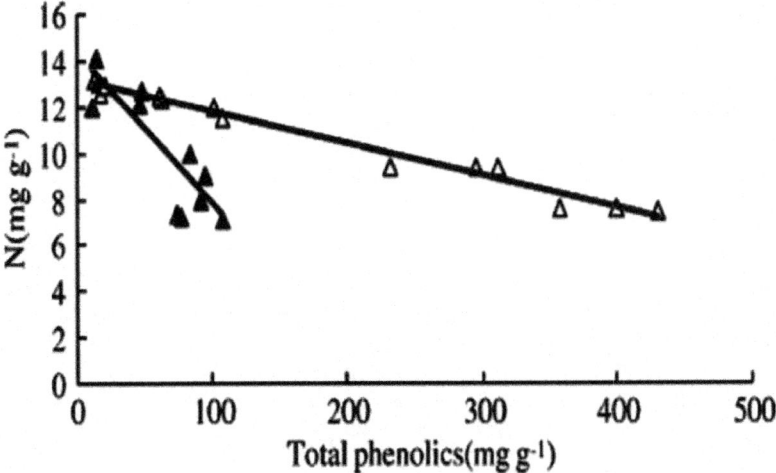

Figure. 5. Relationship between total phenolics and N contents during leaf decomposition of Bruguiera gymnorrhiza (Bg) and Kandelia candel (Kc). Symbols are: black triangle for Bg leaf; white triangle for Kc leaf (Adapted from Lin et al., 2006).

Tannins

Tannin is the fourth most abundant biochemical substance in vascular plant tissue after cellulose, hemicellulose and lignin (Kraus et al., 2003). Leaves and bark may contain up to 40% tannin by dry weight (Matthews et al., 1997; Kraus et al., 2003) and in leaves and needles tannin concentrations can exceed lignin levels (Benner et al., 1990). Tannin reduce herbivore load either directly through toxicity or growth alteration or indirectly through reduction of palatability (Feeny 1970).

Chemistry and Occurrence of Tannins

Tannins are heterogenous group of phenolics compounds derived from flavonoids and gallic acid. Bate-Smith & Swain (1962) defined tannin as water-soluble polyphenolic compounds ranging in molecular weight from 500 to 3000 Daltons that have the ability to precipitate alkaloids, gelatins and other proteins. Haslam (1998) has substituted the term "polyphenol" for "tannin", in an attempt to emphasize the multiplicity of phenolic group's characteristic of these compounds. Haslam also notes that molecular weights as high as 20,000 Daltons have been reported, and that tannins complex not only with proteins and alkaloids but also with certain polysaccharides. However, tannins found in higher plants are divided into two major classes termed Proanthocyanidins or Condensed tannins and hydrolyzable tannins (Fig. 6).

a

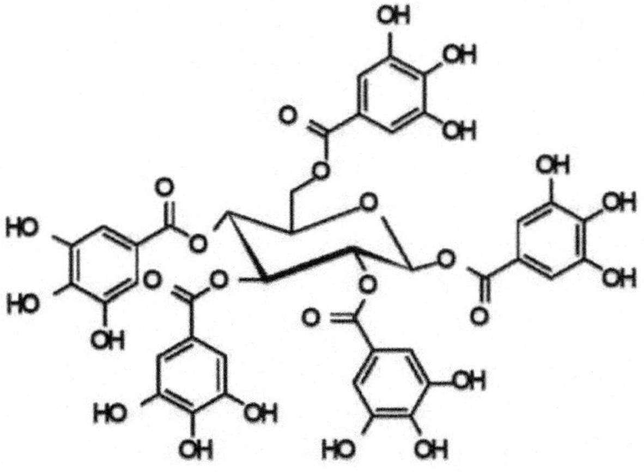

β-1,2,3,4,6-pentagalloyl-O-D-glucose

b

Figure. 6. Structures of tannins; a. Flavan-3-ols (+)-catechin and (-)-epigallocatechin, examples of monomeric precursors that polymerize to form macromolecular products such as linear proanthocyanidins composed of monomeric flavanoid units connected by C4-C8 linkages. b. hydrolyzable tannins β-1,2,3,4,6-pentagalloyl-O-D-glucose.

Tannins are distributed in species throughout the plant kingdom. They are commonly found in both gymnosperms as well as angiosperms. Tannins are mainly physically located in the vacuoles or surface wax of plants. Because tannins are complex and energetically costly molecules to synthesize, their widespread occurrence and abundance suggests that tannins play an important role in plant function and evolution (Zucker 1983). Tannins occur in plant leaves, roots, wood, bark, fruits and buds (Peters & Constabel, 2002; Ossipov et al., 2003). Tannin distribution in plant tissues appears to vary from species to species. In leaf tissues, tannins have been reported to occur preferentially in the epiderm, hypoderm, periderm, mesophyll, companion cells and vascular tissues, as well as throughout the leaf tissue (Grundhöfer et al., 2001). In roots, anatomical studies have identified a 'condensed tannin zone' in pine and eucalyptus located between the growing tip and the more developed cork zone (Peterson et al., 1999; Enstone et al., 2001). Hydrolyzable tannins have a more restricted occurrence than condensed tannin, being found in only 15 of the 40 orders of dicotyledons (Hättenschwiler & Vitousek 2000).

Effects of Tannin on Litter Decomposition

N and lignin concentration or C: N and lignin: N ratios are often used to predict rates of litter decomposition. However, a number of studies have shown that tannin and/or polyphenol content is a better predictor of decomposition, net N mineralization and N immobilization (Palm & Sanchez, 1991; Gallardo & Merino, 1992; Driebe & Whitham, 2000; Kraus et al., 2003). Coq et al., (2010) mentioned that litter decomposition in tropical rainforest correlated well with condensed tannin concentration. They concluded that leaf litter tannins play a key role in decomposition and nutrient cycling in the tropical rainforest.

In the past decades, many studies have shown that tannins are involved in defense mechanisms of plants against attack by bacteria, fungi, and herbivores (Zucker 1983; Scalbert 1991). There is not much knowledge about the mechanisms of action of the tannin (Zucker 1983; Scalbert 1991) even though modern analytical methods have improved the analysis of these complex structures (Mole & Watermann 1987; Schofield et al., 2001). Proposals on the mechanism of action include tannins forming stable complexes with plant proteins to make the tissue unattractive and difficult to digest (Schofield et al., 2001) and tannins acting like a toxin through highly specific reactions with digestive enzymes or directly at the cell membranes (Zucker 1983) or through depletion of essential iron by complexation (Mila et al., 1996). Leaves with high initial contents of condensed tannins, seem to decompose slowly in both terrestrial (Valachovic et al., 2004) and aquatic ecosystems (Wantzen et al., 2002). Condensed tannin may play an important role in aquatic leaf litter decomposition, as they may deter invertebrate shredders (Wantzen et al., 2002). Condensed tannin deters herbivore feeding by acting as toxins and not as digestion inhibitors by protein precipitation. Other researchers have obtained data that suggest the toxic nature of tannins (Robbins et al., 1987; Provenza et al., 1990; Clausen et al., 1990). Alongi (1987) noticed that if decomposers are inhibited by high contents of tannins in their food, strong effects on litter breakdown would be expected. Handayanto et al., (1997) found a strong negative correlation between N mineralization rates and the protein precipitation capacity of litter material, a measure of tannin reactivity. Litter material high in tannin content is commonly associated with reduced decomposition rates (Gallardo & Merino, 1992; Kalburtji et al., 1999). The convergent evolution of tannin-rich plant communities has occurred on nutrient-poor acidic soils throughout the world. Tannins were once believed to function as anti-herbivore defenses, but more and more ecologists now recognize them as important controllers of decomposition and nitrogen cycling processes. Tannins inhibit soil nitrogen accumulation and the rate of terrestrial and aquatic decomposition (Hissett & Gray 1976). Tannins make plant tissues unpalatable and indigestible for animals. Tannins impede digestion of plant tissues by blocking the action of digestive enzymes, binding to proteins being digested or interfering with protein activity in the gut wall (Howe & Westley 1988; Lambers 1993). Tannins may also reduce insect predation because they increase the leaf toughness (Haslam 1988). Kraus et al., (2003) summarized that tannins may limit litter decomposition in a number of different ways: (1) by themselves being resistant to decomposition (2) by sequestering proteins in protein-tannin complexes that are resistant to decomposition (3) by coating other compounds, such as cellulose, and protecting them from microbial attack (4) by direct toxicity to microbes, and (5) by complexing or deactivating microbial exoenzymes.

The studies by Schimel et al., (1998) in the Alaskan taiga provide some of the most comprehensive examinations of the diversity of phenolics and condensed tannin effects on soil processes.

Secondary succession in these forests starts with Salix/Alnus communities and continues to an Alnus/Populus, a Populus, and finally a Picea alba–dominated community. Populus balsamifera was found to play a key role during succession by the production of polyphenols that interfere with soil processes. Plants from strongly N-limited ecosystems are generally defended by tannins, whereas N-based and structural defenses become more abundant with increases in N supply (Gartlan et al., 1980).

For example, in the savannas of southern Africa, infertile miombo woodlands and savannas on soils derived from highly weathered granites have trees whose leaves are defended by tannins, while on nearby savannas on higher-nutrient soils such as shales and young volcanic soils, plants are defended by spines. The prevalence of chemical defenses depends on ecosystem nutrient supply (Craine et al., 2003). Fig. 7 represents the schematic overview of the effects of quantitative chemicals from leaf litter on various soil processes and its consequences for the nitrogen cycle and successional dynamics in terrestrial ecosystems.

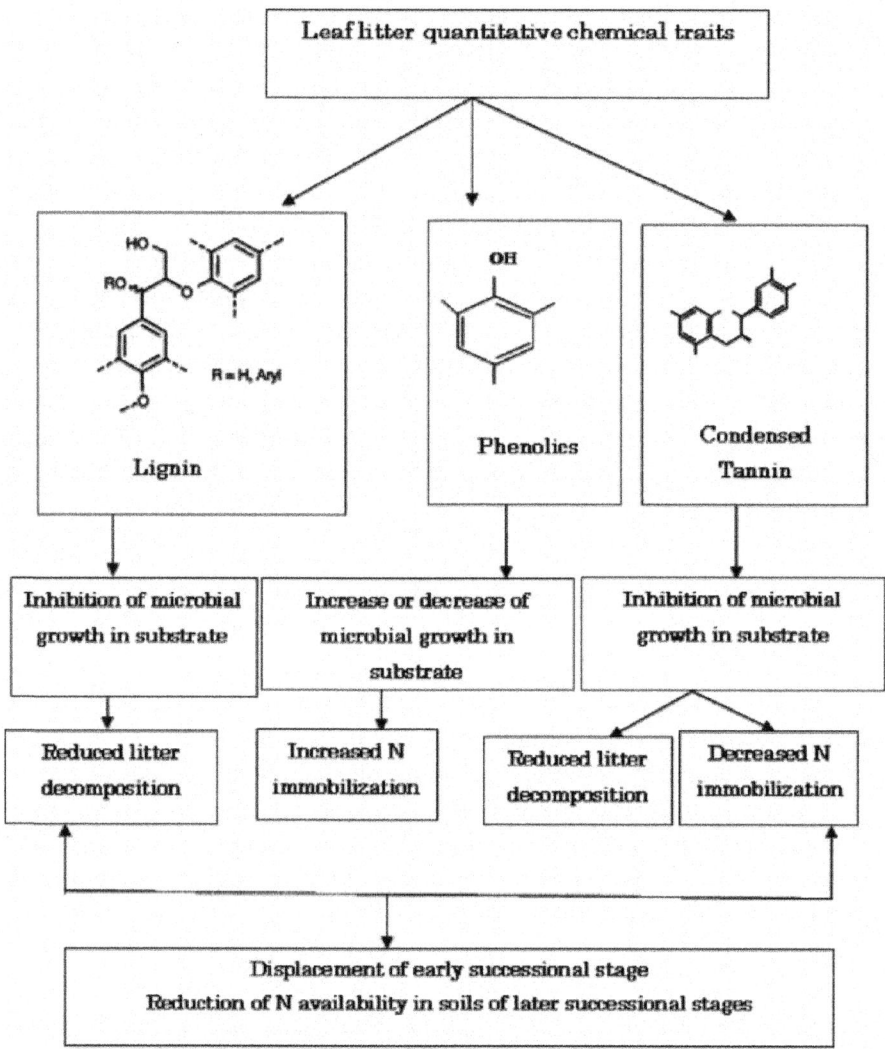

Figure. 7. Schematic reorientation of the effects of quantitative chemicals from leaf litter on various soil processes and its consequences for the nitrogen cycle and successional dynamics in terrestrial ecosystems (modified after Schimel et al., 1998).

LITTER DECOMPOSITION AND ECOSYSTEM FUNCTIONING

Ecosystem is composed of three subsystems i.e., producer- consumer- and decomposer subsystem. Ecosystem functioning is affected not only by the function of each subsystem but also by interactions between them. Quantitative defense is the driving force of ecosystem functioning (fig. 8).

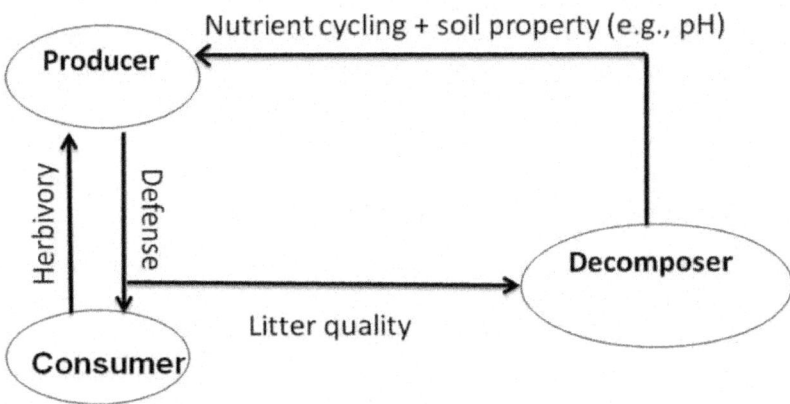

Figure. 8. Ecosystem functioning.

Wardle et al., (1997) demonstrated the importance of tree species composition in determining turnover rate of organic matter and N mineralization by comparing the forests dominated by early successional, fast-growing species and those dominated by slow-growing climax species. This is a verification of the hypothesis in the boreal zone the climax of which is characterized by slow rate of decomposition of hardly decomposable spruce needle litter.

Plant litter decomposition is important to many ecosystem functions such as the formation of soil organic matter, the mineralization of organic nutrients, and the carbon balance (Austin & Ballaré 2010). It is estimated that the nutrients released during litter decomposition can account for 69-87% of the total annual requirement of essential elements for forest plants (Waring & Schlesinger 1985).

Decomposition and nutrient cycling are fundamental to ecosystem biomass production. Most natural ecosystems are nitrogen (N) limited and biomass production is closely correlated with N turnover (Vitousek & Howarth, 1991; Reich et al., 1997). Typically external input of nutrients is very low and efficient recycling of nutrients maintains productivity (Likens et al., 1970). Decomposition of plant litter accounts for the majority of nutrients recycled through ecosystems. Rates of plant litter decomposition are highly dependent on litter quality; high concentration of phenolic compounds, especially lignin, in plant litter has a retarding effect on litter decomposition (Hattenschwiler & Vitousek 2000). At the ecosystem level, chemical defense can influence litter decomposition and nutrient cycling rates (Hattenschwiler & Vitousek 2000; Kraus et al., 2003). Ecosystems dominated by plants with low-lignin concentration often have rapid rates of decomposition and nutrient cycling (Chapin et al., 2003). Simple carbon (C) containing compounds are preferentially metabolized by decomposer microorganisms which results in rapid initial rates of decomposition. More complex C compounds are decomposed more slowly and may take many years to completely breakdown. Phenols and tannins, affect nutrient cycling in soil by inhibiting organic matter degradation, mineralization rates and N availability (Kraus et al., 2003).

Effects of Management on Litter Decomposition

Litter decomposition rate may be regulated by species composition, leaf litter chemistry, management activities, or any of the combinations of them. The forest management activities had significant effects on leaf litter decomposition. They also had significant effects on leaf litter chemistry during decomposition process. Clear cut or selective thinning method for timber harvesting may change litter decomposition by altering microclimatic conditions of the forest floor, leaf litter chemistry and composition of the microbial community (Li et al., 2009; Blanco et al., 2011). In a cold climate, selective thinning can increase the temperatures of the forest floor, and thus increase decomposition rate whereas selective thinning in a warm climate may slow decomposition by reducing the moisture content of surface organic matter (Li et al., 2009; Blanco et al., 2011). Blanco et al., (2011) observed that thinning effects clearly influenced litter chemistry in Continental forest and Mediterranean forest, generally appeared sequentially, first nutrients and cellulose and then total C and lignin.

CONCLUDING REMARKS AND FUTURE DIRECTION

Leaf litter decomposition is the fundamental process of ecosystem functioning. Quantitative chemical is hypothesized to play a key role in regulating litter decomposition and to be important for the production of dissolved organic matter and CO_2. Lignin decomposition may be relatively slower in boreal forest than tropical forests. Results after three years of decomposition experiments of Canadian boreal forest, Moore et al., (1999) concluded that lignin/N ratio, and some other climatic variables were valuable parameters for predicting mass loss.

Plants growing in tropical regions have higher polyphenolics concentrations in their tissues compared to temperate zone species (Coly 1983). Tannins are abundant in tropical tree foliage and have the potential effects on litter decomposition. Wieder et al., (2009) used natural variations in species litter chemistry combined with a through fall removal experiment to understand how climate–chemistry interactions regulate tropical forest litter decomposition.

Their results suggested that widely used predictors of litter decomposition based on chemical quality are still useful in tropical forests and that these wet systems also require an understanding of litter solubility to best prediction rates of decomposition.

The above discussion justifies that quantitative chemical traits can be used as a predictive tool for litter decomposability and ecosystem functioning. The decomposition of plant litter is an essential process in terrestrial ecosystems, resulting in carbon and nutrients being recycled for primary production. While a great deal of research has addressed quantitative chemical defense and their effect on decomposition and ecosystem functioning, there are many areas of quantitative chemical's biogeochemistry that are not known. There is still little information found regarding how different types of quantitative chemicals influence soil organisms, and how these chemicals' biodegradation affects the soil quality. An understanding of the role of quantitative chemicals in plant litter decomposition will allow for more accurate predictions of carbon dynamics in terrestrial ecosystems. When we can make a relationship between quantitative chemicals and decomposition then we can easily predict ecosystem functioning, which is important for conservation and restoration management of endangered ecosystems. Hence, it is imperative that future research focuses more attention on quantitative chemicals' biogeochemistry and their effects on litter decomposition, CO_2 emission and soil quality. There is a specific need to understand the role of lignin and of lignified cellulose, and their interactions during the late stages of decomposition. How do different types of lignin building blocks influence litter decomposition processes? What is the fate of quantitative chemicals after litter decomposition? Identifying and quantifying links between quantitative chemical defense traits and litter decomposability would enhance our understanding of ecosystem functioning and will provide us with a predictive tool for modeling decomposition rates under different vegetation types.

ACKNOWLEDGEMENTS

We are thankful to Biogeochemistry, and Ecology journal editors for giving permission to use the figure in this manuscript.

REFERENCES

1.   Alongi D.M. (1987) The influence of mangrove-derived tannins on interidal meiobenthos in tropical estuaries. Oecologia, 71, 537–540.

2.   Aerts R. (1997) Climate, leaf litter chemistry and leaf litter decomposition in terrestrial ecosystems: a triangular relationship. Oikos, 79:439–449.

3.   Austin A.T., Ballaré C.L. (2010) Dual role of lignin in plant litter decomposition in terrestrial ecosystems. Proceedings of the National Academy of Sciences USA, 107, 4618–4622.

4.   Austin A.T., Vivanco L. (2006) Plant litter decomposition in a semi-arid ecosystem controlled by photodegradation. Nature, 442:555–558.

5.   Badre B., Nobelis P., Tre´molie`res M. (1998) Quantitative study and modeling of the litter decomposition in a European alluvial forest. Is there an influence of overstorey tree species on the decomposition of ivy litter (Hedera helix L.)? Acta Oecologica, 19, 491–500.

6.   Barlocher F., Graca M.A.S. (2005) Total phenolics, Methods to study litter decomposition: a practical guide. Springer, Berlin Heidelberg New York. Bärlocher F., Newell S.Y. (1994) Phenolics and protein affecting palatability of Spartina leaves to the gastropod Littoraria irrorata. P.S.Z.N.I. Marine Ecology, 15, 65–75.

7.   Bärlocher F., Canhoto C., Graça M.A.S. (1995) Fungal colonization of alder and eucalypt leaves in two streams in Central Portugal. Archiv für Hydrobiologie, 133, 457–470.

8.   Barta J., Applova M., Vanek D., Kristufkova M., Santruckova H. (2010) Effect of available P and phenolics on mineral N release in acidified spruce forest: connection with lignindegrading enzymes and bacterial and fungal communities. Biogeochemistry, 97, 71–87.

9.   Bate-Smith E.C., Swain T. (1962) Flavonoid compounds, Comparative Biochemistry, Vol. 3A. Academic Press, New York. Berg B. (2000) Litter decomposition and organic matter turnover in northern forest soils. Forest Ecology and Management, 133, 13–22.

10. Berg B., McClaugherty C. (2003) Plant litter. Decomposition. Humus Formation. Carbon Sequestration. Springer-Verlag Heidelberg, Berlin, Germany. Berg B., Ekbohm G., McClaugherty C. (1984) Lignin and holocellulose relations during long term decomposition of some forest litters: Long-term decomposition in a Scots-pine forest. IV. Canadian Journal of Botany, 62, 2540–2550.

11. Berg B., McClaugherty C., Johansson M. (1993) Litter mass-loss rates in late stages of decomposition at some climatically and nutritionally different pine sites. Long-term decomposition in a Scots pine forest VIII. Canadian Journal of Botany, 71, 680–692.

12. Bernhard-Reversat F., Loumeto, J.J. (2002) The litter system in African forest tree plantations. In: Reddy, M.V. (Ed.), Management of Tropical Plantation–Forests and their Soil Litter System: Litter, Biota and Soil–Nutrient Dynamics. Science Publishers, Inc., Plymouth, UK, pp. 11–39.

13. Blanco J.A., Imbert J.B., Castillo F.J. (2011) Thinning affects Pinus sylvestris needle decomposition rates and chemistry differently depending on site conditions, Biogeochemistry, DOI 10.1007/s10533–010–9518–2.

14. Cameron G.N., LaPoint T.W. (1978) Effects of tannins on the decomposition of Chinese tallow leaves by terrestrial and aquatic invertebrates. Oecologia, 32, 349–366.

15. Canhoto C.M., Graca M.A.S. (1996) Decomposition of Eucalyptus globulus leaves and three native leaf species (Alnus glutinosa, Castanea sativa, Quercus faginea) in a Portuguese low order stream. Hydrobiologia, 333, 79–85.

16. Canhoto C.M., Graca M.A.S. (1999) Leaf barriers to fungal colonization and shredders (Tipula lateralis) consumption of decomposing Eucalyptus globulus. Microbial Ecology, 37, 163—172.

17. Chapin F.S.III., Matson P.A., Mooney H.A. (2003) Principles of terrestrial ecosystem ecology. Springer-Verlag, New York, USA. Clausen T.P., Provenza., F.D., Burritt E.A., Reichardt P.B., Bryant J.P. (1990) Ecological implications of condensed tannin structure: a case study. Journal of Chemical Ecology, 16, 2381–2392.

18. Coley P.D. (1983) Herbivory and defensive characteristics of tree species in a lowland tropical forest .Ecological Monograph, 53, 209–233.

19. Coq S., Jean-Marc S., Meudec E., Cheynier V., Hättenschwiler S. (2010) Interspecific variation in leaf litter tannins drives decomposition in a humid tropical forest in French Guiana. Ecology, 91, 2080–2091.

20. Craine J., Bond W., Piter L., Reich B., Ollinger. S. (2003) The resource economics of chemical and structural defenses across nitrogen supply gradients. Oecologia, 442,547–556.

21. De Santo AV., De Marco A., Fierro A., Berg B., Rutigliano F.A. 2009. Factors regulating litter mass loss and lignin degradation in late decomposition stages, Plant Soil, 318,217–228.

22. Devi A.S., Yadava P.S. (2007) Wood and leaf litter decomposition of Dipterocarpus tuberculatus Roxb. in a tropical deciduous forest of Manipur, Northeast India. Current Science, 93, 243–246.

23. Driebe E. M. Whitham T.G. (2000) Cottonwood hybridization affects tannin and nitrogen content of leaf litter and alters decomposition. Oecologia, 123:99–107.

24. Enstone D.E., Peterson C.A., Hallgren S.W. (2001) Anatomy of seedling tap roots of loblolly pine (Pinus taeda L.). Trees, 15, 98–111. FAO. (2011) Forest management, harvesting, and silviculture, http://www.fao.

org/DOCREP/003/X4109E/X4109E03.htm.

25. Feeny P. (1970) Seasonal changes in oak leaf tannins and nutrients as a cause of spring feeding by winter moth caterpillars. Ecology, 51, 565–581.

26. Fogel R., Cromack K.Jr. (1997) Effect of habitat and substrate quality on Douglas-fir litter decomposition in western Oregon. Canadian Journal of Botany, 55, 1632–1640.

27. Gallardo A., Merino J. (1992) Nitrogen immobilization in leaf litter at two Mediterranean ecosystems of SW Spain. Biogeochemistry, 15, 213–228.

28. Gartlan J.S., Waterman P.G., McKey D.B., Mbi C.N., Struhsaker T.T. (1980) A comparative study of the phytochemistry of two African rainforests. Biochemical Systematics and Ecology, 8, 401–422.

29. Gessner M.O. (1991). Differences in processing dynamics of fresh and dried leaf litter in a stream ecosystem. Freshwater Biology, 26, 387–398.

30. Gorbacheva T.T., Kikuchi R., (2006) Plant-to-soil pathways in the subarctic – qualitative and quantitative changes of different vegetative fluxes, Environmental Biotechnology, 2: 26–30.

31. Graca M.A.S., Barlocher, F. (1998). Proteolytic gut enzymes in Tipula caloptera – Interaction with phenolics. Aquatic Insects, 21, 11–18.

32. Graça M.A.S., Newell S.Y., Kneib, R.T. (2000) Grazing rates of organic living fungal biomass of decaying Spartina alterniflora by three species of salt-marsh invertebrates. Marine Biology, 136, 281–289.

33. Grundhöfer P., Niemetz R., Schilling B., Gross G.G. (2001) Biosynthesis and subcellular distribution of hydrolyzable tannins. Phytochemistry, 57, 915–927.

34. Handayanto E., Giller K. E., Cadisch G. (1997) Regulating N release from legume tree prunings by mixing residues of different quality. Soil Biology and Biochemistry, 29, 1417–1429.

35. Harbone J.B. (1980) Plant phenolics. In: E.A. Bell & B.V. Charlwood, (Eds). Encyclopedia of plant physiology, secondary plant products Vol. 8, Springer-Verlag, Berlin Heidelberg, New York:329–395.

36. Harborne J.B. (1997) Role of phenolic secondary metabolites in plants and their degradation in the nature. In: G. Cadisch & K.E. Giller (Eds). Driven by nature: Plant litter quality and decomposition. CAB International, Wallingford: 67–74.

37. Harrison A. F. (1971) The inhibitory effect of oak leaf litter tannins on

the growth of fungi, in relation to litter decomposition. Soil Biology and Biochemistry, 3: 167–172.

38. Haslam E. (1998) Practical polyphenolics: from structure to molecular recognition and physiological function, Cambridge University Press: Cambridge, UK. Hättenschwiler S., Vitousek P.M. (2000) The role of polyphenols in terrestrial ecosystem nutrient cycling. Tree, 15, 238–243.

39. Hissett R., Gray T.R.G. (1976) Microsites and time changes in soil microbe ecology. In: J.M. Anderson & A. Macfadyen (Eds.) The role of terrestrial and aquatic organisms in decomposition processes. Blackwell, London: 23–40.

40. Hobbie S.E. (1996) Temperature and plant species control over litter decomposition in Alaskan tundra. Ecological Monograph, 66,503–522.

41. Hobbie S.E., Gough L. 2004. Litter decomposition in moist acidic and non-acidic tundra with different glacial histories, Oecologia, 140, 113–124.

42. Howe H.F., Westley L.C. (1988) Ecological relationships of Plants and Animals. Oxford University Press, New York.

43. Hutzler P., Fischbach R.J., Heller W., Jungblut T.P., Reuber S., Schmitz R., Veit M., Weissenböck G., Schnitzler J.P. (1998) Tissue localization of phenolic compounds in plants by confocal laser scanning microscopy. Journal of Experimental Botany, 49, 953–965.

44. Kalburtji K.L., Mosjidis J.A., Mamolos A.P. (1999) Litter dynamics of low and high tannin Sericea lespedesa plants under field conditions. Plant Soil, 208, 217–281.

45. Kitayama K., Suzuki S., Hori M., Takyu M., Aiba S.H., Lee N.M., Kikuzawa K. (2004). On the relationships between leaf-litter lignin and net primary productivity in tropical rain forests. Oecologia, 140, 335–339.

46. Kraus T. E.C., Yu Z., Preston C.M., Dahlgren R.A., Zasoski R.J. (2003) Linking chemical reactivity and protein precipitation to structural characteristics of foliar tannins. Journal of Chemical Ecology, 29, 703–73.

47. Laishram I.D., Yadava P.S. (1988) Lignin and nitrogen in the decomposition of leaf litter in a subtropical forest ecosystem at Shiroi hills in north-eastern India. Plant Soil, 106, 59-64.

48. Lambers H. (1993) Rising CO_2, secondary plant metabolism, plant-herbivore interactions and litter decomposition. Vegetatio, 104/105, 263–271.

49. Lambers H., Poorter H. (1992) Inherent variation in growth rate between higher plants: A search for physiological causes and ecological consequences. Advances in Ecological Research, 23, 187-261.

50. Lavelle E., Blanchart E., Martin A., and Martin S. (1993) A hierarchical model for decomposition in terrestrial ecosystems: application to soils of the humid tropics, Biotropica, 25, 130–150.

51. Lee D.W., Gould K.S. (2002) Anthocyanins in leaves and other vegetative organs: An introduction. Advances in Ecological Research, 37, 1-16.

52. Lewis N.G. (1999) A 20th century roller coaster ride: a short account of lignifications. Current Opinion in Plant Biology, 2, 153–162.

53. Lewis N.G., Yamamoto E. (1990) Lignin: occurrence, biogenesis and biodegradation. Annual Review of Plant Physiology and Plant Molecular Biology, 41, 455-496.

54. Li Q, Moorhead DL., DeForest J.L., Henderson R., Chen J., Jensen R. (2009) Mixed litter decomposition in a managed Missouri Ozark forest ecosystem. Forest Ecology and Management, 257, 688–694.

55. Likens G.E., Bormann F.H., Johnson N.M., Fisher D.W., Pierce R.S. (1970) Effects of forest cutting and herbicide treatment on nutrient budgets in the Hubbard Brook watershed-ecosystem. Ecological Monographs, 40, 23–47.

56. Lill J.T., Marquis R.J. (2001) The effects of leaf quality on herbivore performance and attack from natural enemies. Oecologia, 126, 418–428.

57. Lin Y.M., Liu J.W., Xiang P., Lin P., Ye G.F., Sternberg L., da S.L. (2006) Tannin dynamics of propagules and leaves of Kandelia candel and Bruguiera gymnorrhiza in the Jiulong River Estuary, Fujian, China. Biogeochemistry, 78, 343–359.

58. Lindroth R.L., Batzli G.O. (1984) Plant phenolics as chemical defenses: effects of natural phenolics on survival and growth of prairie voles (Microtus ochrogaster). Journal of Chemical Ecology, 10, 229–244.

59. Matthews S., Mila I., Scalbert A., Donnelly D.M.X. (1997) Extractable and non-extractable proanthocyanidins in barks. Phytochemistry, 45, 405-410.

60. McKey D. 1979. The distribution of secondary compounds within plants. In: G.A. Rosenthal & D.H. Janzen, (Eds.) Herbivores, their Interaction with SA Secondary Plant Metabolites. Academic Press, New York : 56–134.

61. Meentemeyer V. (1978) Macroclimate and lignin control of litter decomposition rates. Ecology, 56, 465–472.

62. Melillo J.M., Aber J.D., Muratore J.F. (1982) Nitrogen and lignin control of hardwood leaf litter decomposition dynamics. Ecology, 63, 621–626.

63. Mila I., Scalbert A., Expert D. (1996) Iron withholding by plant polyphenols and resistance to pathogens and rots. Phytochemistry, 42, 1551–1555.

64. Mole S., Watermann P.G. (1987) A critical analysis of techniques for measuring tannins in ecological studies. Oecologia, 72, 137–147.

65. Moore T.R., Trofymow J.A., Taylor B., Prescott C., Camire´ C., Duschene L., Fyles J., Kozak L., Kranabetter M., Morrison I., Siltanen M., Smith S., Titus B., Visser S., Wein R., Zoltai S. (1999) Litter decomposition rates in Canadian forests. Global Change Biology, 5, 75–82.

66. Nina S., Lerdau M. (2003) The evolution of function in plant secondary metabolites. International Journal of Plant Science, 164, 93–102.

67. Osono T., Takeda T. (2005) Limit values for decomposition and convergence process of lignocellulose fraction in decomposing leaf litter of 14 tree species in a cool temperate forest. Ecological Research, 20, 51–58.

68. Ossipov V., Salminen J-P., Ossipova V., Haukioja E., Pihlaja K. (2003) Gallic acid and hydrolyzsable tannins are formed in birch leaves from an intermediate compound of the shikimate pathway. Biochemical Systematics and Ecology, 31, 3–16.

69. Paixao, N; Perestrelo, R; Marques, J.C., Camara J.C. (2007) Relationship between antioxidant capacity and total phenolic content of red rose and white wines, Food Chemistry, 105, 204–214.

70. Palm C.A., Sanchez P.A. (1991) Nitrogen release from the leaves of some tropical legumes as affected by their lignin and polyphenolic contents. Soil Biology & Biochemistry, 23, 83–88.

71. Palm C.A., Sanchez P.A. (1990) Decomposition and nutrient release patterns of the leaves of three tropical legumes. Biotropica, 22, 330–338.

72. Pereira A.P., Graça, M.A.S., Molles M. (1998) Leaf decomposition in relation to litter physicochemical properties, fungal biomass, arthopod colonization, and geographical origin of plant species. Pedobiologia, 42, 316–327.

73. Peters D.J., Constabel C.P. (2002) Molecular analysis of herbivore induced condensed tannin synthesis: cloning and expression of dihydroflavonol reductase from trembling aspen (Populus tremuloides). Plant Journal, 32, 701–712.

74. Peterson C.A., Enstone D.E., Taylor J.H. (1999) Pine root structure and its potential significance for root function. Plant Soil, 217, 205–213.

75. Powers J.S, Montgomery R.A, Adair E.C, Brearley F.Q, DeWalt S.J, Castanho C.T, Chave J, Deinert E, Ganzhorn J.U, Gilbert M.E et al., (2009) Decomposition in tropical forests: a pan-tropical study of the effects of litter type, litter placement and mesofaunal exclusion across a precipitation gradient. Journal of Ecology, 97, 801–811.

76. Provenza F.D., Burritt E.A., Clausen T.P., Bryant J.P., Reichardt P.B., Distel R.A. (1990) Conditioned taste aversions: A mechanism for goats to avoid condensed tannins in blackbrush. American Naturalist, 136, 810-828.

77. Rahman M.M., Motiur M.R., Yoneyama, A. (2011) Lignin chemistry and its effects on litter decomposition in forest ecosystem. Chemistry & Biodiversity, DOI: 10.1002/cbdv.201100198.

78. Raich J.W., Russell A.E., Ricardo B.A. (2007) Lignin and enhanced litter turnover in tree plantations of lowland Costa Rica. Forest Ecology and Management, 239, 128–135.

79. Reich P.B., Grigal D.F., Aber J.D., Gower S.T. (1997) Nitrogen mineralization and productivity in 50 hardwood and conifer stands on diverse soils. Ecology, 78, 335–347.

80. Robbins C.T., Hagerman A.E., Austin P.J., McArthur C., Hanley T.A. (1991) Variation in mammalian physiological responses to a condensed tannin and its ecological implications. Journal of Mammology, 72, 480–486.

81. Rouifed S., Handa I.T., David J.F., Ha"ttenschwiler S. (2010) The importance of biotic factors in predicting global change effects on decomposition of temperate forest leaf litter. Oecologia, 163, 247–256.

82. Salusso M.M. (2000) Biodegradation of subtropical forest woods from north-west Argentina by Pleurotus laciniatocrenatus. New Zealand Journal of Botany, 38, 721–724.

83. Scalbert A. (1991) Antimicrobial properties of tannins. Phytochemistry, 30, 3875–3883.

84. Schimel J.P., Cates R.G., Ruess R. (1998) The role of balsam poplar secondary chemicals in controlling soil nutrient dynamics through succession in the Alaskan taiga. Biogeochemistry, 42, 221–34.

85. Schofield P., Mbugua D.M., Pell A.N. (2001) Analysis of condensed tannins: a review. Animal Feed Science and Technology, 91, 21–40.

86. Schweitzer J.A., Bailey J.K., Rehill B.J., Hart S.C., Lindroth R.L, Keim

P., Whitham T.G. (2004) Genetically based trait in dominant tree affects ecosystem processes. Ecology Letters, 7, 127–34.

87. Singh J.S., Gupta S.R. (1977) Plant decomposition and soil respiration in terrestrial ecosystems. Botanical Review, 43, 449-528.

88. Smith J., Potts S.G., Woodcock B.A. Eggleton P. (2009) The impact of two arable field margin management schemes on litter decomposition. Applied Soil Ecology, 41, 90–97.

89. Sterjiades R., Erikson K.E.L. (1993) Biodegradation of lignins. In: Scalbert A. (Eds.) Polyphenolic Phenomena, INRA Editions, Paris: 115–126.

90. Suberkropp K., Godshalk G.L., Klug M.J. (1976) Changes in the chemical composition of leaves during processing in a woodland stream. Ecology, 57, 720–727.

91. Swift M.J., Heal O.W., Anderson J.M. (1979) Decomposition in Terrestrial Ecosystems, Studies in Ecology 5. Blackwell, Oxford. Valachovic Y.S., Caldwell B.A., Cromack K., Griffiths R.P. (2004) Leaf litter chemistry controls on decomposition of Pacific Northwest trees and woody shrubs. Canadian Journal of Forest Research, 34, 2131–214.

92. Vitousek P.M., Turner D.R., Parton W.J., Sanford R.L.Jr. (1994) Litter decomposition on the Mauna Loa environmental matrix Hawaii: patterns, mechanisms, and models. Ecology, 75, 418–429.

93. Vanderbilt K.L., White C.S., Hopkins O., Craig J.A. (2008) Aboveground decomposition in arid environments: results of a long-term study in central New Mexico. Journal of Arid Environments, 72, 696–709.

94. Vitousek P.M., Howarth R.W. (1991) Nitrogen limitation on land and in the sea: how can it occur? Biogeochemistry, 13, 87–115.

95. Vivanco L., Austin A. (2006) Intrinsic effects of species on leaf litter and root decomposition: a comparison of temperate grasses from North and South America. Oecologia, 150, 97–107.

96. Wantzen K. M., Wagner R., Suetfeld R., Junk W.J. (2002) How do plant herbivore interactions of trees influence coarse detritus processing by shredders in aquatic ecosystems of different latitudes? Verhandlungen Internationale Vereinigung für Theoretische und Angewandte Limnologie, 28, 815–821.

97. Wardle D.A., Zackrisson O., Hornberg G., Gallet C. (1997) The influence of island area on ecosystem properties. Science, 277, 1296–1299.

98. Waring R.H., Schlesinger W.H. (1985) Forest ecosystems: concepts and management. Orlando, FL, Academic Press. Waterman P.G., Mole S.

(1994) Analysis of phenolic metabolites. London, UK.

99. Wieder W.R., Cleveland C.C., Townsend A.R. (2009) Controls over leaf litter decomposition in wet tropical forests. Ecology, 90, 3333–3341.

100. Xuefeng L., Shijie H., Yan H. (2007) Indirect effects of precipitation variation on the decomposition process of Mongolian oak (Quercus mongolica) leaf litter. Frontiers of Forestry in China, 2: 417–423.

101. Zhang D., Hui D., Luo Y., Zhou G. (2008) Rates of litter decomposition in terrestrial ecosystems: global patterns and controlling factors. Journal of Plant Ecology, 1, 1–9.

102. Zucker W.V. (1983) Tannins: does structure determine function? An ecological perspective. American Naturalist, 121, 335–365.

Chapter 7

TANDEM MASS SPECTROMETRY FOR SIMULTANEOUS QUALITATIVE AND QUANTITATIVE ANALYSIS OF PROTEIN

Lay-Harn Gam

School of Pharmaceutical Sciences, Universiti Sains Malaysia, Malaysia

INTRODUCTION

Tandem mass spectrometry has long been recognized as a technique for qualitative analysis of proteins via amino acid sequencing. The use of tandem mass spectrometry in quantitative analysis for proteins is much limited. Nevertheless, application of chromatographic separation prior to mass spectrometry analysis can be used as a device for quantitative analysis of proteins. Following chromatographic separation, specific compound in a complex mixture can be determined with minimal interference; this is possible by monitoring only the selected m/z ratio of the compound, which is the characteristic of the compound of interest, an approach know as selected ion monitoring (SIM). The application of SIM using a combined GC-MS instrument was first demonstrated by Sweeley et al. in 1966.

On the other hand, quantitative data can also be obtained by repetitive scanning of compounds during elution of a sample from the gas chromatograph (GC-MS) or liquid chromatograph (LC-MS). The detection limits for such techniques are generally much poorer than those of SIM; this is because the instrument spends very little time at each m/z ratio during scanning. Such techniques can either scan the entire mass range of analysis or only scan a limited mass range with a greater sensitivity; subsequently the quantitative data can be calculated from the peak area or peak height displayed by the extracted chromatograms of the selected masses (Shoemaker & Elliott, 1991). The selected masses can be a molecular ion or a fragment ion provided it is sufficiently intense. Repetitive scanning together with automated data processing (library search of the recorded spectra) for biological samples has been used to measure a large number of biological compounds in complex samples. Such technique was found to be reliable, accurate and considerably more cost-effective than operator-mediated methods (Slivon et al., 1985).

Compared to a number of analytical techniques, tandem mass spectrometry is a technique which is able to provide more reliable data due to its high specificity and sensitivity. The major disadvantages of mass spectrometry are the high capital costs and the relatively low sample throughput. Therefore, the use of mass spectrometry in quantitative analysis is preferably dedicated to sensitivity and specificity rather than throughput. Since the original publication (Sweeley et al. 1966), technology and methodology on mass spectrometry have progressed and the applications of SIM for quantitative analysis, particularly in the analyses of small compounds have increased rapidly. Nevertheless, Bellar & Budde (1988) and Eichelberger, et al. (1983) cited that although improved sensitivity and precision is available with SIM, the loss of qualitative information is significant and usually unacceptable. This is especially true for quantitative analysis of minute protein in biological samples. Although coupled with effective sample cleanups, it is common for such mass spectrometry analysis to generate large amount of data caused by interferences from unrelated compounds derived from the complex sample matrix. In this condition, false identification of protein in SIM quantitative analysis will concomitantly increase.

Other quantitative analyses available are selected reaction monitoring (SRM) and multiple reaction monitoring (MRM). Both SRM and MRM were based on the similar principle, where the target precusor ion will be isolated and subsequently one of the fragment ions (SRM) or multiple fragment ions (MRM) will be monitored and quantified. The specificity of SRM and MRM is better than that of SIM. However, in terms of quantification of peptides and proteins, the methods' specificity vary depending on the nature of the peptide or the protein. This is especially true when the protein or peptide is belonged to a family of proteins, where the proteins in the family shared a great extend of similarity in their amino acid sequence. Furthermore, it is relatively common that the target peptide or protein be found in other forms than its intact form in biological matrix, where these other forms of proteins or peptides are the result of partial digestion of the protein or peptide through the activity of proteases in biological matrix. In these instances, the qualitative data of the target protein or peptide becomes important feature to discriminate the targeted peptides from these other forms of proteins or peptides. An illustration of the complexity of protein or peptide analysis will be demonstrated in the quantification analysis of human chorionic gonadotropin (hCG), a glycoprotein belongs to gonadotropin family. Besides the closely resemblance of hCG and LH, there are also the presence of nicked hCG in the biological matrix. In these circumstances, a complete qualitative data to indicate the identity of the target peptide will surely lead to higher confident in protein or peptides quantification. Other method for quantitative analysis of protein is by ELISA, where the amount of the protein present in a solution of biological matrix is measured by the degree of antibody-antigen interaction that is expressed in the intensity of color developed by the antibody-tag enzymatic reaction with its substrate. ELISA is a device for partial quantitative analysis. Furthermore, ELISA method may lead to false positive result caused by the cross reaction of the antibody with other unrelated proteins or compounds as the assay depends solely on the specificity of the antibody used.

An approach for simultaneous qualitative and quantitative analysis of proteins was developed. In this method, qualitative and quantitative analysis of proteins can be conducted in a single tandem mass spectrometry analysis. The method not only provides unambiguous identification of protein via amino acid sequencing, at the same time, quantitative data can be generated from the same tandem mass spectrometry data. This method is recommended as it can achieve a very low limit of quantification.

THE CONCEPT OF SIMULTANEOUS QUALITATIVE AND QUANTITATIVE ANALYSIS OF PROTEINS

The main problem encounters in the quantification of minute protein in a biological sample matrix is the isolation of the target proteins from the sample matrix. One of the most direct approaches for isolation of protein of interest is by using immunoaffinity extraction of the targeted protein through specific antibody-antigen interaction (Gam et al, 2006), this approach is possible if the antibody for the protein is available. Although Such technique can efficiently isolate target protein from the complex sample matrix, co-extraction of other unrelated proteins cannot be avoided. This may result from the non-specific interaction of the antibody with other proteins or may also result from the non-specific binding of other proteins with the antibody coating materials. Besides, it is also common that non-protein materials being extracted in such procedure. All these unrelated proteins or compounds will interfere with the subsequence analysis of the target protein using tandem mass spectrometry.

Mass spectrometer is an instrument with high sensitive but low selectivity, when analysis is carried out in the positive ion mode, the instrument will register all the positive charged ions that enter the detector. Therefore, quantification of minute protein extracted from complex biological matrix using mass spectrometer poses a great challenge. It is common that the target ion cannot be detected as it was suppressed or masked by other ions from the impurities in the sample. In this scenario, quantification of target ions cannot be achieved as it cannot fulfill the signal to noise ratio of greater than 3. One way to solve such problem is to have a very clean sample, a demand that is hard to fulfill as the target protein is only present in trace quantity in the complex biological sample. The alternative way of solving this problem is to get rid of all the unrelated ions during mass spectrometry analysis making the spectrum to be very clean for quantification analysis. One way of getting rid of the unrelated ions is by filtering them out from the spectrum. Based on this understanding, we have developed a quantification method for minute protein in a complex mixture. The first issue to solve is how to filter the unrelated ions out from the target ion spectrum, surely we cannot place a mass filter at the inlet of sample nebulizer, where it will also filter off our target protein. The other filtering device that possible is by using the MS/MS scan, in this device, only the target ions that were exceeded the threshold programmed in MS scan will be isolated and excited to MS/MS scan. In a way, the target ion is being filtered from other unrelated ions. In the analysis of protein using ion trap mass spectrometer, it is not advisable to analyze intact protein, where identification of intact protein is based solely on its deconvoluted molecular weight. Furthermore, analysis of intact glycoprotein by ion trap mass spectrometer possess additional problem where deconvolution of protein molecular weight may not be able to carry out successfully as the ionization of glycoprotein is inconsistent due to the presence of sugar components, which caused variation in the ionization of intact glycoprotein.

One of the advantages of using ion trap mass spectrometer is its ability to ionize peptide into multiple charged ions and to perform MS/MS scan, however it is not possible to perform MS/MS scan on most of the intact proteins, especially the high molecular weight proteins as the data generated cannot be interpreted. It is commonly understood that MS/MS scan is best performed on tryptic digested peptides, where the length of the peptides digested by trypsin is manageable by MS/MS scan of most types of mass spectrometer. Furthermore, the site of trypsin digestion is either Arginine or Lysine, the basic amino acids that will favor the formation of double charged ion to the peptides in positive ion mode scanning. The formation of double charged ion is an added advantage for collision induced dissociation (CID) in the MS/MS scan, according to the mobile proton hypothesis (McCormack et al, 1993), kinetic energy from collision induced dissociation will be converted to vibrational energy that releases through fragmentation reactions directed by the site of protonated amide bond. In this hypothesis, fragmentation of double charged peptide under the collision induced dissociation will result in detection of both y and b ions from the fragmentation, this is because each of the product ions carry a proton to themselves. When comparing to single charged ions, only one of the product ions resulted from the fragmentation will carry a proton and thus carry a charge for detection. The generation of comprehensive product ions by the double charged peptide ion gives a higher confident in determining its amino acid sequence, where the more y and b product ions identified, the more significance the identity of the peptide. In the currently developed method, both the qualitative and quantitative analyses of protein are based on the data of MS/MS scan, which displays product ions of the selected peptide.

In order to give significant qualitative value of a protein, the peptide marker of the target protein must be uniquely different from peptides of other proteins. In another word, the amino acid sequence of the peptide marker must be specific for the target protein. Furthermore, upon collision induced dissociation, such peptide must produce comprehensive spectrum of y and b product ions which give a definite identity to the target protein.

In terms of quantitative analysis of protein, the same peptide marker will be used. Similar with qualitative analysis, the peptide marker will be subjected to collision induced dissociation to generate y and b product ions, as mentioned earlier, these product ions are used to confirm the identity of target protein. Once the identity of the protein is confirmed, subsequence extracted product ions chromatogram will be performed. When performing extracted product ions chromatogram, two orders of mass filters were in place, the first mass filter is peptide marker mass filter that will single out the molecular ion for the peptide marker, while the second mass filter is quantitative ion (or selected product ions) mass filter that will single out the quantitative ion. In doing so, this method create a high selectivity to our currently developed quantitative method. This is because the first mass filter isolates the double charged m/z value of the peptide marker, subsequently the second mass filter isolates the selected quantitative ion/s that was generated from the peptide marker that has undergone the first mass filter. Quantitative ion/s selected must be the most stable and intense product ion/s (either y or b ions) from the fragmentation of the peptide marker. Quantitative ion can be a single product ion or can be a sum of a few product ions. In the event of the summation of a few product ions, limit of quantification (LOQ) of the method will concomitantly reduce, this is because the total peak areas of product ions will certainly enhance the signal to noise ratio. However, the usefulness of selected product ions to act as quantitative ions must be validated, where its peak area under extracted product ions chromatogram must be proportional to the concentration of the target protein. In the other word, the intensity of collision energy plays an important role in this currently developed method, where the collision energy used must be kept constant throughout the analysis.

The beauty of this method is it is able to simultaneously provide qualitative and quantitative data on trace amount of protein in a complex mixture. The two layers of filtering event will ensure only the correct ions being quantified. This method is very useful for quantification of trace amount of protein in a complex biological sample. It is not impossible that many compounds with similar masses can be filtered in by the first layer of mass filter, nevertheless, these similar masses from the impurities of the sample cannot generate the product ions spectrum similar to that of target peptide marker, as they do not contain the amino acid sequence of the peptide marker that is unique to the target protein. Therefore, these unrelated masses cannot pass the second layer of mass filter.

As a result of the selectivity of the two layers mass filter described above, the extracted product ion chromatogram produced is usually free from any back-ground noises, giving rise to a very low limit of quantification of the method and therefore trace protein in complex sample matrix can be quantified. Furthermore, the method gives a very high confident level in terms of qualitative information of the target protein, where false positive data are completely omitted. In the course of my study, I was given a task to develop a quantitative method for human chorionic gonadotropin (hCG) in human urine. hCG is a glycoprotein misused by male athletes to induce endogenous secretion of testosterone. hCG is present as trace component in urine, a complex biological matrix. In order to concentrate and purify the glycoprotein from urine, I have implemented immunoaffinity purification technique to extract hCG from urine matrix. This purification technique did not produce pure hCG as expected, the numerous amount of contaminants can be visualized when subjecting the extracted hCG to mass spectrometry analysis.

MATERIALS AND METHODS

Immunoaffinity Protein Purification

Approximately 11 ml of urine was centrifuged at 1500 rpm for 5 minutes to precipitate any particulate matters. A volume of 10 ml of the centrifuged urine was transferred to a clean polypropylene tube for hCG extraction. The immunoaffinity column was first flushed with 6 ml distilled water to remove the storage buffer and then conditioned with 5 ml of 0.01 M PBS at pH 7.2. During this step, the column flow was adjusted to 7-9 drops per minute. A 2 ml volume of the centrifuged urine was loaded onto the immunoaffinity column (column volume was 2 ml). A 20 minutes incubation time was allowed for the antibody-antigen association to take place. The urine was then removed from the column by flushing with 2 ml of 0.01 M PBS pH 7.2. The column was then reloaded with another 2 ml aliquot of urine and incubated for 20 minutes. This process was repeated until all the 10 ml urine had passed through the column. Finally, the column was washed with 15 ml (7 bed volumes) of washing buffer (0.1% (v/v) Tween 20 in 0.1 M PBS, pH 7.2) followed by 2 ml of elution buffer (1 M citric acid adjusted to pH 2.2 with 10 M NaOH). After the first 1 ml of the elution buffer had entered the gel, the collection of the eluate began. When the 2 ml of the elution buffer had fully immersed in the gel, a 5 minutes equilibration time was allowed to enable complete antibody-antigen dissociation to take place. This was followed by 8 ml of elution buffer. A total of 9 ml eluate was collected. Eluate was concentrated and desalted using a protein concentrator column (Jones Chromatography).

Preparation of Tryptic Digestion Product of hCG

Digestion of hCG using Trypsin

Protein sample (hCG) was desalted using protein concentrator column (C18, 2 cm x 4.6mm ID, Jones Chromatography). A syringe pump (Harward Apparatus) was used to pump the protein solution through the column at 1 ml/min. The column was then flushed with 25 ml of deionized distilled H20 and the protein was recovered by eluting with 70% acetonitrile 0.1 % formic acid. The eluted protein was dried under N2 at 37'C.

The dried protein was denatured using denaturing buffer (6M Guanidine HCI, 0.5M Tris, 2mM EDTA pH8.6). A volume of 10 µl of 1 M dithiothrietol was added to the mixture and incubated at 37°C for 30 minutes. After which 25 µl of 1M iodoacetic acid in 1M NAOH was added and the mixture was further incubated for 30 minute at room temperature. The excess reagents were removed from the protein sample by using the protein concentrator column (as mentioned above). The dried hCG was then reconstituted in 50 µl of 50 mM NH4HCO3. A volume of 2 µl (0.25 µg/µl) of trypsin solution was added and the mixture was incubated at room temperature for 20 hours. This was followed by another addition of the same amount of trypsin and the sample then further incubated for 4 hours at room temperature. The digested hCG was lyophilized and stored at -20°C.

HPLC Separation

The tryptic digested hCG was first reconstituted in 25 µl of high purity distilled H2O (Maxima, ELGA); 10 µl of the sample was then injected into the C-18 Vydac column (300 Å, 5 µm, 1 mm X 50 mm). Separation of the peptides was performed using a Hewlett Packard series 1100 HPLC. The flow rate was set at 1 ml/min and further split by a fused silica splitting device to 20 μl/min through-column flow rate. Mobile phase A was 0.05% TFA in H2O and B was 0.05% TFA in ACN. The gradient used was 5-95% B for 20 minutes and held at 95% B constant for 5 minutes. The HPLC was interfaced to an ion trap mass spectrometer (LCQ, ThermoQuest).

Mass Spectrometry

Mass spectrometric analysis was carried out using the ion trap mass spectrometer (LCQ, ThermoQuest). Data dependent experimental method was created for the analysis of tryptic peptides of hCG.

Creating a Data Dependent Experimental Method for Qualitative Analysis

The MS data was acquired at heating capillary temperature 200°C, sheath gas flow rate is 60arb, spray voltage at 4kV, tube lens offset is -60V and the capillary voltage is at 38V. Data dependent experimental method was created for the analysis. The experimental method was consisted of 2 scan events. The first scan event was full scan MS, the second was MS/MS scan, which were dependent on the results of the full scan MS. This linkage is known as data dependent scan. The parameters of data dependent scan were default collision energy of 25, charge state of 2, minimum signal acquired was 1 x 105 counts, isolation width was 2 m/z.

Creating a Data Dependent Experimental Method for Simultaneous Qualitative and Quantitative Analysis of hCG

Doubly charged parent ion for peptide VLQGVLPALPQVVCNYR, [964.7]2+ was programmed into the parent ion list in data dependent scan. The parameters set for data dependent scan (MS/MS scan) were default collision energy = 25, default charge state = 2, minimum signal acquired = 1 x 104 counts, and the isolation width = 2 m/z.

A DEMONSTRATION OF THE METHOD BY USING HUMAN CHORIONIC GONADOTROPIN HORMONE

Human charionic gonadotropin or hCG is a hormone misuses by male athlete to induce endogenous production of testosterone (Boer et al, 1991). It will be interesting to use hCG for the demonstration of this analysis method, this is because the hormone belongs to a family of gonadotropin. Other hormones in the same family are follicle stimulating hormone (FSH), lutropin hormone

(LH) and thyroid stimulating hormone (TSH) (Canfield et al, 1976). All these hormones share a similar characteristic in molecular structure, they are glycoprotein hormones that made up of one alpha-subunit and one beta-subunit. The alpha-subunit of the hormones is identical (Vaitukaitis et al, 1976), hence, it cannot serve as marker for hCG. On the other hand, the beta-subunits are basically similar with only minor differences in certain amino acid residues. This is especially true between the beta-subunits of hCG and LH (Figure 1). Due to the high resemblance of the hormones, they share relatively similar electrophoretic mobility in gel electrophoresis separation. Therefore, qualitative and quantitative analysis of hCG possess extra challenges.

Human Chorionic gonadotropin (hCG) is synthesized by the trophoblast cells of the placenta (Canfield, et al., 1971). The hormone is release in the first few weeks of pregnancy. Between the 7th to 12th week of pregnancy, the plasma level of hCG rises to extremely high levels where approximately 11,000 − 289,000 mIU/ml of hCG were released. This is followed by a decline during the last two trimester (Braunstein, et al., 1978).The molecular weight of intact hCG, αhCG (a-subunit) and βhCG (b-subunit) are approximately 36.7, 14.5 and 22.2 kDa, respectively. It has been estimated that 30% of the total weight of hCG is contributed by the carbohydrate content (Canfield, et al., 1976) and these carbohydrates account for the heterogeneity properties of hCG. Besides the intact hCG, αhCG and βhCG, the other commonly found hCG fragments are b-core fragment and nicked hCG fragment. These fragments are formed by proteolytic degradation of hCG in kidney. The b-core fragment composes of two polypeptides linked by a disulfide bond. The molecular weight of the purified b-core fragment is between 12-16kDa. Reduction of b-core fragment resulted in its dissociation into two fragments between 8-12kDa and 5-6kDa molecular weights (Endo, et al., 1992).

S K E P L R P R - C R P I N g A T L A V E K –E G C P V C I T V Ng T T I C A G Y C

P T M T R –V L Q G V L P A L P Q V V C N Y R – D V R –F E S I R –L P G C P R-

G V N P V V S Y A V A L S C Q C A L C R –R –S T T D C G G P K –D H P L T C D

D P R –F Q D S S S Sg K – A P P P Sg L P S P Sg R – L P G P Sg D T P I L P Q

Figure. 1. Amino acid sequence for hCG b-subunit, (-) indicates the site of trypsin digestion. The amino acid residues which are different from those in LH are in italic and red. The site of glycosylation is indicated by g.

In our method, hCG was subjected to reduction, alkylation and digestion using trypsin enzyme prior to tandem mass spectrometry analysis. The expected peptide fragments derived from the digestion were listed in Table

1. A total of 16 peptides and glycopeptides fragment were expected from the tryptic digestion of hCG, these peptides were (S K E P L R P R), (C R P I Ng A T L A V E K), (E G C P V C I T V Ng T T I C A G Y C P T M T R), (V L Q G V L P A L P Q V V C N Y R), (D V R), (F E S I R), (L P G C P R), (G V N P V V S Y A V A L S C Q C A L C R), (R), (S T T D C G G P K), (D H P L T C D D P R), (F Q D S S S Sg K), (A P P P Sg L P S P Sg R), (L P G P Sg D T P I L P Q). (note: amino acid residue in red is different from that of LH while g indicated the site of glycosylation). Amongst these peptides, only 10 peptides show at least one amino acid residue different from that of LH, the difference in amino acid residues can be used to distinguish hCG from LH in tandem mass spectrometry analysis.

Table 1. Predicted tryptic digested βhCG fragments

Position no.	[M + H]+	Selected data dependent parent ions	Sequence
1-2	234.1	-	SK
3-8	767.4	[384.5]²⁺	EPLRPR
1-8	983.2	[491.8]²⁺	SKEPLRPR
9-20	Glycopeptide	-	aCPINATLAVEK
21-43	Glycopeptide	-	aEGCPVCITVNTTICAGYTCPT MTR
44-60	1928.4	[964.7]²⁺	VLQGVLPALPQVVCNYR
61-63	389.2	[389.2]⁺	DVR
64-68	651.3	[326.1]²⁺	FESIR
69-74	700.8	[350.5]²⁺	LPGCPR
75-94	2228.6	[743.5]³⁺	GVNPVVSYAVALSCQCALCR
95	175.2	-	R
96-104	924.0	[462.5]²⁺	STTDCGGPK
105-114	1227.3	[614.1]²⁺	DHPLTCDDPR
115-122	Glycopeptide	-	bFQDSSSSK
123-133	Glycopeptide	-	bAPPPSLPSLSR
134-145	Glycopeptide		bLPGPSDTPILPQ

a shows N-link glycopeptide at the bold amino acid

b shows O-link glycopeptide at the bold amino acid/s

Selection of Peptide Marker

In the selection of suitable peptide marker for simultaneous qualitative and quantitative analysis of protein, three criteria are implemented:

  a. The amino acid sequence of the peptide marker must be unique to the protein. This is important as the marker will be used in the qualitative analysis to differentiate the protein of interest from other unrelated proteins.

b. The length of the peptide marker must be suitably long. The long marker will lead to generation of a more comprehensive spectrum of product ions which will give higher confident level in the qualitative analysis. Moreover, the generation of greater number of product ions will also lead to greater choice of quantifying ions to be used in subsequent quantitative analysis.

c. The degree of ionization of the peptide marker. Only the peptide marker that can be easily ionized by the ionization mode of mass spectrometer to produce high abundant molecular ions will be selected as peptide marker. This is important in the analysis of protein from complex biological sample, where interferences from the matrix and chemical noise may mask the low abundance ions.

The b-subunit of hCG has an extension of 30 amino acid residues at its C-terminus compared to LH. This distinctive region of hCG is used as the antigenic epitope for raising of hCG-specific antibodies. This region is highly glycosylated and following trypsin digestion, a total of 3 glycopeptides, F Q D S S S Sg K , A P P P Sg L P S P Sg R , L P G P Sg D T P I L P Q are expected to be formed from this particular region. However, we did not detect any of these glycosylated peptides in MS scan, the possible reason is the presence of carbohydrate in the peptides suppressed the ions signals of glycopeptides and therefore reduced the detection of these glycopeptides.

On the other hand, the molecular ions of two other glycopeptides C R P I Ng A T L A V E K and E G C P V C I T V Ng T T I C A G Y C P T M T R derived from the b-subunit of hCG were detected in different glycoforms, which indicates more than one type of sugar moieties were present in each individual glycopeptide. However, when subjecting these molecular ions to MS/MS scan, there is no useful MS/MS data to indicate the identity of these glycopeptides. Once the sugar moieties were removed (deglycosylation), four out of the five deglycosylated glycopeptides derived from the b-subunit of hCG were detected (Gam et al, 2006). Nevertheless, the deglycosylation process was time consuming and also resulted in sample lost. Hence, the glycopeptides do not serve as suitable marker for hCG. In terms of peptides, only the peptides that are differ in at least one amino acid residue were targeted in the analysis. This is possible as a different in one amino acid residue will be resulted in a different product ions masses upon tandem mass spectrometry analysis. An example is for a peptide sequence of V L Q G V L P A L P Q V V C N Y R of hCG. The amino acid residues in red are the different amino acid residues than those of LH, the expected product ions series of these closely resemblance peptides of hCG and LH is listed in Table 2.

It is clear that out of 32 product ions expected from the peptides, there were only three identical product ions between hCG and LH, namely b1 +, b2 +, b3 + product ions. The extensive difference in the product ions series between the seemingly similar peptides of hCG and LH was great, which allow distinctive identification of these two closely resemblance hormones. Due to the high similarity between the two hormones, false identification of LH as hCG in ELISA was reported as a result of cross reaction of hCG antibodies with LH (Bottger et al, 1993).

Peptide S K E P L R P R was formed from an incomplete digestion of b-subunit at the 2nd to 3rd amino acid residue. It is reasonable to suggest that the formation of this peptide was not due to insufficient enzyme used but rather it was due to the presence of a carboxyl side chain group (glutamic acid, E) on the digestion site, which remarkably reduced the rate of hydrolysis. The use of this peptide as hCG marker may not be suitable as its formation may vary due to incomplete trypsin digestion. In general, taking into the consideration of all the peptides and glycopeptides derived from hCG, only 4 peptides fulfilled the first two criteria (A and B) of peptide marker, these peptides were (V L Q G V L P A L P Q V V C N Y R), (G V N P V V S Y A V A L S C Q C A L C R) (S T T D C G G P K), (D H P L T C D D P R).

Table 2. The predicted product ions of V L Q G V L P A L P Q V V C N Y R and V L Q A V L P P L P Q V V C T Y R of hCG and LH, respectively

y-fragment ions	M/z LH	M/z hCG	b-fragment ions	M/z LH	M/z hCG
y_1^+	175.1	182.1	b_1^+	100.1	100.1
y_2^+	338.1	337.4	b_2^+	213.2	213.3
y_3^+	439.2	451.5	b_3^+	341.2	341.4
y_4^+	560.2	613.7	b_4^+	412.3	398.5
y_5^+	699.3	712.9	b_5^+	511.3	497.6
y_6^+	780.3	812.0	b_6^+	624.4	610.8
y_7^+	926.4	940.1	b_7^+	721.4	707.9
y_8^+	1023.5	1037.2	b_8^+	818.5	779.0
y_9^+	1136.5	1150.4	b_9^+	931.6	892.1
y_{10}^+	1233.6	1221.5	b_{10}^+	1028.6	989.3
y_{11}^+	1330.6	1318.6	b_{11}^+	1156.7	1117.4
y_{12}^+	1443.7	1431.7	b_{12}^+	1255.8	1216.5
y_{13}^+	1542.8	1530.9	b_{13}^+	1300.8	1315.6
y_{14}^+	1613.8	1587.9	b_{14}^+	1515.8	1476.8
y_{15}^+	1683.9	1716.1	b_{15}^+	1616.9	1590.9
y_{16}^+	1855.0	1829.1	b_{16}^+	1779.9	1754.1

Figure 4 shows the total ion chromatogram (TIC) of the tryptic digested peptides of hCG, a total of 3 out of 4 potential peptide markers were detected, where V L Q G V L P A L P Q V V C N Y R were presented as the most abundant ion followed by G V N P V V S Y A V A L S C Q C A L C R as the second most abundant ion in the analysis. Nevertheless, the detection of G V N P V V S Y A V A L S C Q C A L C R peptide by mass spectrometry is anomalous. In some analyses, the peptide cannot be detected. The rationale for this irregularity in the detection of this peptide is not understood, therefore the use of this peptide as hCG marker is not indicated. Hence, in terms of the ease of peptide ionization (criteria C), V L Q G V L P A L P Q V V C N Y R was found to be the most suitable peptide marker for hCG.

The molecular ion of V L Q G V L P A L P Q V V C N Y R peptide was [964.7]2+ while the corresponding peptide of LH has an amino acid sequence of V L Q A V L P P L P Q V V C T Y R and the doubly charged ion was [977.6]2+. This LH peptide differs from the hCG peptide by 3 amino acid residues (red). With these minor differences, a completely different set of parent ions and product ion spectra were expected (Table 2). These differences can be used to distinguish hCG from LH. Therefore, this peptide is unique to hCG for tandem mass spectrometric qualitative and quantitative analysis. Furthermore, the long amino acid sequence (17 amino acid residues) of the peptide increases the method specificity by preventing false identification of the peptide. In addition, the high m/z ratio of its parent ion avoids interference from chemical noise, which normally found at low m/z ratios. All these features of the peptide fulfilled the requirement of peptide marker. Hence, this peptide was selected as the peptide marker for identification and quantification of hCG.

Nicking of hCG occurs at position between 47-48 amino acid residue of the b-subunit. After tryptic digestion, the nicked hCG peptide which comprises amino acid residue 48 to 60 was formed. This amino acid sequence is within the amino acid sequence of V L Q G V L P A L P Q V V C N Y R peptide (from amino acid residue 44 to 60). For this reason, the nicked hCG peptide produces the y+ product ions (y1 + to y12+) which are identical to some of the y+ product ions of V L Q G V L P A L P Q V V C N Y R peptide. The similarity in the y+ ions of the two peptides did not interfere with the quantitative result. This is because the specificity of the data dependent mass spectrometric method is able to eliminate the interference of nicked hCG peptide as the two peptides have distinctly different parent ion masses ([964.7]2+ and [765.9]2+).

Figure 2 (lower panel) shows the MS/MS spectrum of V L Q G V L P A L P Q V V C N Y R peptide, where the fragmentation of the peptide produced a comprehensive product ions spectrum, which provides a convincing identification of the peptide, and subsequently hCG. Once the targeted peptide markers were identified, the subsequently analysis of protein using tandem mass spectrometry can be carried out. Since this method only emphasizes tandem mass spectrometry analysis and its data interpretation, any peptides chromatography separation device can be applied. In our laboratory, we used either HPLC or 2D-LC dependent on the quantity of the protein being analyzed, in 2D-LC, we included an enrichment column prior to chromatographic separation by reversed-phase column suitable for separation of peptides. This device helps to concentrate the peptides and therefore increase the sensitivity of the analysis method. The parameter for HPLC separation method for hCG analysis was 20 µl/min through-column flow rate at a gradient of 5-95% B for 20 minutes and held at 95% B constant for 5 minutes. Mobile phase A was 0.05% TFA in H2O and B was 0.05% TFA in ACN. The column used was C-18 Vydac column (300 Å, 5 µm, 1 mm X 50 mm). As for the parameters for tandem mass spectrometry analysis, it is advice to apply the optimum parameters for respective mass spectrometers. In our laboratory, an ESI – ion trap mass spectrometer (LCQ, ThermoQuest) is used, where data dependent experiment method consisting of MS and MS/MS scan was created. The peptide markers ions were programmed and MS data were acquired at heating capillary temperature 200°C, sheath gas flow rate was 60 arb, spray voltage at 4 kV, tube lens offset was -60 V and capillary voltage was at 38 V. MS/MS scan was conducted at default collision energy of 25, charge state of 2, minimum signal acquired was 1x105 counts, isolation width was 2 m/z.

Quantitative and Qualitative Analysis of hCG using Selected Peptide Marker

The qualitative and quantitative experimental method involved the introduction of the marker to the tandem mass spectrometric analysis. This is to obtain product ions spectrum that reveals the amino acid sequence of the peptide marker, which will then confirm the presence of hCG. The experiment was carried out in the data dependent acquisition mode, where the parent ion mass of the marker was programmed in the parent ion list. The programmed software will run in such a way that as the parent ion is detected in the full scan MS, they will be selectively excited for the MS/MS scan. The resulting product ions produced by CID fragmentation of the parent ion will be shown as the MS/MS spectrum that provides characteristic of the amino acid sequence of the peptide.

The determination of amino acid sequence of the peptide marker was included in hCG confirmatory analysis to avoid the possibility of producing false positive results, which is more likely to happen in selected ion monitoring (SIM) data acquisition method. SIM depends solely on the parent ion mass and the retention time of the compound of interest, which results in the loss of qualitative information of the compound (Gam et al 2003).

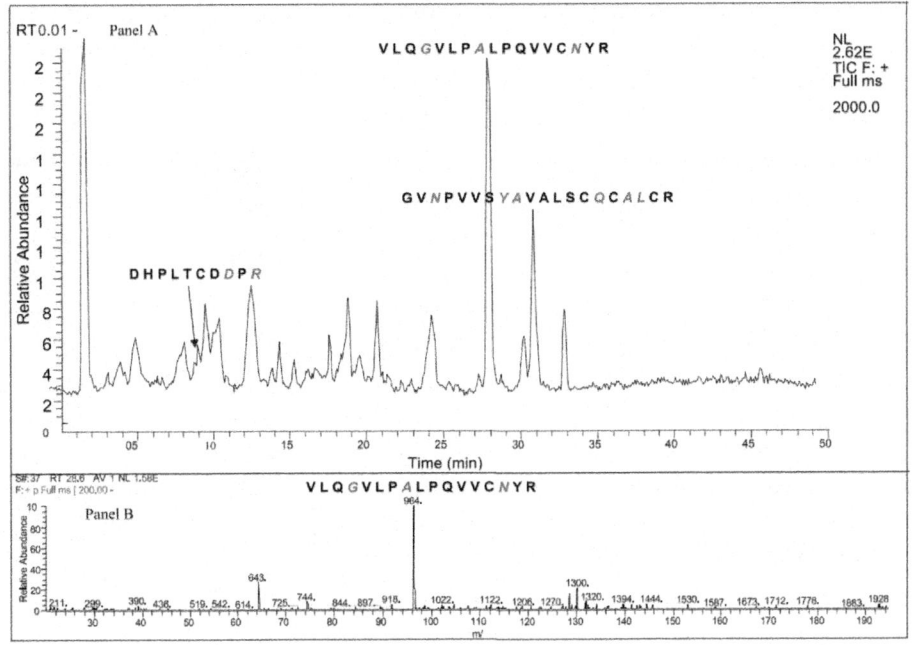

Figure. 2. Panel A; Based Peak Chromatogram for tryptic digested peptides. The y-axis is relative abundance and x-axis is retention time. Panel B; spectrum for peptide VLQGVLPAQVVCNYR. The y-axis is relative abundance and x-axis is mass to charge ratio.

Full scan MS/MS not only gives enough information for the qualitative data of hCG but at the same time, the intensity of the product ions of the peptide marker were used to quantify concentrations of hCG. The quantitative method using product ions spectrum allows relatively low quantification limit as compared to SIM method. This is because the MS/MS experiment is a technique that will minimize or eliminate all chemical and background noises.

For quantification purpose, three most abundant product ions of [964.7]2+, namely b6 +, b9 + and y11+ with the m/z ratios of [610.3]+, [891.5]+ and [1317.8]+ respectively were selected as the quantitative markers. These ions can be evaluated individually by peak area display in the extracted product ion chromatogram or by the summation areas of the three product ions (Figure 3). The summation of three product ions increases the total peak area and therefore greatly reduced the quantification limit as the signal to noise ratio is tremendously increased.

At 5 mIU/mL which approximately equal to 1 pg/mL of hCG in urine matrix, the signal to noise ration for each product ions chromatogram of the peptide marker were exceeded one hundred (Figure 3, upper panel). The high signal to noise ratio reveals that this method would be able to detect a much lower concentrations of hCG. This detection limit of hCG (5 mIU/mL) using our current method is superior than SIM method where a detection limit of 25 mIU/mL was reported (Liu & Bowers 1997).

At 5 mIU/ml hCG, the parent ions of the peptide marker was indistinguishable from the background noise in full scan MS. Nevertheless, as long as the parent ion intensity surpasses the threshold set, the ion will be isolated and excited to data dependent scan to generate MS/MS data. The MS/MS data obtained not only gives identification to the peptide; furthermore, the product ions were used to quantify the protein concentration.

At 5 mIU/ml hCG concentration, our currently developed method yields minimal, if there is any background interference. Quantitative analysis of hCG using SIM method at this concentration is not possible due to the same reason discussed above, where the parent ion was indistinguishable from the background noise, in this situation, the signal to noise ratio valid for quantification analysis could not be established.

Using this approach, we are able to conduct simultaneous qualitative and quantitative analysis on protein. The qualitative data (MS/MS scan data) confirmed the identity of the protein via its unique peptide marker while the product ions (quantifying ions) of the peptide marker were subjected to product ion extracted chromatogram to generate quantitative data of the protein. This approach avoids false quantification of ions, which is possible in SIM.

Method Validation

A standard curve was constructed using hCG at 5 mIU/mL (1pg/mL), 8 mIU/mL (1.6 pg/mL), 10 mIU/mL (2 pg/mL), 15 mIU/ml (3 pg/mL), 20 mIU/mL (4 pg/mL) and 30 mIU/mL (6 pg/mL) concentrations. The protein was subjected to tryptic digestion and analyzed by HPLC/MS/MS according to the method described. In which, the peptide marker was eluted from the column at 12 minutes retention time.

Each of the standard points was performed in triplicate. Tables 3 to 6 show the peak areas (triplicate) of the extracted product ions chromatograms for the chosen quantifying ions. The reliability of each product ions as a quantitative marker was measured by their coefficient of variance (r^2) values. The coefficient of variance for $[610.0]^+$, $[891.8]^+$, $[1317.8]^+$ and the summation of $[610.0]^+$ + $[891.8]^+$ + $[1317.8]^+$ were 0.998 (Figure 4), 0.993 (Figure 5), 0.997 (Figure 6) and 0.995 (Figure 7), respectively. Thus, it is obvious that the intensity of the product ions formed correlates well with the concentrations of hCG. The reliability of using the selected product ions as the quantitative markers is remarkable as the precision (C.V) of each standard point in triplicate is always < 10% (Tables 3 to 6). The linearity of the standard curve was only obtained in a narrow range of hCG concentration between 5 mIU/ml to 30 mIU/ml. Above this range, the curve deviates from linearity.

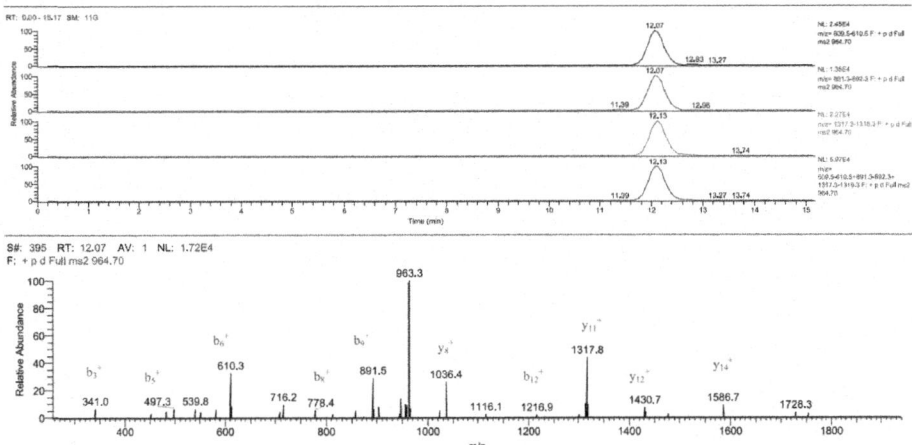

Figure. 3. Data for quantitative and qualitative analysis of hCG at 5 mIU/mL. Upper panel: extracted product ions chromatogram; Lower panel: MS/MS spectrum of the selected peptide marker (Gam et al, 2003).

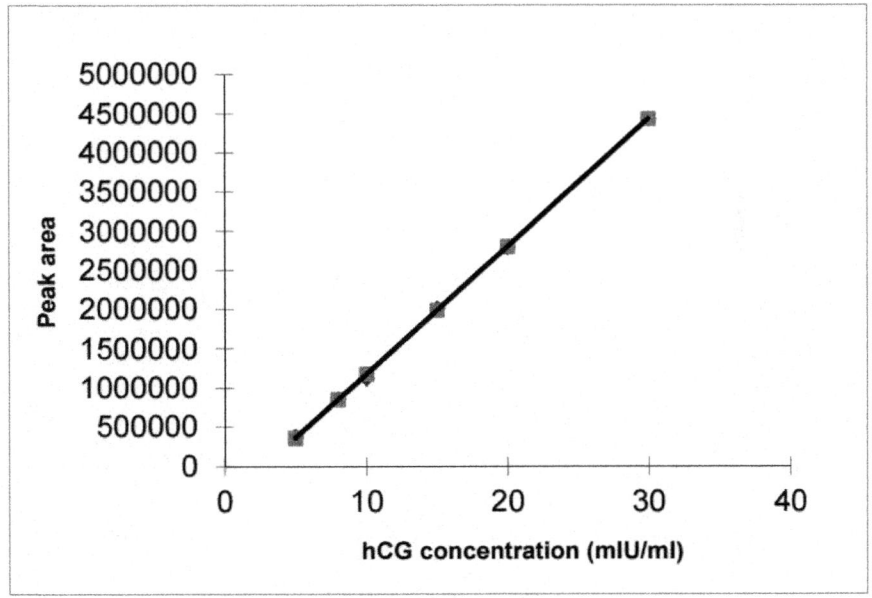

Figure. 4. The standard curve constructed using the b6 + product ion. Intercept = -457221, X variable = 1628623, r2 = 0.999647.

Table 3. Standard curve data quantify using the b_6^+, $[610.0]^+$ product ion

hCG mIU/ml	Peak area			means	SD	CV
5	364576	395481	367517	375858	17057.52	5 %
8	856239	894572	826756	859189	34004.11	4%
10	1173659	1127543	1055078	1118760	59776.41	5%
15	1995623	1973176	2060943	2009914	45595.35	2%
20	2748931	2865973	2762191	2792365	64090.25	2%
30	4468624	4487696	4341096	4432472	79706.43	2%

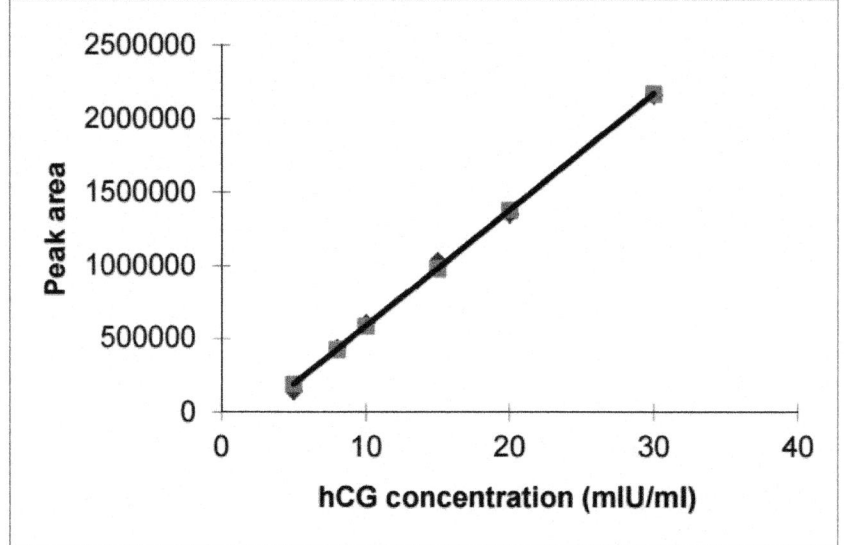

Figurer. 5. The standard curve constructed using the b_9^+ product ion. Intercept = -207664, X variable = 79251.53, r^2 = 0.99765.

Table 4. Standard curve data quantify using the b_9^+, $[891.8]^+$ product ion

hCG mIU/ml	Peak area			means	SD	CV
5	140221	152539	134155	142305	9367.51	7%
8	419835	449395	442145	437125	15406.13	4%
10	623484	601678	594782	606648	4876.21	1%
15	1017695	1013689	1065890	1032425	36911.68	4%
20	1280853	1400054	1365535	1348814	24408.62	2%
30	2198756	2231539	2052210	2160835	12680.80	6%

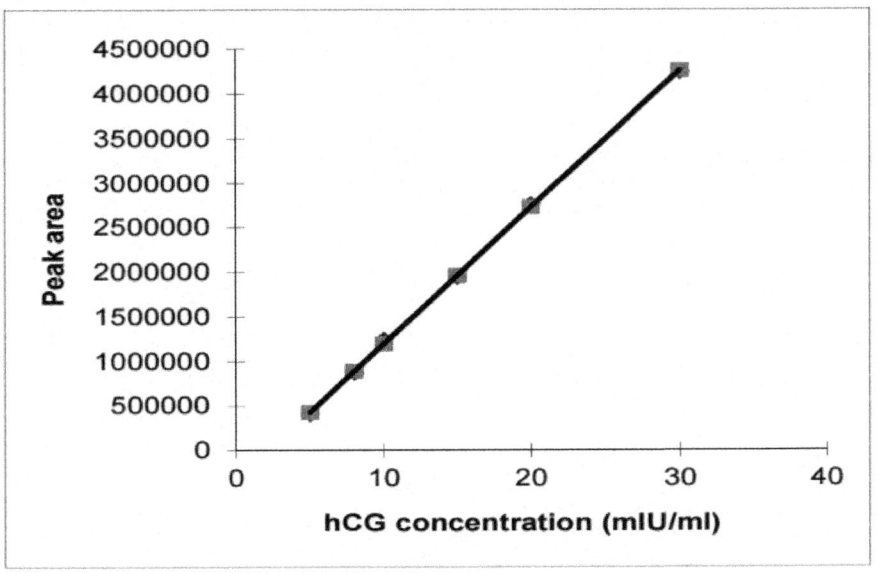

Figure. 6. The standard curve constructed using the $y_{11}{}^+$ product ion. Intercept = -338677, X variable = 153377.7, r^2 = 0.999483.

Table 5. Standard curve data quantify using the $y_{11}{}^+$, $[1317.8]^+$ product ion

hCG mIU/ml	Peak area			means	SD	CV
5	396514	419764	411420	409233	5900.10	1%
8	859176	883765	850385	864442	23603.22	3%
10	1327647	1262137	1141016	1243600	85645.48	7%
15	2032478	1863586	1929342	1941805	46496.51	2%
20	2759796	2789313	2737455	2762188	36669.14	1%
30	4245795	4353626	4132309	4243910	156494.8	4%

FUTURE WORK

I believed the method demonstrated here will be of help to protein chemists whom struggle in protein quantitative analysis, especially for analysis of trace amount of protein in complex biological sample. This method may be useful not only in doping analysis for hCG, it can be applied to other doped proteins such as erythropoietin, growth hormone and ext. This is because an accurate quantitative data (definite amount of doped substances) is needed in doping analysis to differentiate between endogenous and exogenous protein, a fine line between doped and non-dope level. At this time, my works are mainly focused on the identification of biomarkers in diseases, where the biomarkers can be used as diagnostic markers or therapeutic markers for the diseases. The quantitative data of the expression of the biomarkers are important in determining the usefulness of individual biomarker, therefore, this developed method has created a good platform for conducting quantitative analysis on the identified biomarkers.

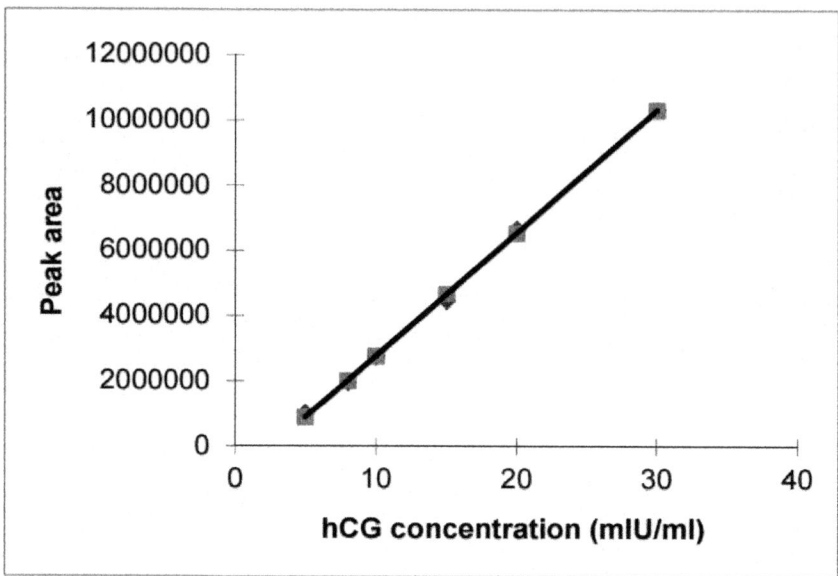

Figure. 7. The standard curve constructed using the summation of three product ions. Intercept = -1000044, X variable = 376994.7, r^2 = 0.998635.

Table 6. Standard curve data quantify using the summation of three product ions: b_6^+, b_9^+ and y_{11}^+

hCG mIU/ml	Peak area			means	SD	CV
5	985703	1083863	998570	1022712	60311.26	6%
8	1903745	2014766	1954676	1957729	42490.05	2%
10	2869837	2729474	2664510	2754607	45963.48	2%
15	4375897	4400165	4563987	4446683	115839.60	3%
20	6542834	6742038	6713948	6666273	19862.62	0%
30	10428756	10514285	10038730	10327257	336238.20	3%

CONCLUSION

Using hCG as an example, the approach for simultaneous qualitative and quantitative analysis of protein by using tandem mass spectrometry has been demostrated. Since the qualitative data for identification of protein was carried out through product ions profiling which revealed the amino acid sequence of the protein, this analysis method give a high confident level of protein identity. Subsequently, the high abundant product ions are selected as quantifying ion for quantitative analysis of the protein. This quantification approach eliminates all background noises. This is because the quantifying ions were chosen from the fragmentation of the selected peptide marker, which was isolated from the remaining ions in MS scan and excited to collision induced dissociation. Using this approach, it is not possible to quantify a false ion, where only the correct parent ion will produce the expected product ions profile and subsequently the product ions were chosen as quantifying ions. Thus, our current method satisfies both the qualitative and quantitative requirements for protein analysis, which normally can be achieved for one aspect but not the other.

ACKNOWLEDGEMENT

This work was carried out to accomplish of my Ph.D degree for a period of three years (1997-2000). I would like to express my gratitude to Universiti Sains Malaysia for providing me scholarship to pursue my study. Furthermore, I thank Prof Aishah Latiff, my Ph.D supervisor and also Doping Control Centre for provided me with support in terms of usage of chemicals and infrastructure for carrying out this research.

REFERENCES

1.   Bellar, T.A. & Budde, W.L. (1988). Interlaboratory comparison of methane electron capture negative ion mass spectra. Anal. Chem. 60:2076-2081.

2.   Bottger, V., Micheel, B., Scharte, G., Wolf, G. & Schmechta, H. (1993). Monoclonal antibodies to hCG and their use in two-site binding enzyme immunoassays. Hybridoma. 12 (1):81-91.

3.   Canfield, R.E., Birken, S., Morse, J. H. & Morgan, F.J. (1976). Human Chorionic Gonadotroin. In: Peptide Hormones. Parsons J.A. eds. The Macmillan press LTP. Pp 299-385.

4.   Canfield, R.E., Morgan, F.J., Kammerman, S., Bell, J. J. & Agosto, F.M. (1971). Studies of human chorionic gonadotropin. Rec. Progr. Horm. Res. 27:121-164.

5.   Eichelberger, J.W., Kerns, E.H., Olynyk, P. & Budde, W.L. (1983). Precision and accuracy in the determination of organics in water by fused silica capillary column gas chromatography/mass spectrometry and packed column gas chromatography/ mass spectrometry. Anal. Chem. 55:1471-1478.

6.   Endo, T. Nishimura, R. Saito S., Kanazawa, K., Namura, K., Katsuro, M.,

Shii, K., Mukhopaddyan, S., Baba, S. & Kobato, A. (1992). Carbohydrate structure of b-core fragment of hCG isolated from pregnant individual. Endocrinol. 130:2052-2058.

7.   Gam LH and Latiff A. (2006) Tandem mass spectrometic analysis of glycopeptides derived from the tryptic digestion of human chorionic gonadotropin (hCG). Malaysian Journal of Science. 25(2): 87-95.

8.   Gam LH, Tham SK and Aishah L. (2003), Immunoaffinity extraction and tandem mass spectrometric analysis of human chorionic gonadotropin in doping analysis. Journal of Chromatrography B, 79: 187-196.

9.   Liu, C.L. & Bowers, L.D. (1997). Mass spectrometric characterization of the b-subunit of human chorionic gonadotropin. J. Mass Spectrom. 32: 33-42.

10.   McCormack, AL, Jones, JL and Wysocki, VH. (1992). Surface-induced dissociation of multiply charged protonated peptides, J Am. Soc. Mass Spectrom. 3:859-862

11.   Shoemaker, J. D. & Elliott, W.H. (1991). Automated screening of urine samples for carbohydrate, organic and amino acids after treatment with urease. J. Chromatogr. 562: 125-129.

12.   Sweeley, C.C., Elliott, W.M., Fries, I. & Ryhage, R. (1966). Mass spectrometric determination of unresolved components in gas chromatography effluents. Anal. Chem. 38:1549- 1953.

13.   Vaitukaitis, J.L., Ross, G.T., Braunstein, G.D. & Rayford, P.L. (1976). Gonadotropins and their subunits: basic and clinical studies. Recent Prog. Horm. Res. 32:289-298.

Chapter 8

QUANTITATIVE MEASUREMENTS OF X-RAY INTENSITY

Michael J. Haugh[1] and Marilyn Schneider[2]

[1] National Security Technologies, LLC, USA

[2] Lawrence Livermore National Laboratory, USA

INTRODUCTION

This chapter describes the characterization of several X-ray sources and their use in calibrating different types of X-ray cameras at National Security Technologies, LLC (NSTec). The cameras are employed in experimental plasma studies at Lawrence Livermore National Laboratory (LLNL), including the National Ignition Facility (NIF). The sources provide X-rays in the energy range from several hundred eV to 110 keV. The key to this effort is measuring the X-ray beam intensity accurately and traceable to international standards. This is accomplished using photodiodes of several types that are calibrated using radioactive sources and a synchrotron source using methods and materials that are traceable to the U.S. National Institute of Standards and Technology (NIST). The accreditation procedures are described.

The chapter begins with an introduction to the fundamental concepts of X-ray physics. The types of X-ray sources that are used for device calibration are described. The next section describes the photodiode types that are used for measuring X-ray intensity: power measuring photodiodes, energy dispersive photodiodes, and cameras comprising photodiodes as pixel elements. Following their description, the methods used to calibrate the primary detectors, the power measuring photodiodes and the energy dispersive photodiodes, as well as the method used to get traceability to international standards are described. The X-ray source beams can then be measured using the primary detectors. The final section then describes the use of the calibrated X-ray beams to calibrate X-ray cameras.

Many of the references are web sites that provide databases, explanations of the data and how it was generated, and data calculations for specific cases. Several general reference books related to the major topics are included. Papers expanding some subjects are cited.

BRIEF INTRODUCTION TO X-RAYS: CHARACTERISTIC SPECTRAL LINES AND BREMSSTRAHLUNG

Characteristic X-Ray Spectral Lines from Atoms

The electronic structure of an atom, using Ag as an example and shown in Fig. 1, is:

$$^{47}\text{Ag: } 1s^2\, 2s^2\, 2p^6\, 3s^2\, 3p^6\, 3d^{10}\, 4s^2\, 4p^6\, 4d^{10}\, 5s^1 \tag{1}$$

Shown in Fig. 1 is the energy level diagram for the lowest four quantum numbers of the Ag ion, i.e., one of the 1s electrons, has been removed. How the electron is removed is covered in the next section.

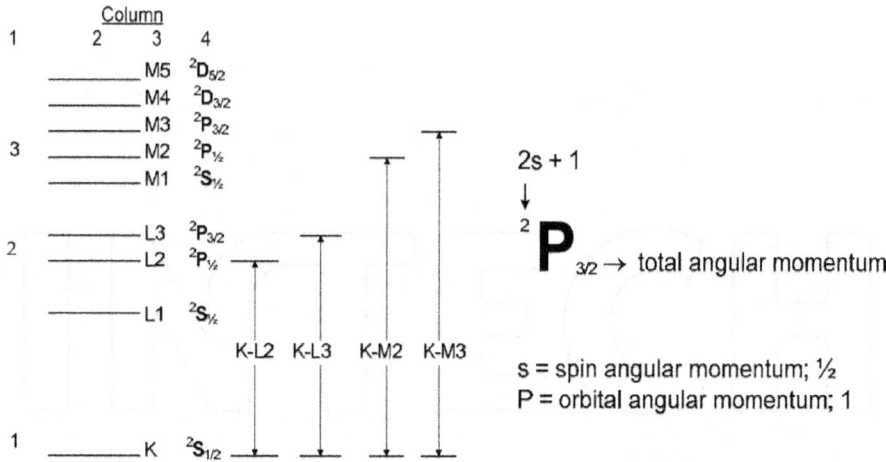

Figure 1. The energy level diagram for the lowest levels of the singly ionized Ag atom.

Column 1 is the principle quantum number, column 2 indicates an energy level, column 3 is the IUPAC designation for the state, and column 4 is the state electronic structure.

We use the Russell-Saunders angular momentum coupling scheme (Herzberg, 1945) to describe the electronic structure of each state. The notation is illustrated in Fig. 1. The superscript "s" denotes $2s + 1$ where "s" is the total spin of the state. In this case, it is a doublet state since there is an unpaired electron (one electron has been removed). The upper case letter indicates the orbital angular momentum (S means zero angular momentum; P means one unit angular momentum; D means two units, etc.) The subscript indicates the combination of orbital angular momentum and spin angular momentum. This description is somewhat simplified but it gives insight into transition probabilities and what X-ray transitions are expected along with their relative intensities.

When an electron has been removed from the lowest energy level, the $1^2S_{1/2}$ state, an electron can drop from a higher level with the simultaneous emission of a photon. The relative energies of these states can be obtained from the binding energies of the electrons in each state, and these are given in Table 1. A good source for this information is the Center for X-Ray Optics web site of the Lawrence Berkeley National Laboratory (CXRO reference).

The readily observed X-ray spectral lines to the electron deficient K level are shown in Fig. 1 column 4 and the energies and relative probabilities are given in Table 2.

Both the Siegbahn and the newer International Union of Pure and Applied Chemistry (IUPAC) notations for the transitions are given. The IUPAC notation is a bit less obscure but the Siegbahn notation is still more popular in current literature. In the IUPAC notation, the number refers to the order of the energies in the shell as shown in Fig. 1, so that L1 refers to the n=2 $s_{1/2}$, L2 refers to the $p_{1/2}$, and so on. Note that the spectral emission energy can be estimated by taking the difference between the corresponding binding energy given in Table 1. This estimate is reasonably accurate for K transitions, but care should be taken when using this estimate for higher energy level transitions. If the electron is removed from the n=2 shell, the spectral emission is referred to as an L line. The set of easily observed emission lines is given in Table 3.

Table 1. Electron Binding Energies for the Ag Atom

Energy level	IUPAC	Energy, eV
K1s	K 1	25514
L2s	L1	3806
L2$_{p1/}$2	L2	3524
L2$_{p3/}$2	L3	3351
M3s	M1	719.0
M3$_{p1/}$2	M2	603.8
M3$_{p3/}$2	M3	573.0
M3$_{d3/}$2	M4	374.0

M3$_{d5/}$2	M5	368.3
N4s	N1	97.0
N4$_{p1/}$2	N2	63.7
N4$_{p3/}$2	N3	58.3

Table 2. The K Type X-ray Spectral Lines for the Ag Ion

Siegbahn Designation	IUPAC Designation	Spectral Line Energy, eV	Relative Intensity
Kα2	K-L2	21990.3	53
Kα1	K-L3	22162.92	100
Kβ3	K-M2	24911.5	9
Kβ1	K-M3	24942.4	16
Kβ2	K-N2,3	25456.4	4

Table 3. The L Type X-ray Spectral Lines for the Ag Ion

Siegbahn Designation	IUPAC Designation	Spectral Line Energy, eV	Relative Intensity
Lα2	L3-M4	2978.21	11
Lα1	L3-M5	2984.31	100
Lβ1	L2-M4	3150.94	56
Lβ2,15	L3-N5,6	3347.81	13
Lγ1	L2-N5	3519.6	6

Spectral Line Widths, Lifetimes, and Competing Processes

The X-ray spectral lines have a narrow width relative to the photon energy. The line widths for several fluorescence transitions in the Ag singly ionized atom are given in Table 4. One can estimate the fluorescence lifetime for the line width using the uncertainty relation given in Equation (1):

$$\Delta E^* \Delta t \geq h \quad (2)$$

ΔE =fluorescence line width, eV

Δt =lifetime of the fluorescence state, sec

h = 4.135x10^{-15} eV•s, Planck's constant

The calculated lifetimes for the Ag transitions are given in Table 4. The fluorescence process can also be treated as a rate of decay from the higher energy state to the lower energy state with the rate constant given as the reciprocal of the lifetime. The rate constants for the two excited states Ag^+ ion decay is given in Table 4.

Table 4. Line widths and fluorescence lifetimes for several Ag transitions

Ag Transition	Line Width, eV	Lifetime, sec	Decay Rate Constant, $se^{c-}1$
K-L2	8.9	4.65×10^{-16}	2.15×10^{15}
K-L3	8.6	4.81×10^{-16}	2.08×10^{15}
L3-M4	2.2	1.88×10^{-15}	5.32×10^{14}
L3-M5	2.34	1.76×10^{-15}	5.66×10^{14}
L2-M4	2.4	1.72×10^{-15}	5.80×10^{14}

There are other processes that compete with the fluorescence process. The process that most affects the X-ray fluorescence is called the Auger effect after Pierre Auger, although it was first discovered and published a year earlier by Lise Meitner. The Auger effect describes the transfer of energy that can occur when a vacant state is filled by an electron from the next higher state, but the energy for this transition is transferred to an electron in a higher state which is ejected from the ion, carrying the excess energy as kinetic energy. For example, consider an ion with a hole in the K shell that is filled by an electron from the L1 state. For the Auger process, the energy from the K-L1 transition is transferred to the L2 electron which is ejected from the ion with kinetic energy equal to the difference between the energy for the K-L1 transition and the binding energy of the L2 electron. The rate constant for this Auger process is about 1×10^{15} sec^{-1}. Comparing this rate to the fluorescence rate for the Ag K transitions, we note that the Auger rate is smaller so that the fluorescence yield for that condition will be about 85% of the of the total rate for filling the hole in the K shell. For K transition energies near 9keV (atomic number near 30), the fluorescence decay rate and the Auger rate are about equal and the yields will be about 50%. For lower atomic numbers, the fluorescence yield will be lower than 50% and conversely, the yield will be larger than 50 % for atomic numbers larger than 30. Recall that the line width of the transition is determined by the total rate of the excited state decay. For the Ag^+ K transition the fluorescence rate is dominant. For L transitions, the Auger rate dominates up to the atomic number 100. It is only at this value of Z that the fluorescence yield is 50%. So the line widths for L transitions are determined by the Auger process and never drop below 2 or 3 eV. Refer to graphs of the relative yield as a function of atomic number (Podgorsak, 2010). There are several other internal conversion processes that compete with fluorescence but their rates are much lower and they will not be discussed here. The Auger effect is used for chemical analysis by measuring the kinetic energy of the Auger electron, a technique called Auger electron spectroscopy. The other competing processes also have a niche in analytical chemistry.

Electron Impact to Produce X-Rays

The X-rays used in our measurements are primarily produced by the impact of electrons on solid materials. When an electron moving at a high velocity enters a solid material it deposits its energy in the solid in a variety of ways. Most of the energy ends up heating the anode but our interest is in that small percent of interactions that produce X-rays. Our strongest interest is in the collisions that remove an inner electron from the target material and produce the characteristic X-ray spectral lines from the atom. In fact, the X-radiation produced by the interactions of the electron with the solid material is a small fraction of the electron's energy loss processes. As can be seen in the NIST ESTAR tables (ESTAR), the radiation yield from Ti for electron impact in the energy range of 110 keV is less than 0.5%, and the majority of that radiation is bremmstrahlung. Bremmstrahlung is the spectrum of X-rays produced by the deceleration of electrons. Fig. 2 shows a typical emission produced by electron impact for a Ti anode target. The transition energies are given in Table 5.

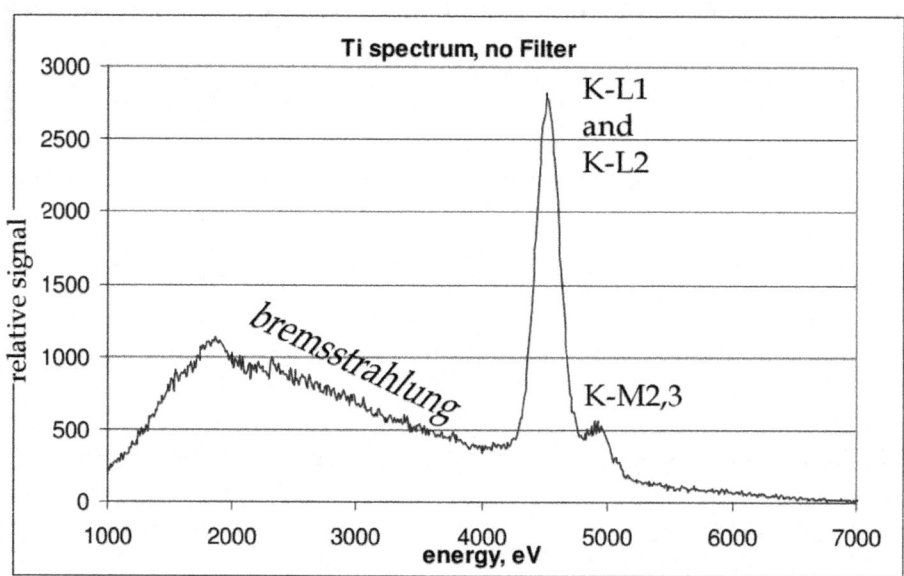

Figure 2. This is a spectrum of the X-rays produced when an accelerated beam of electrons strikes a Ti anode.

Table 5. Transition energies are taken from the NIST X-ray database

Transition	K-L2	K-L3	K-M2,3	K-M4,5
Energy, eV	4504.92	4510.899	4931.83	4962.27

Ti K X-ray Emission Lines

The anode was at 8,000 V and the heated filament electron source was near ground potential. The energy dispersive detector that was used to take this spectrum has a resolution near 240 eV. The tall, narrow band near 4500 eV comprises the K-L3 and the K-L2 Ti spectral lines. The spectral lines of Ti are approximately 2 eV wide. The bremsstrahlung is the broad band ranging from less than 1000 eV to 7000 eV and peaking near 2000 eV. The count of photons in the bremsstrahlung is 1000 times larger than the counts in the spectral lines.

The characteristic radiation depends on the anode material properties and the energy of the impacting electron. We have observed that the intensity of emission for characteristic lines follows this equation:

$$I_c = k \, (V_e - V_b)^n \qquad (3)$$

I_c =intensity of the characteristic X-ray line

k =proportionality constant

V_e=accelerating voltage of the electron

V_b=binding energy of the bound electron

n=a number somewhat greater than 2

Using this introduction to the basics of X-ray emission, the sources used to produce X-ray emission are presented in some detail in the following section.

X-RAY SOURCES

The Diode Source

NSTec laboratories have four X-ray sources that cover the X-ray spectral energy range from 50 eV to 110 keV. All the primary X-ray sources are the diode type; electrons are emitted from a heated tungsten filament, and then accelerated by an electric field to strike an anode. Two sources use a secondary beam that is generated when the primary beam strikes a sheet of material that fluoresces.

The diode sources produce spectral lines that are characteristic of the anode material and a broad spectrum of radiation known as bremsstrahlung, peaking near one-third of the accelerating voltage. A typical diode source is shown in Fig. 3. The filament is heated by an independent electrical circuit that is near ground potential. The anode is maintained at a high positive voltage so that the electrons emitted from the filament are accelerated and strike the anode at the energy determined by the voltage difference. The electric field is shaped using guide wires. X-rays are emitted in all directions and some exit the aperture, as shown in Fig. 3, and enter into the sample chamber. The anode is water-cooled so that a high beam current can be tolerated, thus giving a strong X-ray intensity. This intensity allows collimation of the X-ray beam with a pair of slits, as well as isolation of individual spectral lines using a diffraction crystal. The narrow band X-ray source can measure sample properties such as filter transmission, crystal reflectivity, and sensor efficiency. The source and sample chamber are in vacuum. The voltage supply is 20 kV, making the highest available spectral line nearly 17 keV (Zr K spectral lines).

The other diode source uses anodes that are cooled only by thermal conduction through the mechanical connections. This limits its operation to 10 W and 10 kV, with a usable spectral range from 700 eV to 8400 eV. It is often used to measure the absolute efficiency of X-ray cameras and the sensitivity variation across the sensor pixels.

The third source covers the spectral energy range from 8 to 111 keV. It uses X-rays from a diode source to produce fluorescent X-rays from a fluorescer material. This source is progressing toward NVLAP accreditation and will be described in detail in the next section. The fourth source is currently being built and will cover the X-ray spectral region from 50 eV to several keV, also operating on the fluorescer principle.

The NSTec X-ray sources are used to calibrate and characterize components or complete systems that are used in the study of plasmas and similar efforts. A large component of our present calibration efforts is for diagnostics that are used on the NIF target diagnostics.

Reducing the Band Width of the Source: Filters, Grazing Incidence Mirrors, and Diffraction Crystals

The emission from a diode source produced by the impact of the electrons on the anode has a broad band of bremsstrahlung and the characteristic spectral lines from the anode composition as was shown in Fig. 2. The large amount of bremsstrahlung X-rays does not allow one to use the raw emission from the diode source to accomplish calibrations such as measuring the energy dependence of a detector's sensitivity. There are several methods for reducing the spectral band width of the raw diode emission:

- using thin sheets of solid materials that can act as high pass filters;
- using a high pass filter combined with a grazing incidence mirror to make a band pass filter;
- using fluorescers that produce only the spectral lines of the fluorescer sheet; and
- using diffraction crystals to reflect only the X-rays that meet the Bragg angle requirements.

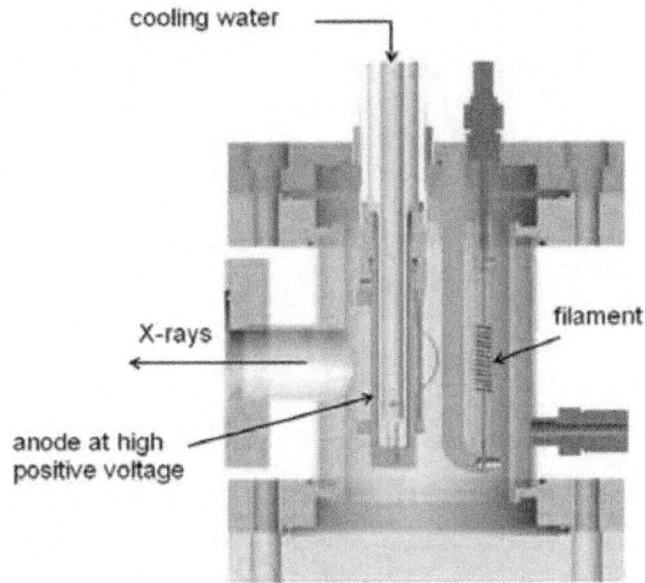

Figure 3. Example of an X-ray diode.

Filters

Thin sheets of solid material absorb X-rays and the transmission of the sheets depends upon the X-ray energy, the material thickness, and the atomic number Z of the material. Gases can also absorb X-rays but are not practical as filters for the applications described in this chapter. The transmission of a Ti sheet that is 25 μm thick is shown as Fig. 4(a).

The X-rays are absorbed in the Ti until the X-ray energy gets above 3000 eV. At the binding energy of the Ti 1s electron, 4966 eV, referred to as the K edge, the X-rays are again strongly absorbed. The sheet begins transmitting X-rays again when the X-ray energy rises above 6000 eV. This is the typical behavior of the X-ray transmission for solid materials. The transmission of materials for X-rays up to 30 keV is readily obtained using the CXRO web site. For higher energies, one can obtain absorption cross sections in the NIST tables.

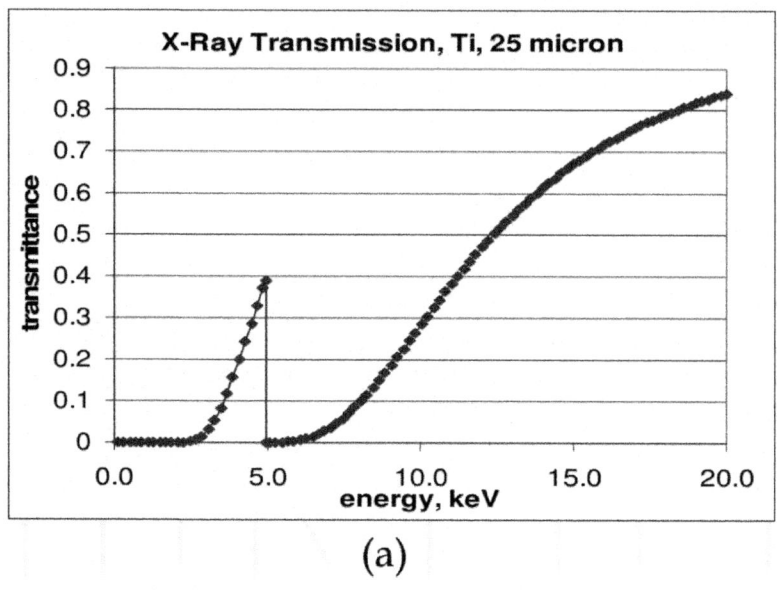

(a)

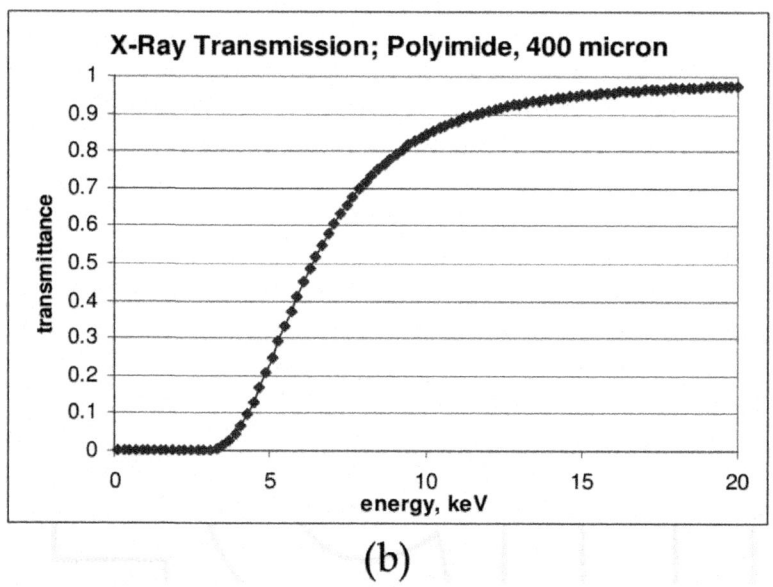

(b)

Figure 4. Graphs showing the X-ray transmission of (a) a 25 μm thick sheet of Ti, Z=22, and (b) a 400 μm thick sheet of polyimide.

The transmission characteristics shown in Fig. 4(a) can be used to make a band pass filter for transmission of the Ti K lines when the electron accelerating voltage is at 8000 eV or lower and the Ti filter is sufficiently thick. This application will be discussed in more detail in the description of camera calibrations. High pass filters can be made from low Z materials and plastics are the most convenient. The transmission of 400 μm thick polyimide is shown in Fig. 4(b). The DuPont version of this material is called Kapton and the material is reasonably resistant to X-ray damage. The X-ray energy at 50% transmission is near 6 keV and the range of X-ray energy for the transmission range from 10% to 90% is 6 keV. This is a very broad cut off for the high pass filter.

Grazing Incidence Mirror

In materials, the index of refraction for X-rays is complex, with a magnitude slightly less than 1. The consequence of this is that an X-ray beam incident from vacuum onto a material is mainly absorbed, unless it is incident at a shallow (grazing) angle to the surface. Since the vacuum is the more optically dense region, the X-ray experiences "total internal reflection" and is specularly reflected. This forms the basis of grazing incidence X-ray mirrors. These mirrors reflect X-rays at the specular angle for angles less than a few degrees. As the mirror is rotated with respect to the direction of the X-ray beam, at some angle the reflected intensity will start to decrease and will eventually go to zero reflected intensity. The angle at which the X-ray intensity drops to 50% of the reflection at very low energies is referred to as the maximum reflection angle. The maximum reflection angle is a function of the X-ray energy, the mirror composition, and the mirror roughness. Calculated reflectivity curves for various materials and surface roughness can be obtained from the CXRO web site. A typical measured grazing incidence reflectivity curve is shown in Fig. 5 (green scatter). The corresponding calculated reflectivity curve is shown in red in Fig. 5. Given their angular dependence, grazing incidence mirrors are often used as low pass filters. The combination of a grazing incidence mirror with an appropriate thin sheet filter described previously forms a band pass filter.

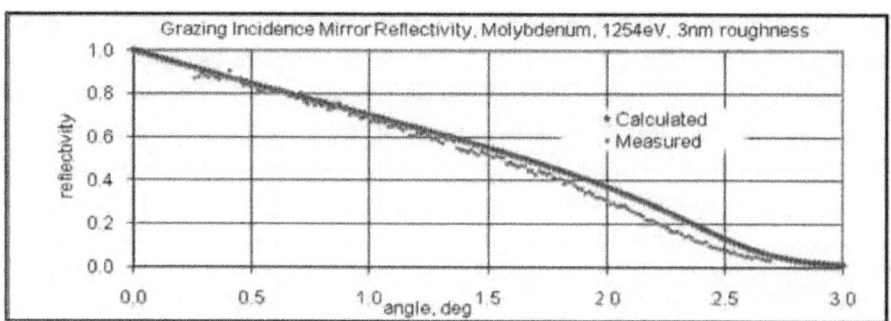

Figure 5. This graph is a comparison of the measured reflectivity curve for the molybdenum grazing incidence mirror at an X-ray energy of 1254 eV with the calculated reflectivity with a surface roughness of 3 nm rms.

The reflectivity curve for a grazing incidence mirror is affected by materials adsorbed on the surface. Water vapor and oxygen can significantly affect the reflectivity curve. For this reason, the grazing incidence mirror reflectivity curve is usually calibrated before it is used in experimental applications. This can be done using the NSTec sources. The synchrotron at Brookhaven is also used for these calibrations.

Diffraction Crystal

Crystals are often used to isolate individual spectral lines from a diode source. They are used in plasma diagnostics as components of a spectrograph. The crystal reflectivity follows the Bragg law for the location of the maximum reflection as a function of X-ray energy:

$$n(12398.425/E) = 2d\sin\Theta \qquad (4)$$

n=an integer equal to the diffraction order

E=X-ray energy, eV

d=distance between the crystal planes, Å

Θ=angle between the X-ray beam and the crystal plane

For n=1 and a given Θ, only the X-rays having the energy E given by the Bragg law will be reflected. For a monochromatic plane wave the Bragg reflection curve has a finite width. Theoretical calculations of the reflection curves for many crystals can be obtained at the Argonne web site (Stepanov, 1997&2009). Real crystals can approach this theoretical width if properly made. Two of the NSTec sources have the ability to measure the reflectivity curve of flat and curved crystals such as those made of mica. (Haugh & Stewart, 2010) The use and calibration of crystals is not covered in this chapter.

The Manson Type Diode Source: An X-Ray System Used For Calibration

One of the NSTec diode type X-ray sources that is used for testing and calibrations generates X-rays in the energy range from 400 eV to 9 keV. We refer to this as the Manson source since this was the manufacturer. The source is not water cooled, and the power is limited to 10 W to avoid melting the anodes. The filament is shaped to a point near the anode. This produces a small spot, approximately 1 mm diameter, where the electrons impact the anode. This small X-ray emission spot acts as a point source providing a flat X-ray intensity in the sample region allowing us to do radiographic type measurements and to measure the sensitivity variation across the sensor array of a camera.

Fig. 6 shows a schematic diagram of the NSTec Manson system, looking down on it from above. The Manson comprises three compartments: the source chamber and two testing chambers which are the rectangular boxes in the figure. The two test chambers are connected to the main chamber by stainless steel vacuum components that include an isolation gate valve and a mechanical shutter. The diagnostic that is shown attached to the top arm in the figure is at vacuum. Components, such as filters, can be mounted inside the chamber

Each test chamber has its own vacuum pump and controls and can be isolated from the source chamber by a gate valve, then brought to atmosphere. Test chambers have photodiode and an energy dispersive detectors for measuring X-ray flux and the X-ray spectrum, mounted on push rods so that they can be moved into or out of the beam.

The X-ray beam paths that are used for testing are shown in red in Fig. 6. Filter 1, shown in the source chamber, is used to isolate a narrow wavelength band of X-rays. These filters are mounted in a vertical stalk that holds up to three filters. A light blocker prevents visible light emitted by the filament from entering the test chamber which would overwhelm the detectors and CCD.

The Manson system is a multi-anode device, holding up to six different anodes on a hexagonal mounting bracket. Two X-ray beams are isolated from the anode emission for use in the test chambers. A typical X-ray emission produced by the impact of electrons with a metal anode was shown in Fig. 2. The Ti spectrum that is observed when a 100 μm thick Ti filter is placed between the X-ray source and the detector as a band pass filter is shown in Fig. 7. See also Fig. 4(a) for the spectral characteristic of a thin sheet of Ti. Comparing the unfiltered Ti spectrum shown in Fig. 2 with the filtered spectrum shown in Fig. 7, we can see that the transmission is now limited to the spectral energy range between 4000 eV and 4966 eV, the latter being the K edge of Ti. The spectral content now includes the Ti K lines and the bremsstrahlung within the energy range given.

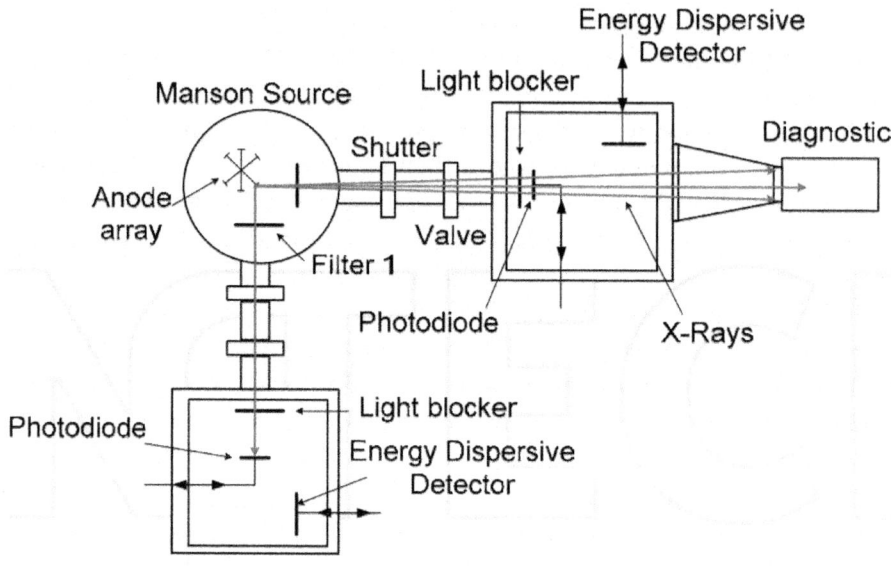

Figure 6. Manson Schematic. The diagnostic being calibrated is shown directly attached to the chamber at the end of the upper arm. The red lines are the X-ray beam path.

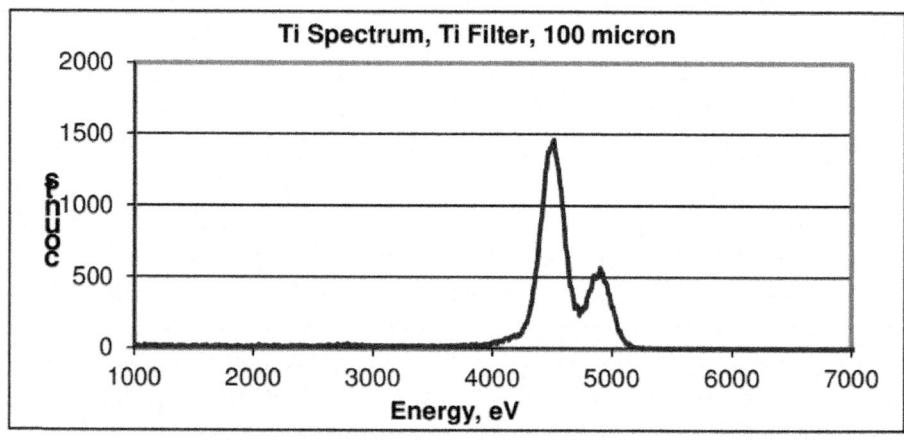

Figure 7. The spectrum of Ti X-rays shown in Fig. 2 using a Ti filter 100 micron thick to limit the spectral bandwidth.

Fluorescer Source

The High Energy X-ray system (HEX) uses a diode type source to produce monochromatic X-rays. X-rays from the diode (a commercial 160 kV X-ray tube) excite characteristic X-ray lines in the fluorescer foil. The X-ray tube and the fluorescing targets are enclosed in a lead box. An exit collimator in the lead box shapes the X-rays into a beam. The fluorescer operation is illustrated by Fig. 8. For this example, the fluorescing material is a thin lead (Pb) sheet, with a thickness of approximately 250 μm, and the filter is a thin platinum (Pt) sheet. Table 6 gives the properties of the fluorescer and filter. The high energy X-ray lines are transmitted by the filter but the low energy lines are stopped by the filter.

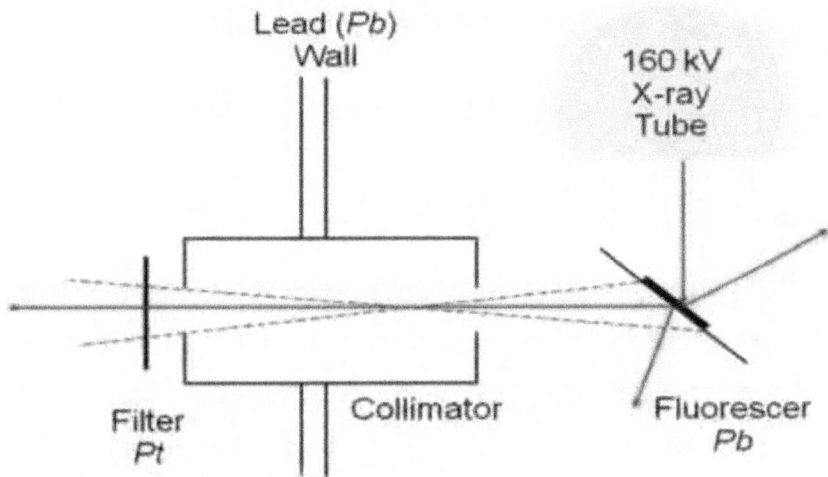

Figure 8. Illustration of fluorescence principle.

Table 6. X-ray Fluorescence

Filter	Fluorescer
Platinum (Pt), 50 μm thick	Lead (Pb)
Transmission	**Spectral Lines, keV**
0.72	73, 75
0.40	85
2 x 1⁰·5	10.4, 10.6, 12.6, 14.8
0	2.3

This method provides a reasonably narrow spectral energy that can be used to calibrate detectors at a range of well defined energies. The resulting spectrum from the arrangement is shown in Fig. 9.

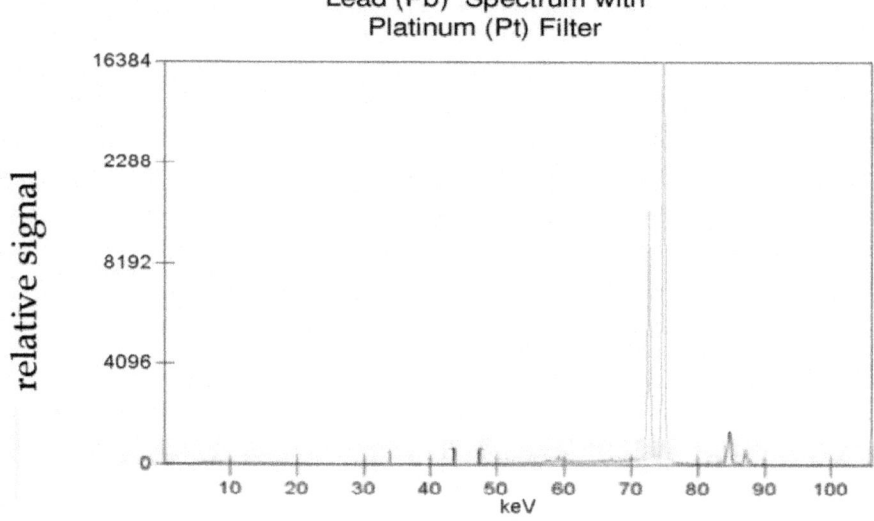

Figure 9. Pb Spectrum, Pt Filter.

The arrangement of the components is shown in Fig. 10(a) and (b).

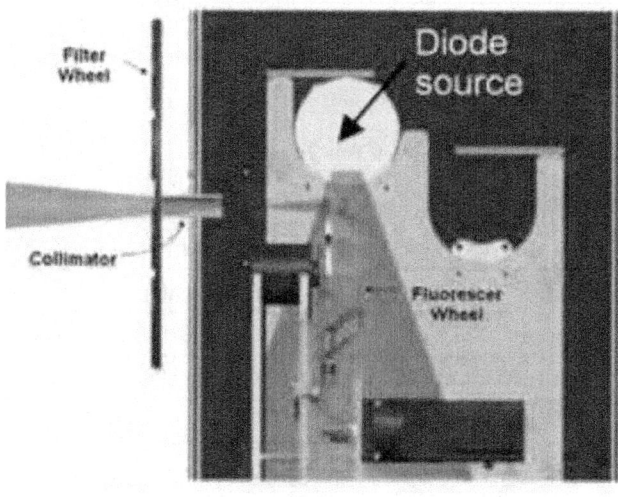

(a)

Pb chamber containing the diode
source and the fluorescer wheel Filter Wheel

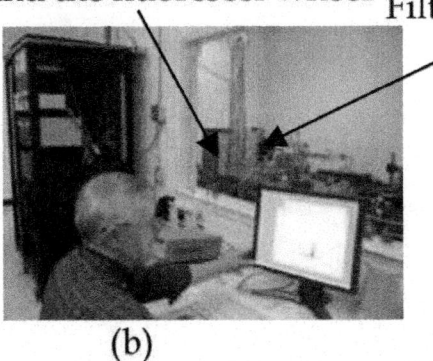

(b)

Figure 10. a) HEX source component inside the Pb chamber and (b) a view of the control room looking through a window at the HEX optical table.

The end of the commercial X-ray tube is shown in yellow. The pink trapezoid that starts at the tube represents the primary X-ray beam. The fluorescers are mounted on the motorized wheel in the rectangles shown on the wheel. The fluorescer emits in all directions, but the X-ray beam is defined by the collimator inserted into the wall of the lead box, and the beam path is illustrated by the pink triangles. There is a filter wheel mounted downstream from the collimator, and it is also motorized. The fluorescer and the filter can be set from the computer in an adjacent control room, as shown in Fig. 10(b). The fluorescer is usually a thin sheet made of elemental metal, but metal compounds are sometimes used. The maximum intensities obtained when an 11.5 mm diameter collimator is used are on the order of 1×10^6 photons per cm^2 per second, at one meter from the fluorescer, depending on the fluorescer material. The spectral lines used range from 8 keV to 115 keV.

Remote adjustment of the fluorescer wheel and the filter wheel is done through the control room computer. Data from the detectors and devices being calibrated are received in the control room.

The HEX source sits at the end of an optical table as shown in Fig. 11. The sample and the detectors are mounted on the optical table. The control room is separated from the HEX laboratory, as seen in Fig. 10(b).

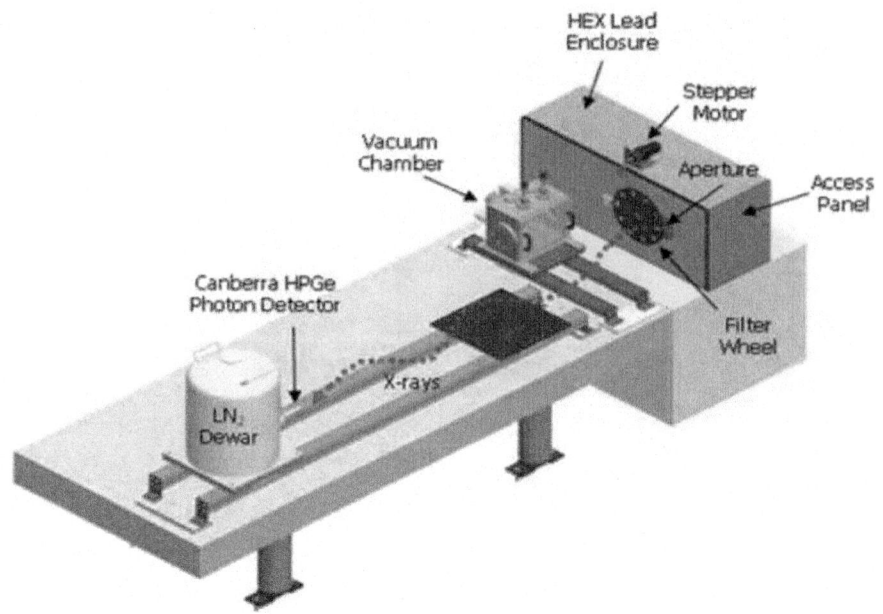

Figure 11. HEX Layout.

SOLID STATE DETECTORS

Introduction to Semiconductor Detectors

In the field of X-ray measurements, the term solid state detector usually refers to semiconductor type detectors. There are other X-ray detectors that are solid materials, such as scintillator –photomultiplier combinations, but for this work the term solid state detectors will include only semiconductor detectors. There are 3 types of semiconductor materials used for the work described in this chapter: silicon (Si), germanium (Ge), and cadmium telluride (CdTe). A semiconductor is defined as a material that has a small band gap (which can be manipulated) between the valence electrons and the conduction band, on the order of 1 eV to several eV. A metal has electrons populating the conduction band at any temperature, and an insulator has a large gap, on the order of 10 eV and higher. At normal temperatures, the semiconductor will have some population of electrons in the conduction band, and corresponding holes (positive charge where the electron vacated) in the valence band. The ratio of electrons in the conduction band to those in the valence band is given by the Boltzmann probability relation:

$$`N/N_0 = CT^{3/2} \, e^{-(E/kT)} \tag{5}$$

N=population of electrons in the conduction band

N_0=population of electrons in the valence band

E=band gap energy, eV

k=Boltzmann constant, 8.617343×10^{-2} ev/K

T=Absolute temperature, K

C=A material property

    Consider an X-ray photon incident on the semiconductor. It has an energy that is many times that of the band gap. It interacts with the semiconductor material, primarily through the photoelectric effect, to produce energetic free electrons. These then produce electron-hole pairs. The number of electron-hole pairs produced is proportional to the energy of the X-ray photon. In general it takes several eV to produce an electron-hole pair, the exact energy depending on the semiconductor material. The number of electron-hole pairs produced by an X-ray photon depends upon the material and the X-ray photon energy as given by:

$$N_{e-h} = eV/\varepsilon \tag{6}$$

N_{e-h}=number of e-h pairs produced

eV=X-ray photon energy, eV

ε=energy required to form an e-h pair in the semiconductor material

ε is usually referred to as the "ionization energy." Table 7 gives the ionization energy and the semiconductor band gap for the three materials we are discussing. The electron-hole pair formed will drift apart. But if a voltage is applied, the electrons will move in the direction opposite to the electric field, and the holes will move in the direction of the electric field, so the charges can be collected and measured.

    The room temperature population of electron-hole pairs of the semiconductor (Eq. 4) will be noise in the measurements. Details of semiconductor behavior can be found in Knoll (2001), which describes the physics and behavior of semiconductor detectors.

Table 7. Properties of semiconductors discussed in this section (*Quaranta, 1969)

Semiconduc-tor material	ε, Energy required for one hole/pair production, eV	Semiconductor band gap, eV
Si	3.63	1.116
Ge	2.96	0.665
CdTe	4.43*	1.44

There are several ways that the signal is lost and the solid state detector output fails to give the full measure of the X-ray intensity. Electrons produced by the X-ray interaction can recombine with another hole and electron lost from the charge collection process. Charge can be trapped by impurities or lattice defects and enhance the recombination process. For a properly designed semiconductor detector, this problem can be minimized. A semi-conductor typically has a coating on the surface that protects the active semiconductor material from damage by the environment. This surface coating can absorb some of the X-ray beam and this will not generate a current out of the detector. This front surface absorption will have greater impact for the lower energy X-ray photons. For the higher energy photons, at some energy they will begin to pass through the detector's active material and will not be detected. Another signal loss mechanism occurs when the X-ray energy rises above the K edge of the semiconductor active material. The semiconductor can fluoresce and this energy is lost from the detector.

Photodiode Used to Measure X-Ray Power

Calibrated photodiodes are used to measure the X-ray power of X-ray sources. The sources are used to calibrate X-ray detectors. This description of photodiode operation will use the photodiode design that was used to measure the efficiency of X-ray CCD cameras and other imaging and non-imaging detectors. The photodiode is made from Si, 55 μm thick. If all of the X-ray photon energy is deposited in the active Si, and 100% of the e-h pairs formed are collected, the diode current i is given inEquation (6) as:

$$i = F \cdot (eV/\varepsilon) \cdot (1.6 \times 10^{-19}) \text{ amp} \qquad (7)$$

1.6×10^{-19} coulomb = Charge on the electron

F=Rate of X-ray photons hitting the photodiode, number/s

ε =Energy needed to form an electron/hole pair

eV =Photon energy

Applying Equation (6) to a Si diode with a 1 cm^2 area, with an incident beam of Ti Kα1 (4510.84 eV) X-rays, with a photon intensity of 2×10^5 photons•cm⁻

$^2 \cdot s^{-1}$, yields a current of 39.8×10^{-12} A. This can be readily measured using a commercial picoammeter.

The silicon photodiode in use was designed to measure X-ray intensity with nearly 100% efficiency for X-ray energies up to about 5 keV (IRD). Several of these photodiodes have been calibrated from 1000 eV to 60 keV at the Physikalisch-Technische Bundensanstalt (PTB Reference, Gottwald, 2006)). Within the measurement uncertainty, which is near 1%, the silicon photodiode is 100% effective up to the Si K edge, where it dropped several percent. The efficiency rises back to near 100% by 3 keV. The 55 µm thick Si photodiode begins to transmit near 5 keV X-ray energy, and at higher energies the diode efficiency follows the Si transmission curve.

Energy Dispersive Photodiodes

Some photodiodes are designed to measure the energy of the X-ray photons. The types of semiconductor materials that are commonly used include Si, Ge, and CdTe. A bias voltage is applied to the semiconductor and the electric fields generated require cooling of the detector. The voltage pulses produced by an individual X-rays are amplified and then counted according to pulse height. Pulses that have heights within a certain range are effectively assigned to channels according to the average pulse height by a processor referred to as a multi-channel analyzer. This produces an energy spectrum of the detected X-ray photons. The resolving power of the energy dispersive detector is generally limited to several hundred eV.

The detector sensitivity falls off at lower energies due to absorption at the front surface by a "dead" layer of the sensor and/or a window separating the vacuum chamber containing the detector from the environment. It falls off at higher energy when the photons begin to be transmitted by the sensor. The Ge and CdTe detectors are designed to operate optimally in the 10 keV to slightly over 100 keV. The X-ray photon interacts with the sensor material in ways other than forming electron-hole pairs, which can reduce its sensitivity. If a photon has sufficient energy, it can knock out a 1s electron of the sensor material, a higher state electron can transition down into the 1s vacancy. A second photon having an energy equal to the energy between these two states can then be emitted. The sensor then produces a corresponding pulse of electron-hole pairs that is smaller than the base peak by the energy difference between the incoming X-ray photon and the binding energy of the sensor material 1s electron. This is referred to as an "escape peak,"and effectively this means the incoming X-ray photon is not counted. There are other losses that are introduced by the detector fabrication details.

The Ge detector made by Canberra uses a high purity Ge disk, 8 mm diameter and 5 mm thick, that is cooled to liquid nitrogen temperature. The sensor is in a vacuum chamber that has a 4 mil thick beryllium (Be) window for X-ray beam entry. Ge has an escape peak near 11.1 keV. Fig. 12(a) shows a spectrum from the radioactive isotope of americium (Am) having an atomic mass of 241 (Am-241). This source emits gamma radiation (X-rays that are produced by nuclear transitions) at 59.5 keV and 26.4 keV, and X-rays that are produced by electronic transitions from the Am decay daughter neptunium (Np) ion are seen in the 13 keV to 22 keV range.

The CdTe sensor is a 5 mm square that is 1 mm thick. It is cooled sufficiently using a thermoelectric cooler so that a 400 V bias voltage can be applied without electrical breakdown. It also operates in vacuum with a 0.001 inch Be window. The energy required to form an electron-hole pair for this material is 4.43 eV. (Herzberg, 1945) Escape peaks occur near 27 keV and 23 keV. Electron-hole pairs formed near the back contact of the detector cause fluctuations in pulse height, and they are not seen as belonging to the true peak. This is a loss in sensitivity. The Am-241 spectrum for the CdTe detector is given in Fig. 12(b). Note that there are several peaks in the 30 keV to 40 keV energy range that are not in the spectrum when the Ge sensor is used. A small escape peak is seen near 32 keV, and a larger escape peak is seen near 36 keV.

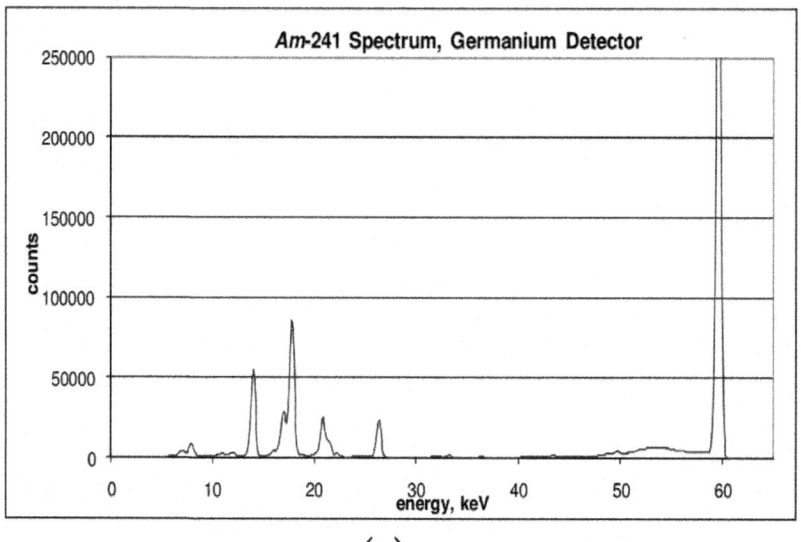

(a)

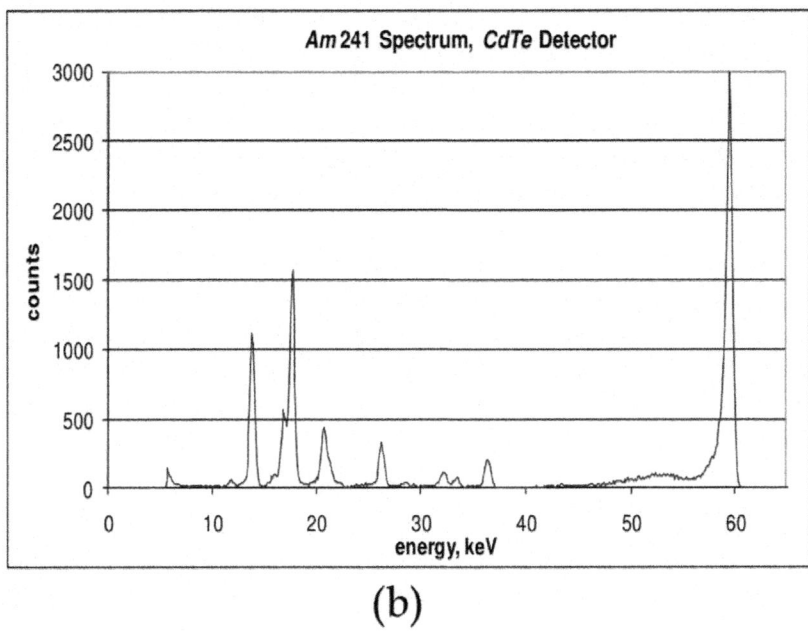

(b)

Figure 12. Am-241 spectrum. (a) Ge detector, showing γ-rays (gamma rays) at 59 keV and 26 keV, X-rays from Np from 22 keV to 14 keV, and escape peaks below 14 keV. (b) CdTe detector, showing X-rays and γ-rays as in (a), and escape peaks in the 30 - 40 keV range.

Imaging Devices

X-ray imaging is used to obtain geometric and qualitative intensity information in many applications such as tooth damage, stellar luminosity, and temperature and density of laser-produced plasmas. In the past few decades, several types of solid state detectors have replaced film in most imaging applications. These include charge-coupled devices (CCDs), charge-injection devices (CIDs), image plates (photostimulable phosphor plates - PSP), and other devices and sensor materials. After a brief description of image plates, the work described in this chapter will be limited to the CCD and CID Si-based sensors.

Image Plates

Image plates use principle of photostimulated luminescence (PSL) to to read the X-ray image after the image plate has been exposed. The exposed image is read and digitized with an image plate reader and stored in the computer as an image file. The image plate contains very small crystals (~5 μm) of barium fluorobromide phosphor with a trace amount of europium as a color center (Maddox, 2011). When exposed to X-rays, an atom of Eu^{2+} is converted to Eu^{3+} and the free electron is trapped in the barium fluorobromide lattice, creating a metastable state. The metastable electrons are freed by stimulation from the laser in the image plate reader, and they recombine with the Eu^{3+} to give off a blue-violet fluorescence that is recorded by the image plate reader with a suitable detector. The digitized image can then be analyzed for geometric information and quantitative intensity information. We now use image plates extensively for characterizing the X-ray beam in a variety of arrangements. We have also calibrated image plates for quantitative intensity measurements that are then applied to plasma studies at LLNL (Maddox, 2011). LLNL and other groups have also evaluated their performance for medical imaging (AAPM, 2006). Image plates will not be considered further in this chapter since the subject of the book is photodiodes.

Silicon Based Cameras

The basic sensing function is the same for all Si-based sensors, and a camera operates in a manner similar to the principles described for the X-ray photon detectors described earlier. The photons interact with the Si semiconductor to produce electron hole pairs. The camera types vary in the method in that the charge is moved from the sensor pixel and eventually produces a digital output proportional to the charge formed. Janesick (2000) is a good source for the details of camera readout methods.

The charge coupled device (CCD) transfers the charge from a row of the sensor pixels to an adjacent row until it reaches the "read-out" row. The charge is read from each pixel sequentially in that "read-out" row as voltage. The voltage is then converted to digital counts. A major advantage of the CCD camera is low electronic noise because it is usually cooled to minimize the dark current from thermally-produced electron-hole pairs (Eq. 4). The major contribution to the noise is then from the read-out. The read-out noise is quite low, typically no more than a few counts. The dominant noise contribution for X-ray applications is then what was produced by the arrival time statistics of the photons themselves, a Poisson distribution. The standard deviation per pixel is then the square root of the number of X-ray photons the pixel absorbed during the exposure time.

An alternative electronic method for reading the charge stored in the individual pixels is the CID. Every pixel can be individually addressed using indexing row and column electrodes. A displacement current proportional to the stored signal charge is read when stored "packets" are shifted between capacitors within individually selected pixels. The displacement current is amplified and converted to a voltage, and fed to the outside world as a digitized signal. The CID technology offers certain advantages over CCD technology such as cost, reduced heat generation, and resistance to blooming. The disadvantage is the high readout noise. Improvements continue in this area but for X-ray photon intensity measurements, the readout noise far exceeds the photon statistical noise for a single image. When calibrating the CID, this can be mitigated by multiple imaging.

Most CCD cameras are front illuminated, the front side being where the gates are located. This region is not active to photon detection. This dead region will stop low energy X-rays and the front illuminated CCD cameras lose sensitivity as the X-ray energy drops below 2000 eV. Below 1000 eV they are not useful. For this reason many X-ray CCD cameras are back illuminated and back thinned and are sensitive even into the vacuum ultraviolet. The Si active region is typically 15 μm to 30 μm. The Si thickness thinning does reduce the sensitivity at the high energy side. Camera efficiency measurements for both front illuminated and back illuminated cameras are given in Section 6.

CALIBRATING PRIMARY DETECTORS USING RADIOAC-TIVE SOURCES AND A SYNCHROTRON SOURCE

The Calibration Concept Using Radioactive Sources

Radioactive sources provide a variety of spectral lines at well-defined energies, as was described in the energy dispersive detector section (section 4.3). The photon output is directly proportional to the activity of the radioactive source, and the activity measurement is traceable to NIST. The uncertainty for the activity is provided by the vendor. The activity R at time t is given by:

$$R = R_0 \exp((t \ln 2)/\tau_{1/2}) \qquad (8)$$

R_0 activity at time $t=0$ (disintegrations/sec)

$\tau_{1/2}$ radionuclide half life

The radioactive source is placed at some distance from the detector that is sufficiently large so that the source intensity is uniform over the detector area. The power Ω at the detector in photons per second for a selected spectral band is given by:

$$\Omega = R\ B\ (A_d/4\ \pi r^2)\ T \qquad (9)$$

B branching ratio

A_d detector area

r distance between radioactive source and the detector

T X-ray transmittance through r cm of air

The branching ratio is the fraction of nuclear decays that produce the selected spectral band. It has an experimental uncertainty associated with it. If the activity was measured using a gamma emission, and the same gamma is used for the detector calibration, the uncertainty is that given is that given by the vendor in his activity certification. The detector efficiency , is then given by:

$$\eta = S/\Omega \quad (10)$$

S is the photon count per second as measured by the detector. The measurement arrangement is shown in Fig. 13. The instrument to be calibrated, the Ge detector, is seen on the right facing the radioactive source. An optical distance meter is located at the far left of the optical rail and is also at the same height as the source and detector and measures the distance from the radioactive source and to the detector window. The internal distances for source and detector are provided by the manufacturers to an accuracy of 0.5 mm or better. The distance sensor has an accuracy of ± 1 mm and was calibrated at NIST within a month of the detector calibration measurements. Thus the source-to-detector distance accuracy is ± 1 mm.

Measurement Results

Measurement results are given in this section. The spectral lines that were used for these measurements are given in Table 8.

Figure 13. Experimental arrangement for calibrating detectors using radionuclides.

Table 8. Spectral Lines Used for the Calibration of the Ge and CdTe Detectors

Spectral Energy, keV	Spectral Type	Radionuclide
5.97	Mn, K line	Fe55
22.16	Ag, K line	Cd109
26.34	γ	Am241
41.3	Eu, K line	Gd153
59.54	γ	Am241
88.00	γ	Cd109
97.43	γ	Gd153
103.18	γ	Gd153

Quantum efficiency measurements for the Ge detector are shown in Fig. 14(a), and Fig. 14 (b) for the CdTe detector. These measurements show a precision near 3% at the 95% confidence level. The Ge detector shows a peak efficiency near 60 keV and falls off in efficiency at lower and higher spectral energies. The CdTe detector has a peak efficiency near 30 keV and also falls off in efficiency at lower and higher spectral energies.

It has been described earlier that escape peaks occur above the K edges (minimum energy needed to remove a 1s electron) of the detector materials, and this reduces the detector efficiency. This effect has been directly observed in the calibration of a Si photodiode in the vicinity of the Si K edge (1.39 keV). The K edges for Ge, Cd, and Te are 11.1 keV, 26.71 keV, and 31.81 keV respectively. (CXRO) In Fig. 14 (a) and (b), note the large gaps that exist between points. The situation regarding the large gaps can be improved by using more radionuclides but there are availability and economic restrictions that will limit this approach. We plan to fill in these energy gaps by calibrating an Si detector up to 60 keV using a synchrotron as described in the next sub-section. This calibration can then be transferred to the other detectors using the HEX source.

Synchrotron Calibration of a Silicon Photodiode

The silicon photodiode used by NSTec for calibrations of camera efficiency that is described in Section 6 was manufactured by International Radiation Detectors (IRD) who claimed that its measurements are absolute (IRD Reference). This claim was substantiated by sending two photodiodes to the German synchrotron at the PTB. The photodiode efficiency was measured from 1 keV to 60 keV. The measured efficiency was 100% from 1000 eV up to the Si K edge at 1839 eV, then dropped about 3% to 97% due to the Si escape photon. The efficiency rose back to 100% by 3.5 keV. The efficiency begins to fall near 5 keV as the absorption of the Si falls below 100%.

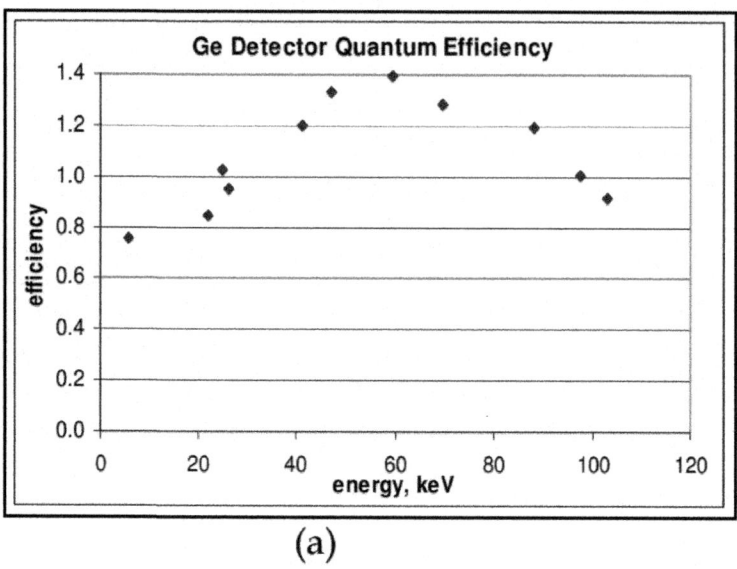

(a)

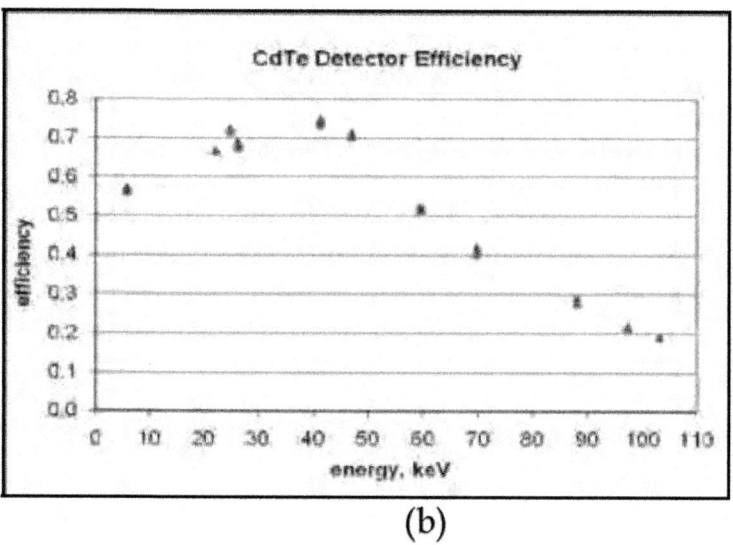

(b)

Figure 14. Measurements of the quantum efficiency of the (a) Ge detector and (b) CdTe detector using the radionuclide sources.

NSTec has performed calibrations using the IRD Si photodiode, accepting the concept that it was an absolute photodiode and using the IRD measured thickness to correct for X-ray transmission at the higher energies. PTB calibration is an internationally accepted standard. The 2 calibrated photodiodes are then utilized as primary standards to cross calibrate other photodiodes which will be our working devices.

Accreditation Procedure for Calibration Laboratories at NSTEC

In order to provide substantiated traceability to international standards that would verify the quality of the NSTec calibration laboratories, we have been working with the National Voluntary Laboratory Accreditation Program (NVLAP). NVLAP is one of several organizations that can provide the certification that calibration procedures meet national and international standards. The NSTec laser timing laboratories have obtained this accreditation. The US National Institute for Standards and Technologies (NIST) does not have standards for X-ray intensity measurements; their concentration has been dosimetry. The calibration methods described previously are the starting point in the NSTec effort to achieve accreditation for the X-ray laboratories. Through the development of an accredited X-ray intensity calibration methodology, the diagnostics and system components for pulsed high energy physics experiments conducted at the national laboratories such as LLNL can now be calibrated traceable to national standards. Since the NSTec had already passed the NVLAP accreditation process on specific measurements in two laser laboratories, the groundwork for the X-ray laboratories was already done. The HEX laboratory would be the first one to be accredited. The required rigor of the quality assurance management system is already in place and approved. The required steps for accreditation are:

- Determination of customer requirements
- Development of calibration procedures and documentation of the procedures
- Traceability
- Evaluation of measurement reproducibility
- Determination of uncertainty
- Generation of calibration certificates.

The main customer requirement is to have the capability to calibrate customer detectors and components to an absolute value with known uncertainties.

The previous sections have described the detector calibration procedure used to calibrate our X-ray source. The radionuclides used are traceable to

NIST, and the vendor of these radionuclides participates in a measurement assurance program with NIST. NIST regularly sends standard radionuclide sources to the vendor to conduct "blind" measurements; the activity of the sources is unknown to the vendor. The vendor then measures the activity and submits the results back to NIST. NIST then checks the measurements to verify that they meet the required accuracy. The distance measurement traceability was described earlier.

As with all traceable measurements, repeatability and reproducibility determination are requirements for accreditation. The variation in repeated measurements quantifies random errors. Only when calibrations using a common procedure and common equipment produce the results that vary within an acceptable range, even with different operators, is the technical process is ready for NVLAP review.

NSTec is in the process of evaluating the overall uncertainty in its calibration of X-ray detectors with radionuclide sources. A list of the factors being considered is given in Table 9. The procedures are being written at the time of this publication.

Table 9. Quantities in the Calibration Procedure Contributing to the Overall Uncertainty

---

Radionuclide signal measurement reproducibility.

Radionuclide activity.

Radionuclide branching ratio, when needed; daughter X-ray emission.

Distance measurement, source to detector.

Detector size.

Measurement duration.

Choice of region of interest (ROI) for counting photons belonging to a spectral line.

Radionuclide decay rate.

Energy accuracy.

Air transmission of the X-ray.

Curve fit to points between the measured points.

CALIBRATING X-RAY CAMERAS USING THE PRIMARY DETECTORS

Measuring the Quantum Efficiency and its Spatial Variation (Flat Field) for an X-Ray CCD Camera Using the Manson Source

The Static X-Ray Imager (SXI) is an X-ray camera used on the NIF target chamber to measure quantities such as laser beam pointing and the sizes of the laser entrance hole in ignition targets (Schneider, 2010). The SXI records a time integrated X-ray image. The sensor is a back thinned CCD chip with 2kx2k pixels, 24 μm square (approximately 50 mm x 50 mm). The use of this camera at NIF requires the knowledge of camera sensitivity S(E) as a function of X-ray energy E.

The X-ray photon interacts with the Si sensor to produce hole-electron pairs that the CCD electronics process to produce the digital signal count S. The number of electron-hole pairs produced by an X-ray photon that interacts with the Si sensor is a function of the photon energy and is slightly dependent upon the temperature (Janesick, 2000). The sensor is cooled to 253K when operating.

A useful model relating the camera signal to fundamental quantities is found in Janesick (2000).

$$S(E) = P \cdot A_{pix} \cdot T \cdot \eta \cdot QE \cdot K^{-1} \text{ counts/pixel} \qquad (11)$$

P =Photon rate at the CCD, photons/cm^2/sec

A_{pix} =Pixel area

T=Exposure time

QE=Quantum efficiency, fraction of photons that interact with the pixel

η=Quantum yield, number of electron-hole pairs produced by the photon

K=Camera gain, electron-hole pairs per count

For the Si CCD, the quantum yield is given by (Knoll, 2001):

$$\eta = E/3.66 \text{ hole pairs/photon} \qquad (12)$$

The camera manufacturer's measurement of the gain constant was used for the calculations of quantum efficiency. The quantity being determined by the calibration is S(E), but the model given by Equation 10 is a valuable check of the calibration procedure, the calibration implementation, and is useful for troubleshooting camera problems. For the Si based cameras, the QE is related to the properties of Si as we shall see in the following results.

The photon intensity P' was measured using a photodiode. We use a photodiode manufactured by IRD, model AXUV100. These detectors are designed with no doped dead region and zero surface recombination, so that they have near theoretical quantum efficiencies for the soft X-ray spectral region. One unit of this model was sent by the photodiode manufacturer for calibration to the synchrotron at the PTB. The results from this calibration showed agreement within 1% from 1000 eV to 1839 eV, and the photodiode gave a 5% lower reading above this energy. The thickness of the Si photodiode is 54.5 μm. The current i from the photodiode is related to the photon intensity as given in Equation 12:

$$i = P' \cdot \eta_{PD} \cdot 1.6 \times 10^{-19} \text{ amp} \tag{13}$$

Here $\eta_{PD} = E/3.62$ and 1.6×10^{-19} is the charge on the elctron in coulombs. The area of the IRD detector is exactly 1 cm^2. The photodiode and the CCD are not at the same distance from the anode, as can be seen in Fig. 6. These distances were measured to an uncertainty of 2 mm. The camera sensitivity ξξ, defined as counts per photon:

$$\xi = S(E)/(P \cdot A_{pix} \cdot T) \tag{14}$$

$$\xi = S(E)/((i/(\eta_{PD} \cdot 1.6 \times 10^{-19})) \cdot (d_{IRD}/d_{CCD})^2 \cdot A_{pix} \cdot T \tag{15}$$

P=Photon intensity at CCD, photons•cm^{-2} •s^{-1}

d_{IRD}=Distance from the photodiode to the anode

d_{CCD} =Distance from the CCD to the anode

Note that P›, the photon intensity at the photodiode in Eq. 12, is given by:

$$P' = P \cdot (d_{IRD}/d_{CCD})^2 \tag{16}$$

Methods for Imaging

The SXI's CCD camera was mounted on the diagnostic arm is shown Fig. 6. There was an extension between the camera and the Manson chamber of sufficient length that the X-ray beam uniformly illuminated the CCD. The camera calibration proceeded by the following steps:

- Locate the bad pixels so that they can be masked out for image analysis;
- Determine the linear range of the camera;
- Measure the camera sensitivity;
- Measure the uniformity of the CCD chip response over the area of the camera.

The cameras had a large number of bad rows and hot pixels. The bad rows were associated with the readout and identified using closed shutter images with a 3 ms exposure time. The hot pixels were identified by taking an image using the Ti anode and no filter, and using the same exposure time that was used for the experiments on the NIF target chamber experiments. A map was made that identified the bad rows and bad pixels.

The photon intensity was measured with the photodiode in arm #1 as seen in Fig. 6. An exposure time was chosen to be as short as possible to give a reasonable signal. Photodiode readings were taken before and after acquiring each CCD image. During imaging, the X-ray beam intensity was monitored continuously for beam fluctuations using the photodiode in arm #2. If there were beam intensity fluctuations observed during imaging, that image was discarded.

Flat field images are images where the CCD is uniformly illuminated in order to measure the uniformity of the camera response over its area. They were taken using the same anode voltage that was used for the camera efficiency measurements and maximum anode current. The exposure time was chosen to produce a signal that was 50% to 60% of saturation. Ten flat field images and ten background images were taken at each photon energy.

Image Analysis

The camera images for the efficiency analysis had the background subtracted and the bad pixels replaced by the average of adjacent pixels. The mean pixel count was determined by randomly selecting 1000 regions 20x20 pixels in size, calculating the mean counts/pixel for each region and calculating the average of the means for each region. This is the signal S for that image. Then, for the flat field images, average all images that have the same exposure time, average the background images, and subtract the average background from the average flat field image.

Camera Sensitivity

The camera sensitivity for one of the SXI cameras is given in Fig. 15(a). The Quantum Efficiency (QE) calculated using Eq. 10 through 14 and camera gain K=7.62 electrons per count is plotted as a function of photon energy in Fig. 15 (b). The data scatter as measured by the standard deviation was 1% or less at each point. The dip near 1800 eV and the fall-off after 2000 eV are properties of Si. Si that is 15 µm thick transmits up to 35% as it approaches the K edge at 1839 eV. It begins transmitting again above 2500 eV and is transmitting 80% at 8 keV. These QE results are similar to that obtained by Poletto (1999). There are two possible causes why the QE does not approach 1 when the photons are completely absorbed:

- There may be absorption at the surface coating of the Si;
- the Quantum Yield may be less than the photon energy divided by 3.66 eV per electron-hole pair.

Analysis of a large number of single photon events could show the relative contribution of each effect.

Flat Field

The flat field source is the 1 mm diameter spot on the anode. The anode is 1405 mm from the CCD. This arrangement would produce a flat field within 1% if there were nothing between the anode and the CCD. There is a light blocker that has an aluminum coating on a polyimide film (Al 1054 Å ±50 Å; polyimide 1081 Å ±100 Å). This item does not affect the flat field within the 1% cited above. The filter can cause a variation in the beam intensity across the CCD if there is sufficient variation in thickness, foreign material, or misalignment with the anode. A comparison of all the flat field images implies that the maximum variation is ±1% peak-to-peak.

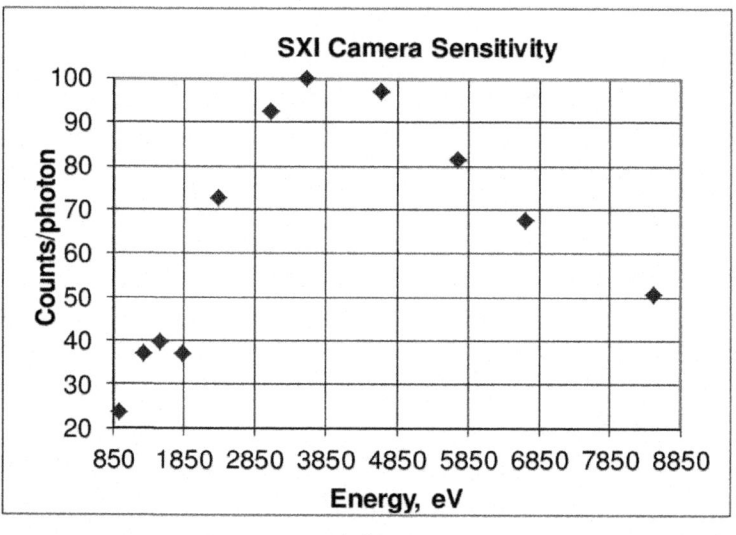

(a)

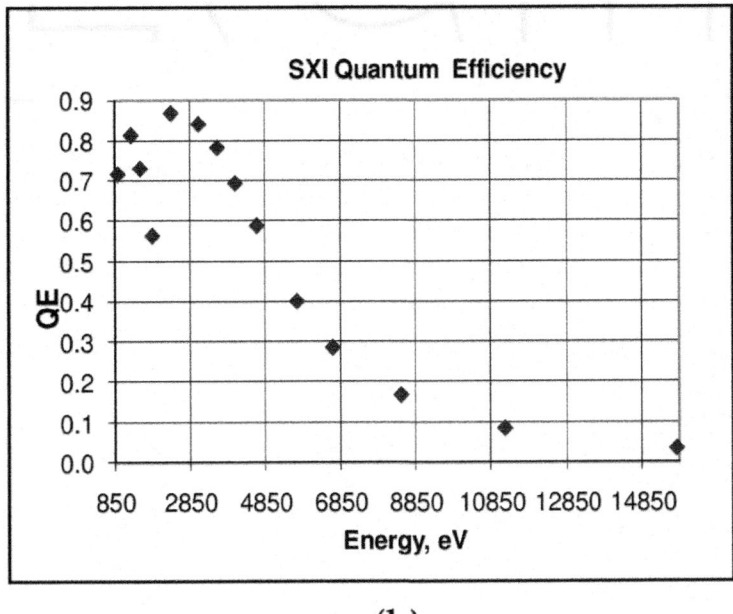

(b)

Figure 15. The SXI (a) camera sensitivity and (b) quantum efficiency as measured by the camera count per pixel for each photon of a given energy. The measurements made at X-ray energies below 8800 eV were done on the Manson. The higher energy measurements were done on the HEX.

Fig. 16(a) shows the flat field image for one of the SXI cameras at the Cu 8470 eV energy band. The image is set at high contrast so that the pixel signal variation shows clearly. A gross pattern is observed with the sensitivity at a maximum near the left center and decreasing slowly going away from the maximum. The image in Fig. 16 (b) is at Ti 4620 eV; it shows the same pattern but decreased magnitude. The pattern continues to decrease in magnitude until it is no longer visible at 3000 eV. Vertical lineouts averaged over a small horizontal width (see band in Fig. 16(b)) for three images at three different X-ray energies are shown in Fig. 17. The lineouts are normalized by dividing by the maximum counts in each image. The maximum sensitivity variation for each of the curves in Fig. 17 is 13% at 8470eV, 6% at 4620eV and 2% at 3580eV.

A flat field image of the Mg 1275 eV band is shown in Fig. 16(c) for comparison to the higher energy flat field images. There is no trace of the sensitivity variation pattern that is seen at higher energies. The 1275 eV lineout in Fig. 17 shows that the maximum variation is less than 1%, which is the measurement limit of our flat field procedure.

This sensitivity variation is a large scale effect; it includes groups of pixels and is probably related to the CCD manufacturing process. Any sensitivity variation of individual pixels is less than the photon noise associated with averaging 10 images.

A different phenomenon was seen at low energies. Small irregular patches having diminished sensitivity were observed that are readily seen in Fig. 18(a). This image shows a portion of the CCD. The effect on sensitivity in these regions also shows an energy dependence. Fig. 8b is a similar image taken at 3080 eV. The irregular patches have now become quite dim compared to what was observed at 1275 eV. At 4500 eV, these paths of low sensitivity have completely disappeared.

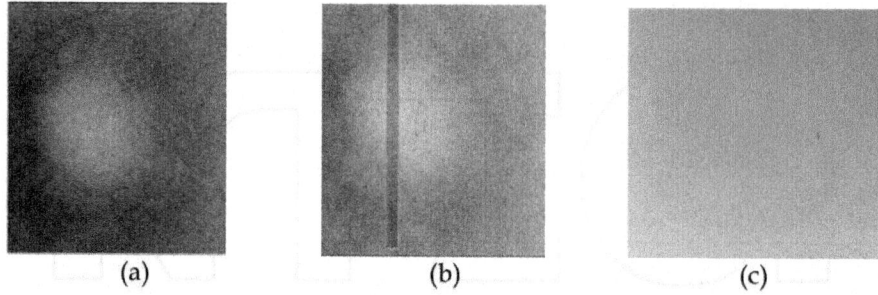

(a)   (b)   (c)

Figure 16. Flat field image for the (a) Cu anode, 8470 eV and (b) Ti anode, 4620 eV, showing the pixel sensitivity variation (Signal range: 5200 to 7200 counts/pixel) The vertical band was the area used to calculate the cross section that is shown in Fig. 17. The same region was used for the cross section at the other energies. (Signal range: 5200 to 7200 counts/pixel) (c) Flat field image for the Mg anode, 1275 eV, showing the pattern observed at the higher energies shown in Fig. 16(a) and (b) has completely gone and the pixel sensitivity is flat.

There are several possible causes for these dark regions. Debris on the CCD surface could absorb X-rays and would be energy dependent, absorbing X-rays less as the energy increased. Damage to the CCD would likely cause an energy dependence that would increase the variance of the defective region from the surrounding pixels as the energy increased. Damage to the surface coating could produce this effect if the coating were thicker in that defective region. When we examined the CCD surface with a magnifying glass it did appear that the coating was deformed. It looked like a manufacturing defect.

It is difficult to correct these images using the normal method of flat field inversion. This could be done if you limit the energy range of the X-ray source. But the characterization always provides the information necessary for the effective use of the X-ray camera.

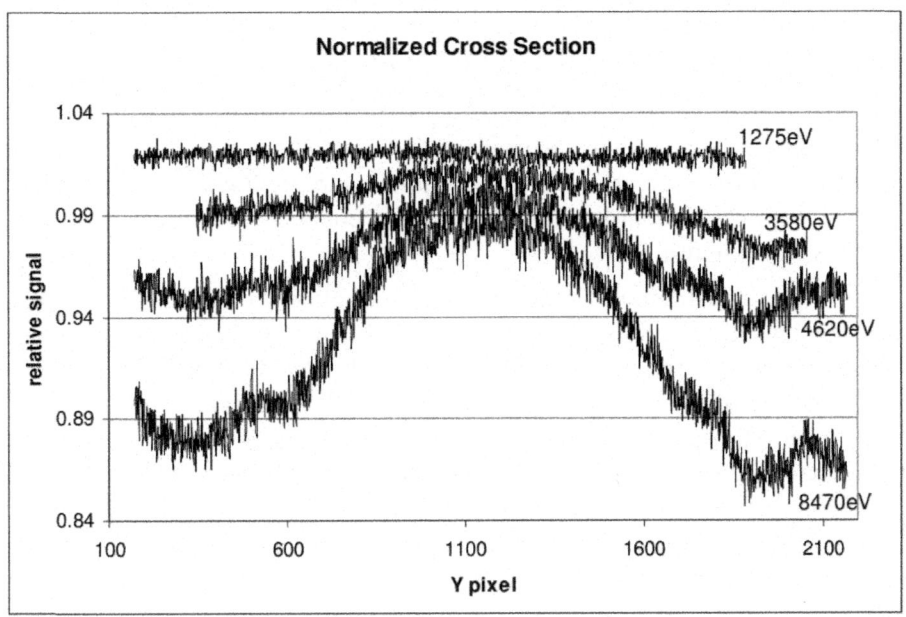

Figure 17. Normalized vertical lineouts from flat field images at several X-ray energies. The lineouts were normalized to the maximum counts in each image. As the X-ray energy increases, the pixel sensitivity shows a greater vatiation.

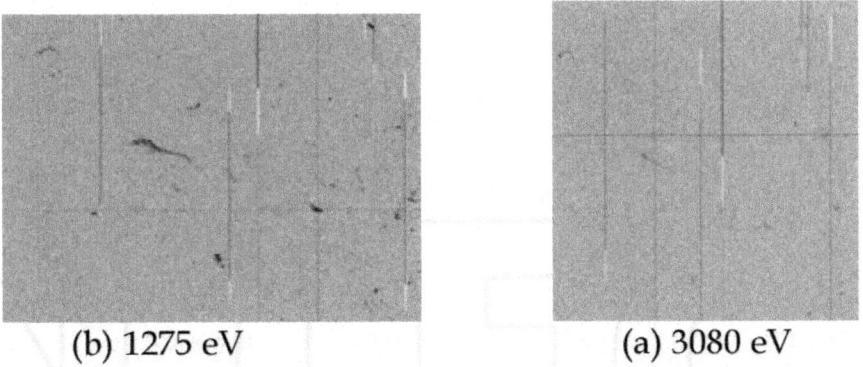

Figure 18. These are the same sections of a flat field image taken at two different energies, (a) 1275 eV and (b) 3080 eV. The sections cover about ¼ of the entire CCD. The dark regions are CCD surface defects causing diminished pixel sensitivity. For the 1275 eV section shown in (a) the blemishes are much darker than in the 3080 eV image shown in (b).

Calibrating a Front Illuminated CCD Camera from 705ev to 22kev Using the Manson and Hex Sources

The SXI camera described above plays a critical role in the NIF operation, but this specific chip is no longer manufactured. There is another chip on the market with this large array, 2kx2k, 24 μm square, and we were requested to test the chip in a standard camera. The major concern regarding this chip was that it is front illuminated.

The QE measurements at X-ray energies below 10 keV were done using the Manson source following the procedures given in 6.1. These measurements are shown in the graph of Fig. 15. Compare this to the results shown in Fig. 19 for the QE of the back illuminated camera. The maximum QE for the front illuminated camera is QE=0.34 near 2300 eV. This is almost a factor of 3 lower than the QE measured for the back illuminated camera. The predominant difference begins to show below 1000 eV. At the Cu L lines, near 930 eV, the QE for the front illuminated camera is down by a factor of 10 from the front illuminated camera. At the Fe L lines near 705 eV, the QE is down by a factor of 100.

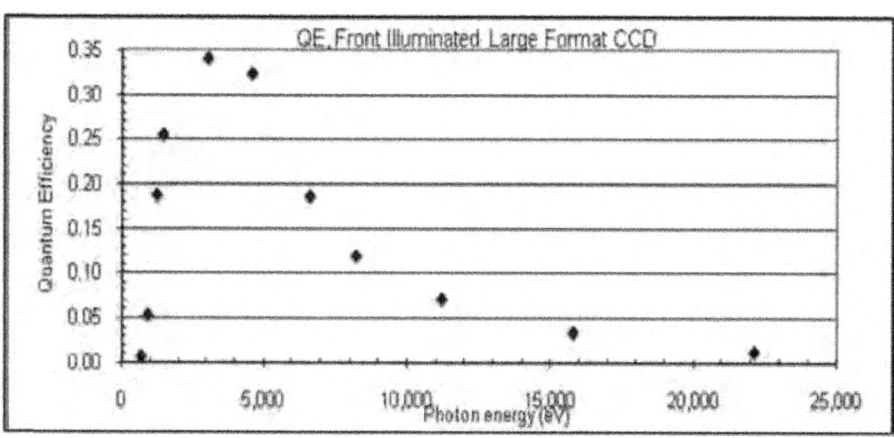

Figure 19. The quantum efficiency measured for a front illuminated CCD sensor.

The measurements at 10 keV and lower energies were done on the Manson. The measurements at higher energies were made using the HEX. Compare this to the QE measurements shown in Fig. 15.

The Manson can only be used effectively up to the Cu K lines. The QE measurements at higher energies have to be done on the HEX. The CCD cameras must be kept in a vacuum since they are cooled and the HEX has a vacuum chamber on a rail as is seen in Fig.11. The chamber is very similar to that shown on the Manson. It differs in having a Be window on the side facing the Hex source. The camera is mounted on the opposite side from the Be window. The HEX fluorescer source is near 10mm diameter rather than the "point" source of the Manson. The X-ray beam is not flat across the entire CCD surface but is flat near the beam center. The camera is moved horizontally and vertically until the X-ray beam is centered on the CCD. The camera is then moved aside on the rail and the CdTe detector is placed at the same distance from the source as was the CCD. The beam center is then determined by moving the detector horizontally and vertically. These are the measurements used in Eq. 10 to determine the QE shown in Fig. 15 and Fig. 19 for the higher X-ray energies. The observation then is that the QE at these energies is the same for the front illuminated and the back illuminated cameras.

Measuring the sensitivity variation on the HEX requires that the X-ray intensity measurement be carefully measured over the entire area and an analytical representation be developed. This functionality is being developed now. We will use both the CdTe detector on a motorized X,Y positioner and image plates to measure the X-ray intensity distribution.

Single Photon Measurements using the Manson Source

Images can be taken at sufficiently short exposure times so that most or all of the incidents recorded by the camera are caused by individual photons. These single photon images provide spectral information. This technique is used for astronomical measurements and laser plasma studies. The image shown inFig. 20(a) was taken on the Manson source using a Ti anode and a Ti filter 100 μm thick. This is the same condition that was used to generate the spectrum shown in Fig. 7 using an energy dispersive detector. The camera used was a silicon CCD type having 1300 pixel x 1340 pixel array and the pixel size was 13 μm square. A background image using the same exposure time and no X-rays has been subtracted from the original X-ray image. The region shown in the figure is a 100 pixel square. There are approximately 95 single photon events in this 10000 pixel area, or about a 1% fill. This is the fill rate typically used in single photon measurements. Note that a significant fraction of the single photon events produce counts in more than one pixel, that is, the production of electron/hole pairs produces by the photon occurs in more than one pixel.

The graph shown in Fig. 20(b) is a histogram of the entire pixel array for the single photon image of the Ti X-rays. This plot shows the number of times a pixel has a given count as a function of counts. The histogram exhibits two peaks and they are above 400 camera counts. The two peaks are the Ti Kα photons occurring at 415 camera counts and the Ti Kβ photons occurring at 454 camera counts. These peaks represent single pixel events where the total number of electron/hole pairs produced by the photon is contained within that single pixel. As stated in the previous paragraph, there are many incidents in the image where the single photon produces counts in multiple pixels. These multi-pixel events produce the rising number of incidents in the graph going toward lower counts. There are no incidents at counts above the K-M band. Compare this spectrum to that shown in Fig. 7 where an energy dispersive Si detector was used. The spectral resolution is nearly the same for each detector. In general then, a camera is an energy dispersive detector when operated in the single photon mode.

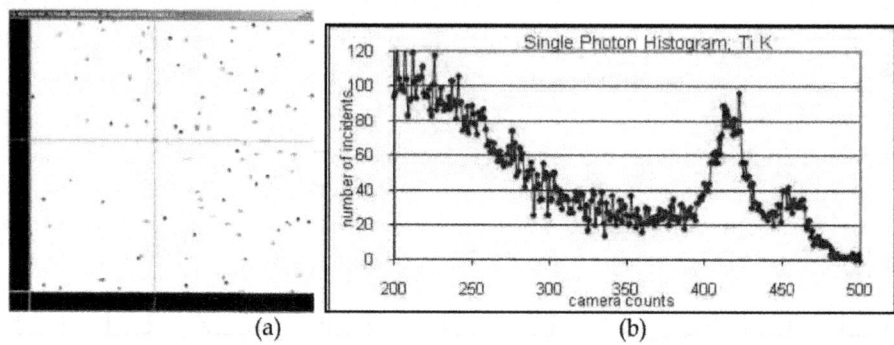

(a)   (b)

Figure 20. a) This image shows single photon incidents on a CCD camera zoomed in to show the individual pixels in a small region of the camera active area. (b) This graph is a CCD active area showing the Ti K-L and K-M spectral bands. Compare this to the spectral scan of the Ti emission using the energy dispersive detector shown in Fig. 3.

The above description also describes a method for calibrating the camera count to spectral energy. As described earlier for the camera efficiency calibration, images are taken with several anode/filter combinations. The camera count for the peak center is then plotted against the literature value for the spectral energy (more precisely, a weighted average of the unresolved spectral lines).

More sophisticated software than a simple histogram can be devised that would capture a large portion of the multi-pixel incidents that are single photon events. This would reduce the noise that is seen in the histogram peaks. The method requires identifying significant pixels by a thresholding technique, then adding the counts of adjacent pixels to the central pixel. This represents a new image that generates a new histogram. The spectral peaks will be better defined because the noise is reduced.

Characterizing and Calibrating an Uncooled X-Ray CID Camera Using the Hex Source

This section describes the characterization of a CID camera that was planned as the detector in a spectrometer system that was to be used on the LLNL NIF target chamber. The initial interest was to measure the emission from highly ionized Ge so the camera was characterized in the 10 keV region using the HEX source (Carbone, 1998 and Marshall, 2001). The fluorescers chosen were Cu, Ge, and Rb giving weighted average for the K-L and K-M transitions of 8.13 keV, 10.01 keV, and 13.58 keV respectively.

The major use for this CID sensor is for dental X-rays. It is relatively cheap and therefore expendable, a desirable property for the NIF application. The camera operates at room temperature normally, which gave a challenging problem to the characterization on HEX. Since the CID operates at room temperature, the dark current can saturate the camera for exposure times less than 10 seconds. This not a problem on NIF since the exposure time can be less than 1 second with sufficient X-rays to provide a bright spectral image.

As indicated in the earlier description of CCD camera calibrations on the HEX, minutes of exposure time are needed to get a satisfactory signal. Preliminary experiments with the CID camera showed that we would be limited to three-second exposure times. It was determined that multiple exposures, on the order of 100 exposures, would be needed to obtain satisfactory photon statistics. The multiple exposures would also allow us to average the readout noise and get to the limit that photon statistics were dominant. A shutter control system was implemented for automatically taking the multiple images. We quickly found that drift in the dark current required us to take background images immediately after the X-ray exposure. The system was designed so that an image was taken with the shutter open to the X-rays, then the next image was taken with the shutter closed. In this way a pair of images were produced, one image exposed to X-rays and the other as a background, that were close enough in time that there was no observable dark current drift. A black Kapton sheet, 50 μm thick, was used to shield the camera from visible light. The same type shield is used for the camera on the NIF target chamber.

The X-ray beam was characterized geometrically using image plates to optimize collimator and distance choices. The intensity distribution was measured using the CdTe energy dispersive detector at multiple locations across the beam. Multiple images were taken with the CID, and then the detector was placed at the same location as the center of the CID had been located to verify that there was no drift in the X-ray source intensity. The multiple images were analyzed by subtracting each background from the previously taken X-ray image and summing the 100 resulting images. The final image then was effectively a 300 second exposure with the background removed. The measurements concentrated on the X-ray beam center for this initial effort. The CID camera efficiency, counts per pixel per photon, could then be calculated using the CdTe intensity measurements.

The results are shown in Fig. 21. The camera response was measured for two CID cameras at three spectral energies over the range of interest. The responses of the two cameras are the same within the experimental uncertainty. The expected response was modeled using the vendor's specification for camera gain and Si thickness and a typical surface coating. This is shown by the blue line in the figure. This did not fit the measurement data so a second model curve is shown using a thinner Si effective thickness.

The CID camera is now considered to be suitable for the spectrometer operation. The spectrometers will be incorporated as part of existing diagnostics at several locations on the NIF target chamber. All cameras will be calibrated using an extension of the procedure. It will extend to lower X-ray energies using the Manson source and measure the sensitivity variation of the CID over the full pixel array.

CONCLUSION

The chapter started with a presentation of basic X-ray physics needed to follow the description of X-ray detector calibration. The X-ray sources used at NSTec for calibrating detectors were described. The operation and characteristics of solid state semiconductor detectors was presented. Single sensor photodiodes, both current detectors and pulse counters, are used to measure the X-ray source beam intensities. The detectors are calibrated using either of 2 procedures: radioactive sources that are NIST traceable; a synchrotron beam that has an internationally accepted beam intensity accuracy. The chapter presented the methods used and the results obtained for calibrating several types of X-ray cameras.

The accreditation procedure for recognition of the X-ray calibration labs as certified to international standards is in process. This requires the full analysis of all uncertainties associated with the detector calibration. The calibrated photodiode has yet to be completed for the synchrotron calibration. It will then be used to better fill the efficiency curves of the energy dispersive photodiodes. There are several agencies around the world that oversee and certify the accreditation. NSTec will be working with one of them to achieve certification. The NSTec X-ray labs will continually improve existing procedures and develop new methods for calibrating X-ray detection systems and components.

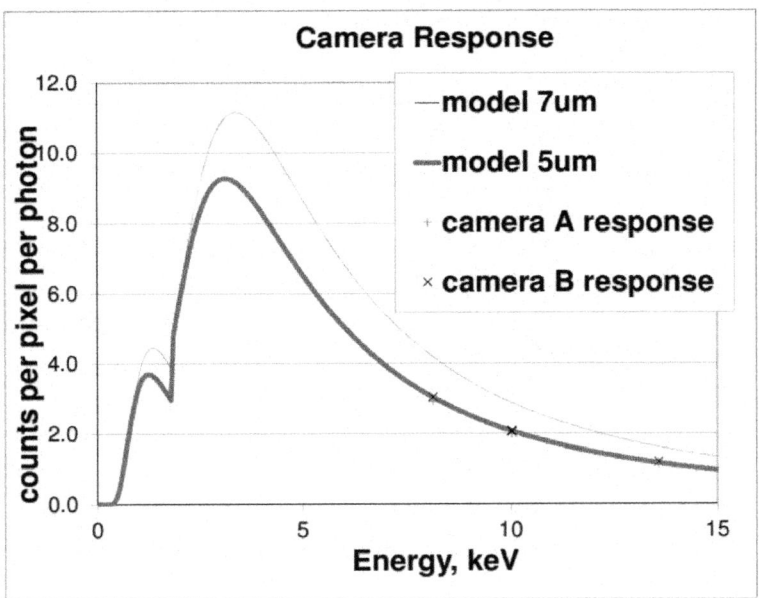

Figure 21. The measurement results for the CID camera efficiency are shown as the crosses and the plus signs. The curves are model calculations for the CID camera response based on camera characteristics described in the text.

ACKNOWLEDGEMENTS

This manuscript has been authored by National Security Technologies, LLC, under Contract No. DE-AC52-06NA25946 with the U.S. Department of Energy. The United States Government and the publisher, by accepting the article for publication, acknowledges that the United States Government retains a non-exclusive, paid-up, irrevocable, world-wide license to publish or reproduce the published form of this manuscript, or allow others to do so, for United States Government purposes. This manuscript was done under the auspices of the U.S. Department of Energy by Lawrence Livermore National Laboratory under Contract DE-AC52-07NA27344.

There were many persons from both NSTec and LLNL involved in developing the X-ray laboratory calibration methods. I particularly thank Susan Cyr for special effort in putting this manuscript together.

REFERENCES

1. American Association of Physicists in Medicine (AAPM) 2006Report 93Acceptance Testing and Quality Control of Photostimulable Storage

Phosphor Imaging Systems, available from http://www.aapm.org/pubs/reports/rpt_93.pdf

2. J. Carbone, A. Zulfiquar, C. Borman, S. Czebiniak, H. Ziegler, 1998Large format CID x-ray image sensors, Proceedings of SPIE 3301, 90 doi:10.1117/12.304550,Solid State Sensor Arrays: Development and Applications II

3. Center for X-Ray Optics (CXRO) (n.d.).X-ray interactions with Matter, available from http://henke.lbl.gov/optical_constants/

4. E. S. T. A. R. Program, (n.d., Available from http://physics.nist.gov/PhysRefData/Star/Text/ESTAR.html

5. A. Gottwald, U. Kroth, M. Krumrey, M. Richter, F. Scholze, G. Ulm, 2006The PTB high accuracy spectral responsivity scale in the VUV and x-ray range, Metrologia 43

6. M. J. Haugh, R. Stewart, 1 EOF 10 EOF 2010 Measuring Curved Crystal Performance for a High Resolution Imaging X-ray Spectrometer, Hindawi Publishing

7. M. J. Haugh, R. Stewart, 2010X-Ray Optics and Instrumentation, Article ID 583620

8. G. Herzberg, 1945 Atomic Spectra and Atomic Structure, Dover

9. International Radiation Detectors (IRD) (n.d.).Available from http://www.ird-inc.com/axuvhighnrg.html

10. J. Janesick, 2000 Scientific Charge-Coupled Devices, SPIE Press, Bellingham, WA

11. G. F. Knoll, 2001 Radiation Detection and Measurement, 3rd edition, John Wiley & Sons

12. B. Maddox, et al. 2011High-energy backlighter spectrum measurements using calibrated image plates, RSI 82, 023111 EOF

13. F. J. Marshall, T. Ohki, D. Mc Innis, Z. Ninkov, J. Carbone, 2001Imaging of laser-plasma x-ray emission with charge-injection devices, Rev. Sci. Instru. 72 713 EOF 716 EOF

14. L. Poletto, A. Boscolo, G. Tondello, 1999Characterization of a Charge-coupled Detector in the 1100 0nm (1 eV to 9 keV) Spectral Range, Applied Optics, 38, 1 Jan 99

15. Bundensanstalt. . P. T. B. Physikalisch-Technische, (n.d., available at http://www.ptb.de/index_en.html

16. E. Podgorsak, 2010 Radiation Physics for Medical Physicists 2nd edition, Springer

17. C. Quaranta, G. Canali, G. Ottavani, K. Zanio, 1969Electron-hole Pair Ionization Energy in CdTe between 85K and 350K, Lettere Al Nuovo Dimento, 4, 908 910

18. M. B. Schneider, O. S. Jones, N. B. Meezan, et al. 2010Images of lthe laser entrance hole from the static X-ray imager at NIF, Rev. Sci. Instru. 81 10E538.

19. S. Stepanov, 1997X-ray Server, available from http://sergey.gmca.aps. anl.gov/

20. S. Stepanov, 2009X0h Program, avalable from http://sergey.gmca.aps. anl.gov/x0h.html

Chapter 9

REVIEW OF AEROSOL OBSERVATIONS BY LIDAR AND CHEMICAL ANALYSIS IN THE STATE OF SÃO PAULO, BRAZIL

Gerhard Held[1], Ana Maria Gomes[1], Andrew G. Allen[2], Arnaldo A. Cardoso[2], Fabio J.S. Lopes[3] and Eduardo Landulfo[3]

[1] Instituto de Pesquisas Meteorológicas, Universidade Estadual Paulista, Bauru, S.P., Brazil

[2] Instituto de Química, Universidade Estadual Paulista, Araraquara, S.P., Brazil

[3] Centro de Lasers e Aplicações, Instituto de Pesquisas Energéticas e Nucleares, Universidade de São Paulo, São Paulo, S.P., Brazil

INTRODUCTION

Large-scale forest fires in the tropics, emitting vast amounts of aerosols and trace gases, drew the attention of scientists around the world in the late 80s and early 90s. A number of international collaborative research projects, such as TRACE-A (Transport and Atmospheric Chemistry near the Equator-Atlantic, [1]) and SAFARI-92 (South African Fire-Atmosphere Research Initiative, [2]), were initiated under the auspices of the International Geosphere-Biosphere Programme to investigate biomass burning emissions and their long-range transport. One of the areas of great interest was the Amazon region (Figure 1), which later led to the creation of the international Large Scale Biosphere-Atmosphere Experiment in Amazonia (LBA, [3]) in Brazil in 1998. Several intense observation campaigns were dedicated, not only to rainfall measurements by radar and storm structure, but also to biomass burning, monitoring of emissions and transport of aerosols and their impact on the vegetation and population of the region. However, monitoring of background concentrations of aerosols, deploying stacked filter units, had already been initiated in 1990 at the "Sierra do Navio" site (Amapá, about 190 km north of the equator) and in Cuiabá (Mato Grosso), a town located in the Brazilian savannah [4]. The location of both sites is shown in Figure 1.

São Paulo is Brazil's most populous State, with approximately 42 million inhabitants (21,5% of Brazil's total population) in an area of 249 000 km². The region is diverse in terms of its geography, natural environment and economy, and can be broadly classified into three main zones. In the southeast, the Atlantic coastal strip is separated from the remainder of the State by the scarp of the Serra do Mar, containing Brazil's largest remaining areas of Atlantic rainforest, a threatened ecosystem that has been largely eliminated in most of the Brazilian States bordering the Atlantic ocean. Located on a plateau above the scarp are the densely populated and heavily industrialized regions of metropolitan São Paulo (RMSP) and its satellite cities. Continuing inland, the largest fraction of the area of the State has an economy mostly based on agro industry. Here has been widespread conversion of natural ecosystems to agriculture. The most important single agricultural activity is sugar cane production, although there are also substantial cattle ranching, citrus cultivation and agro forestry for pulping and construction. In all regions, it is largely local emission sources that determine the chemical composition of the atmospheric aerosol, with a smaller influence of long-range transport of polluted air masses from elsewhere in Brazil.

Figure 1. Brazil, showing the location of São Paulo State in relation to the Amazon region, as well as the background monitoring stations in Sierra do Navio (Amapá) and Cuiabá (Mato Grosso).

In terms of atmospheric quality, suspended aerosol particles are (together with ozone) probably the most important atmospheric pollutant in both São Paulo city and the largely agricultural hinterland of the State. Ozone is generated during reactions involving the nitrogen oxides (NO_x) and volatile organic compounds (VOCs) emitted from vehicles, biomass burning and biogenic sources. The particulates are either emitted directly (in the form of primary aerosols), or are produced during reactions involving gaseous precursors (SO_2, NO_x and hydrocarbons). In large urban areas, such as the Metropolitan Region of São Paulo (RMSP), anthropogenic emissions from vehicles and industrial processes are the dominant contributors to elevated aerosol levels, while biomass burning [5-7] and dust lifted from barren fields (Figure 2) during the dry winter season constitute the principal sources of aerosols in the central and western sectors of the state. The State of São Paulo is the largest producer of sugar cane in Brazil, accounting for about 60% of Brazil's harvest [8], with more than 4,7 million hectares planted in 2010, of which 44% are burnt before harvesting [9]. The sugar cane is mostly harvested from April to November. Although progress is being made in mechanization, large areas are still harvested manually, which requires burning of the crop in sectors of the plantations during the night prior to manual cutting to remove excess foliage. This practice results in large quantities of aerosols and trace gases being emitted into the atmosphere (Figure 2a), not only negatively affecting local towns, but also regions much further downwind [10-12], demonstrating the importance of monitoring aerosols throughout the State.

(a)

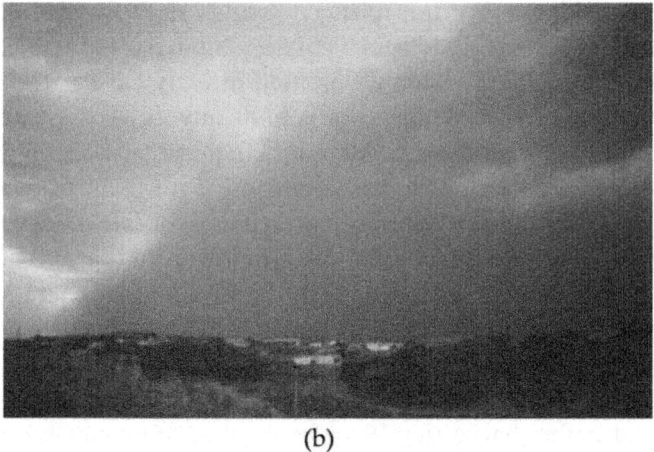

(b)

Figure 2. a) Typical sugar cane fire in central São Paulo State. (b) Dust lifted from freshly cut sugar cane fields by the downdraft of an approaching storm.

Along the São Paulo coast, marine aerosols are modified by the inclusion of pollutants emitted from transport, urban, and industrial sources. There are some areas where levels of anthropogenic pollution are low, and where the aerosol composition can be mainly attributed to natural origins. However, compared to metropolitan São Paulo and the interior of the State, the coastal zone has been much less well studied, with the exception of Cubatão, a heavily industrialized town near the coast close to Santos.

In the State of São Paulo, the first aerosol measurements began in Cubatão [13], and within the metropolitan area of São Paulo, notorious for its traffic emissions [14]. In terms of morphology, São Paulo is among the world's five largest cities, and is sixth largest in terms of population [15], with about 11,3 million inhabitants. The population of the Metropolitan Region of São Paulo (RMSP), which includes peripheral urban areas, reached an estimated 19,9 million persons in 2009 [16]. Human activities including road transport and industry now exert an enormous impact on air quality in the region, and therefore on the health of the population [17]. The total fleet of vehicles (cars, buses, trucks and motorcycles, powered by gasoline, ethanol and diesel) in the State of São Paulo exceeded 12,8 million in 2011, of which about 50% operate within the RMSP [9].

Observations from the Brazilian Lightning Detection Networks (RINDAT [18] and BrasilDAT at ELAT/INPE [19]) have shown a significantly higher lightning frequency over the RMSP and other large urban complexes within the State since the inception of the RINDAT Lightning Network in 1999

[20, 21]. This prompted a study of the impact of anthropogenic emissions on the frequency of lightning [22], showing a distinct increase of cloud-to-ground flashes, not only over the RMSP, but also over other large cities and densely populated or industrialized regions in the State, correlated to the occurrence of heat islands and increased concentrations of PM_{10}.

METEOROLOGY AND CLIMATOLOGY OF THE STATE OF SÃO PAULO

Since the meteorology of a region has a major impact on the dispersion or accumulation of pollutants, a brief characterization of the climate is appropriate. The State of São Paulo is located between the latitudes of about 20 and 25 South (Figure 1), thus falling into the transition zone from a tropical to a subtropical climate, with an annual rainfall total ranging between 1250 and 1650 mm in the interior, increasing to 1850 mm over the narrow coastal strip [23]. The year can be roughly divided into two periods, *viz.*, the rainy season from October to March, when most of the rain is produced by convective storms, and the dry winter months from April to September. During the rainy season, conditions are more representative of the tropical climate, with the occasional occurrence of a South Atlantic Convergence Zone (SACZ), which can be identified from satellite images as a cloud band with orientation northwest to southeast, extending from the southern region of Amazônia into the central region of the South Atlantic Ocean [24]. The SACZ situations can last more or less continuously from 4 days to more than one month and are extremely efficient producers of rain in the form of tropical thunderstorms, with accompanying high humidity. During the relatively dry winter months, the climatic conditions are more typical of the subtropics, with only occasional heavy rainfalls being caused by the passage of baroclinic systems (mostly cold fronts), moving from southwest to northeast across the State, but for the remaining time, the weather is dominated by a high pressure system, resulting in elevated temperatures, with low humidity and high stability in the Planetary Boundary Layer, favoring the accumulation of pollutants in the atmosphere of the region [25].

Sodar observations made during the period of June 2009 to December 2011 showed that strong nocturnal Low-Level-Jets (LLJs) develop on top of the surface radiation inversion, mostly during the relatively dry austral winter months (May – October), when stable conditions prevail [26, 27]. These LLJs generally form during the late evening at altitudes ranging from 250–500 m AGL, with maximum speeds of 12–20 m.s^{-1}. They usually last until 08:00–09:00 Local Time (LT), when the inversion has been eroded by the solar radiation. The frequency of LLJs varied from 3 - 22 days per month, with higher frequencies and greater intensity generally during the winter months. Observations with a sodar were made at three different locations in the central region of the State, *viz.* in Bauru, Rio Claro and Ourinhos. Earlier measurements, deploying tethered balloons and radiosondes in the eastern region of the State, yielded similar results in terms of structure, dynamics, seasonality and development characteristics [28]. LLJs have been observed in many parts of the world and were found to have regional extent. The practical importance of the LLJ lies in the rapid transport of moisture and pollutants in a narrow vertical band above the radiation inversion [29].

GROUND-LEVEL MONITORING OF PARTICULATES

Regular monitoring of air pollutants under the auspices of the Companhia de Tecnologia de Saneamento Ambiental (CETESB), the air quality "watchdog" in the State of São Paulo, started in the 70s, but a fully automatic monitoring network was only installed in 2000. Since then, observations are available in real time [30]. In 2001, 29 automatic stations, the majority in the RMSP, were already in operation [31]. From 2008 onwards, the automatic monitoring network was significantly expanded. In 2011, 42 monitoring stations in 28 towns were in operation, 19 in the RMSP and 23 in the remaining parts of the State [9]. The majority of the stations monitor particulate matter (PM_{10}), NO, NO_2, NO_x and O_3, as well as meteorological parameters, while a few also measure $PM_{2.5}$, SO_2 and CO. The automatic air quality monitoring network is shown in Figure 3. Additionally, CETESB also maintained a network of 41 manual monitoring stations during 2011, where measurements are made of $PM_{2.5}$, PM_{10}, TSP (Total Suspended Particulates), black smoke and SO_2, in various combinations [9]. Aerosol mass concentrations are determined using either β-attenuation instruments (automatic stations) or gravimetric and reflectometric techniques (manual stations).

In accordance with recommendations of the World Health Organization [32], CETESB defines 5 levels of air quality: "Boa" (good), "Regular" (regular), "Inadequada" (insufficient), "Má" (bad) and "Péssimo" (extremely bad), the highest being invoked if one of the monitored pollutants exceeds the pre-defined threshold. The national air quality standards are defined in CONAMA Resolution No. 03/90 (Table 2 in [9]).

PM$_{10}$ and TSP measurements are available since 1984 and 1985, respectively [31], although initially only from very few stations in the interior of the State, but gradually increasing to 41 and 11, respectively, in 2011 [9]. Figure 4 shows the year-to-year variation of annual mean PM$_{10}$ concentrations against the National Air Quality Standard (PQAr) for the RMSP and two sites in Cubatão (Figure 3, Nos. 24 and 25), which is one of the major industrial hubs in Brazil, where one site is located within the industrial suburb (No. 25) and the other in the town centre (No. 24). A significant reduction of mean annual PM$_{10}$ concentrations can be noticed from 1998 onwards, confirming the success of implementation of stringent air quality control measures, administered by CETESB. However, within the industrial suburb, confined in a valley, concentrations are still about twice the PQAr. A detailed description of Cubatão, its industrial activities and their location are found in [33]. More details on current PM$_{10}$ and TSP concentrations are provided in Section 4.3.

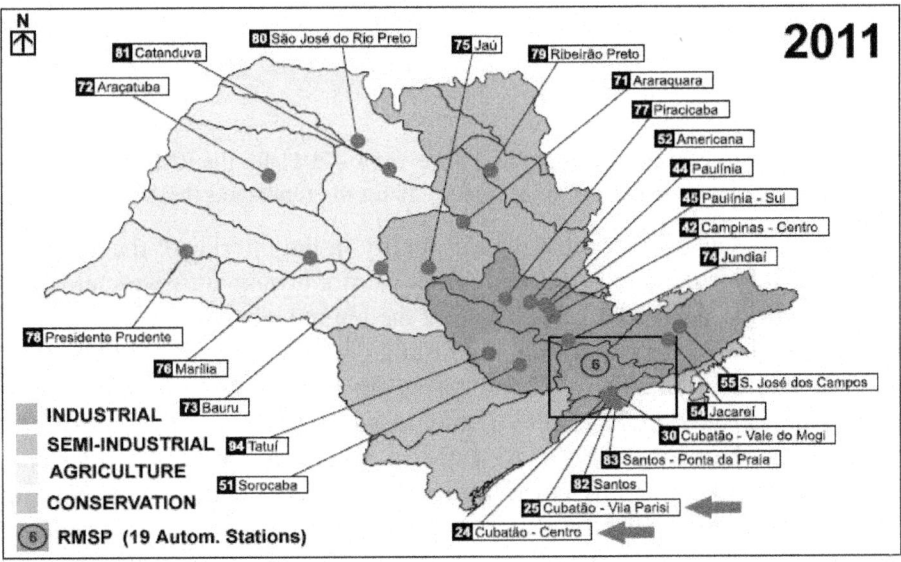

Figure 3. CETESB network of automatic monitoring stations in 2011. The shading indicates the principal land use in four schematic regions of the State, directly related to the type of emissions. Adapted from [9].

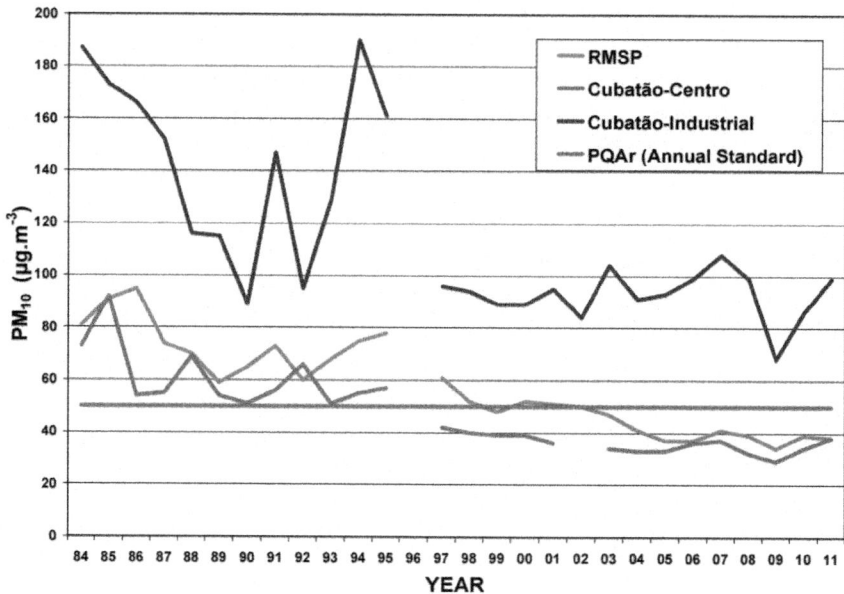

Figure 4. Year to year variation of PM_{10} from 1984 – 2011 for the RMSP and Cubatão. The data were extracted from [9, 31]. $PQAr = 50 \ \mu g.m^{-3}$ represents the Annual Standard.

Figure 5 demonstrates that the air quality in the interior of the State from 2002 to 2011, when a reasonable number of monitoring sites were already in operation [9], was generally well below the annual standard of 50 $\mu g.m^{-3}$ for PM_{10}, with the exception of Santa Gertrudes, just south of Rio Claro (Figure 6), where several large ceramic industries are located, notorious for emitting large quantities of aerosols. At two other monitoring sites, annual means were close to the Annual Standard. At Limeira mixed industrial activities range from metallurgical, through cellulose to ceramics, besides sugar cane and orange production and processing plants. Limeira and Santa Gertrudes are medium-sized industrial towns, about 20 and 40 km northwest of Americana (Figure 3, No. 52). The other site is in Piracicaba (Figure 3, No. 77), which also hosts mixed industrial activities, including a significant petrochemical plant. However, the exceedance in 2011 was most likely caused by major road construction works in the immediate vicinity of the monitoring site [9].

Although annual mean concentrations of PM in the State of São Paulo seem to be quite acceptable, it is obvious that violations of the daily Air Quality Standard do occur occasionally in several towns of the interior and within the RMSP. Comprehensive annual and specialized technical reports and publications on the air quality in the State of São Paulo, including detailed monitoring results, are available online [9].

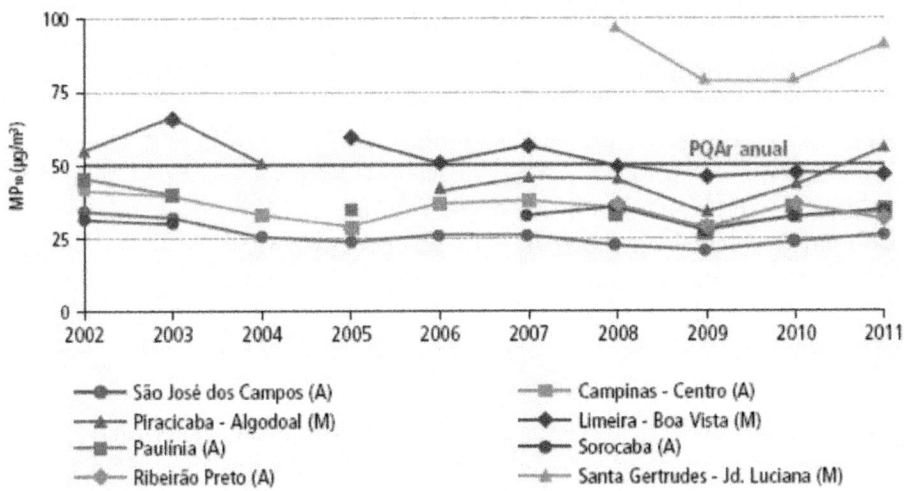

Figure 5. Year to year variation of PM_{10} from 2002 -2011 for monitoring sites in the interior of the State of São Paulo (after [9]). PQAr = 50 µg.m^{-3} represents the Annual Standard.

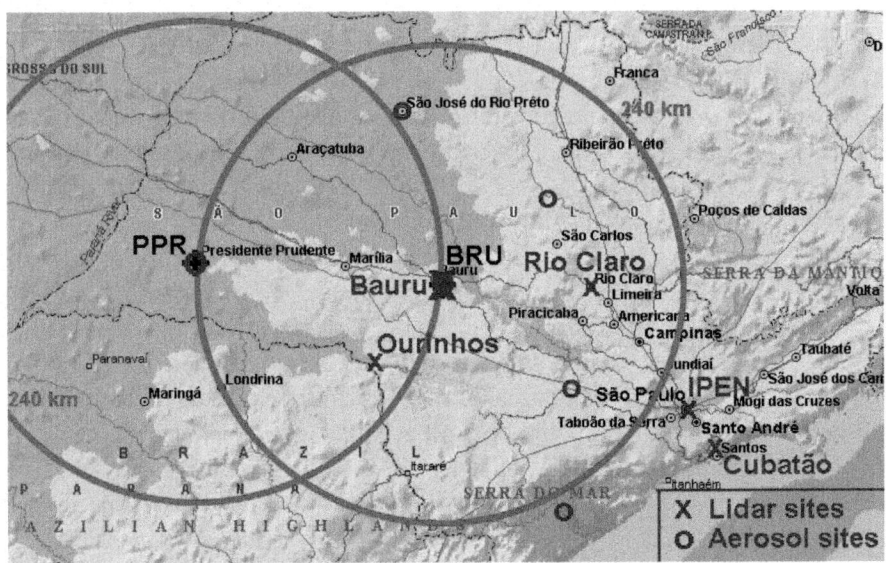

Figure 6. Aerosol monitoring sites in the State of São Paulo (except CETESB network) and 240 km ranges of IPMet's radars in Presidente Prudente (PPR) and Bauru (BRU). Sites from where lidar measurements are available are marked with x.

CHEMICAL COMPOSITION OF AEROSOLS

Agro-Industrial Rural Regions

Seasonal variability in the major soluble ion composition of atmospheric particulate matter in the principal sugar cane growing region of central São Paulo State indicates that pre-harvest burning of sugar cane plants is an important influence on the regional-scale aerosol chemistry [34]. The size-distributed composition of ambient aerosols is used to explore seasonal differences in particle chemistry, and to show that dry deposition fluxes of soluble species, including important plant nutrients, increase during periods of biomass (sugar cane trash) burning [6, 10].

Concentrations of trace gases and aerosols were determined at six measurement sites of a regional network in São Paulo State (blue circles in Figure 6), installed in rural areas including the State's central agricultural zone and the eastern coast [11] as part of an experimental research project to determine the anthropogenic component of nutrient deposition. The measurements were made over 12 months during 2008/2009 (one week of continuous sampling per month). Aerosols were collected onto 47 mm diameter Teflon filters using active samplers, and trace gases (NO_2, NH_3, HNO_3 and SO_2) were sampled using diffusion-based devices. The soluble ions NO_3^-, NH_4^+, PO_4^{3-}, SO_4^{2-}, Cl^-, K^+, Na^+, Mg^{2+} and Ca^{2+} were analyzed in aqueous extracts of the aerosol filters, using ion chromatography. NO_2, HNO_3 and SO_2 were similarly determined as NO_2^-, NO_3^- and SO_4^{2-}, following aqueous extraction of the collection media. NH_3 was determined using a colorimetric technique. Identification and quantification of nutrient sources was achieved using principal component analysis (PCA) followed by multiple linear regression analysis (MLRA) applied to the chemical data. Dry deposition fluxes were estimated using the measured atmospheric concentrations together with dry deposition velocities of gases and aerosols to different surface types, including tropical forest, savannah, sugar cane, pine, eucalyptus, orange, coffee, pasture and water. The annual cycle in deposition, to a sugar cane surface, of reactive nitrogen and sulphur in the gaseous, aerosol and dissolved phases is illustrated in Figure 7.

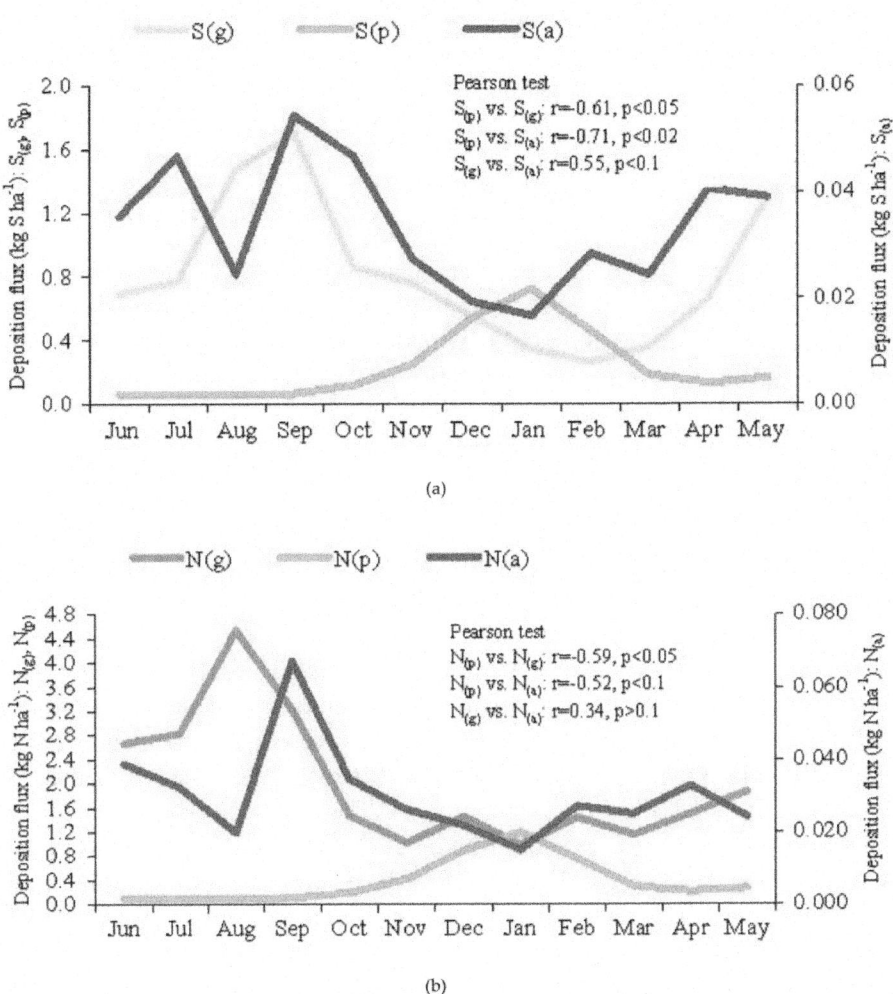

(a)

(b)

Figure 7. Annual cycle in deposition fluxes to a sugar cane surface of: (a) sulphur in gaseous ($S_{(g)}$), aerosol ($S_{(a)}$) and rainwater ($S_{(p)}$) phases; (b) nitrogen in gaseous ($N_{(g)}$), aerosol ($N_{(a)}$) and rainwater ($N_{(p)}$) phases. Primary y-axes: gas and rainwater; secondary y-axes: aerosol. Data for Araraquara.

The sugar cane industry has a major impact on air quality and the characteristics of the atmospheric aerosol. During the dry season (May to October), the burning of the cane, a prerequisite of manual harvesting, has for many years resulted in very large emissions of pollutants, including high carbon content aerosols. These particles contain water-soluble organic carbon (WSOC), anions (sulphates, nitrates and chlorides), cations (potassium, ammonium, calcium, magnesium, sodium), black carbon (BC), insoluble organic carbon and trace metals. Carbonaceous material comprises the bulk of the aerosol mass, especially in fine particles [5-7, 35-39]. In 2004, the annual emission of nitrogen oxides (NO_x) from sugar cane burning in Sao Paulo State was in excess of 45 Gg.N [40]. This is not only indicative of the scale of the emissions, but also of their potential for formation of secondary aerosols (containing nitrates, amongst other components).

In 2011, annual mean PM_{10} concentrations measured at automatic monitoring stations in the agro-industrial interior of São Paulo State were in the range 23-91 g.m^{-3}, with the highest values at locations affected by primary emissions from ceramics industries (Figure 5). At sites in the sugar cane production areas, annual mean PM_{10} concentrations were in the range of 32-41 g.m^{-3} [9]. The data revealed no obvious trends in PM_{10} concentrations during the period 2002-2011 (Figure 5).

A proportion of the primary material emitted during sugar cane fires is in the form of the large ash fragments notorious for causing domestic soiling problems in the region. During 1995-1996, CETESB investigated deposition rates, and measured the concentrations of PAHs, PCBs, dioxins and furans. The sedimented material was collected during the harvest period using plastic funnels lined with polyurethane foam, positioned near to plantations and in the urban area of the city of Araraquara. Samples were also collected in parallel using a high volume filter-based sampler. Levels of PCBs were in the range 4-12 ng.m^{-3}, and showed no association with levels of carbonaceous material derived from the fires. Deposition fluxes of the dioxins and furans were in the range 1-17 pg.m^{-2}.day^{-1}, and were higher in greater proximity to plantations, indicating that sugar cane burning was a source of these compounds. The PAHs were found in two distinct groups. Naphthalene, fluorene, phenanthrene, anthracene, fluoranthene, and pyrene were present at concentrations exceeding 30 ng.m^{-3}, while acenaphthene, chrysene, benzo(a)fluoranthene, benzo(k) fluoranthene, benzo(a)pyrene, dibenzo(a,h)anthracene, benzo(g,h,i)perylene and indeno(1,2,3,c,d)pyrene were found at up to 21 ng.m^{-3}. Concentrations were always higher during the harvest period [41].

The presence of PAHs in ash from sugar cane fires was also reported by Zamperlini et al. [42, 43]. In the PM_{10} fraction, it was found that the most abundant polycyclic aromatic hydrocarbons were phenanthrene and fluoranthrene, and the least abundant was anthracene [44]. Cluster analysis of the total PAH concentrations for each day of sampling, and the corresponding meteorological data, suggested that concentrations of PAHs were independent of climatological conditions or season of the year. Vehicular sources were identified during both dry and wet seasons, although sugar cane burning emissions were the dominant source during the dry season.

Sugar cane burning is a major source of acidic gases that contribute to the formation of secondary aerosols. In Araraquara, Da Rocha et al. [36] reported concentrations of 9,0 ppb (HCOOH), 1,3 ppb (CH_3COOH), 4,9 ppb (SO_2), 0,3 ppb (HCl) and 0,5 ppb (HNO_3). Extremely high concentrations of these gases were measured in the plumes downwind of sugar cane fires: 1160-4230 ppb (HCOOH); 360-1750 ppb (CH_3COOH); 10-630 ppb (SO_2); 4-210 ppb (HCl); and 14-90 ppb (HNO_3). Highest levels of SO_2, HCl and HNO_3 in Araraquara were measured during the harvest period, with peak concentrations in the evening (the time of the fires).

The distribution of soluble ionic material between fine (<3,5 μm) and coarse (>3,5 μm) aerosol fractions was determined by Allen et al. [5], who measured the ions $HCOO^-$, CH_3COO^-, $C_2O_4^{2-}$, SO_4^{2-}, NO_3^-, Cl^-, Na^+, K^+, NH_4^+, Mg^{2+} and Ca^{2+}. The fine and coarse particles showed acidic and basic properties, respectively, and concentrations of all major ions increased significantly during the dry season (Figure 8). Da Rocha et al. [6] collected aerosols in twelve size fractions, and used calculation of ion equivalent balances to show that during burning periods, the smaller particles (Aitken and accumulation modes) were more acidic, containing higher concentrations of SO_4^{2-}, $C_2O_4^{2-}$, NO_3^-, $HCOO^-$, CH_3COO^- and Cl^-, but insufficient NH_4^+ and K^+ to achieve neutrality. Larger particles showed an anion deficit due to the presence of unmeasured ions, and comprised re-suspended dusts modified by accumulation of nitrate, chloride and organic anions. Increases of re-suspended particles during the burning season were attributed to release of earlier deposits from the surfaces of burning vegetation, as well as increased vehicle movement on unsealed roads. During the winter months, the relative contribution of combined emissions from road transport and industry diminished due to increased emissions from biomass combustion and other activities specifically associated with the harvest period.

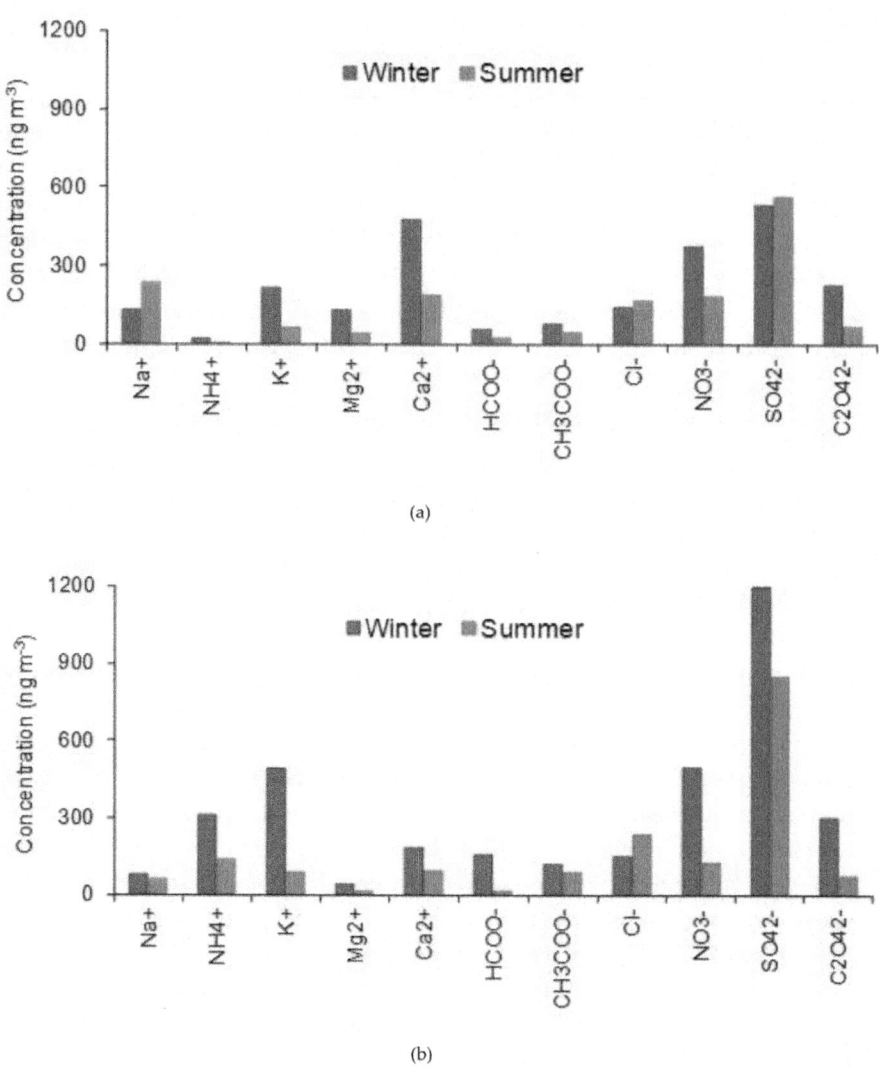

Figure 8. Comparison of aerosol composition in the Araraquara region during winter (biomass burning) and summer (non-burning) periods: (a) Coarse particles; (b) fine particles.

In separate work, biomass-burning aerosols were found to contribute around 60 and 25% of the mass of fine and coarse aerosols, respectively, in the Piracicaba sugar cane growing region [7]. A high proportion of the elements K, S, Cl, Br, Fe and Si in aerosols has been linked to biomass burning [45], indicative of both a combustion component (emissions of K, S, Cl and Br) and a suspended soil dust component (emissions of Fe and Si).

In a study reported in [38], elemental analysis of individual and bulk aerosols collected in rural areas was followed by evaluation of the data using statistical hierarchical clustering, which revealed the contributions of two different types of carbonaceous material (biogenic and carbon-rich) and two aluminosilicate fractions (pure or mixed with carbon). These findings contrasted with the findings of similar work in the atmosphere of São Paulo city, where hierarchical clustering analysis revealed the presence of metal compounds, silicon-rich particles, sulphates, carbonates, chlorides, organics and biogenic particles [46]. This reflects the very different characteristics of the aerosols found in the two regions.

Da Rocha et $al.$ [6] showed that dry deposition fluxes of important plant nutrients increased during the sugar cane burning season. During this period, the fine fraction aerosol was more acidic and contained elevated concentrations of SO_4^{2-}, $C_2O_4^{2-}$, NO_3^-, $HCOO^-$, CH_3COO^- and Cl^-, but insufficient NH_4^+ and K^+ to achieve neutrality. Larger particles consisted of re-suspended dust, modified by inclusion of nitrate, chloride and organic anions. The increases in annual particulate dry deposition fluxes due to higher fluxes during the sugar cane harvest were 44,3% (NH_4^+), 42,1% (K^+), 31,8% (Mg^{2+}), 30,4% ($HCOO^-$), 12,8% (Cl^-), 6,6% (CH_3COO^-), 5,2% (Ca^{2+}), 3,8% (SO_4^{2-}) and 2,3% (NO_3^-). The contributions of dry deposition to total deposition (including precipitation scavenging, excluding gaseous dry deposition) were 31% (Na^+), 8% (NH_4^+), 26% (K^+), 63% (Mg^{2+}), 66% (Ca^{2+}), 32% (Cl^-), 33% (NO_3^-) and 36% (SO_4^{2-}).

Deposition rates of aerosol nutrient species to a range of natural and agricultural surfaces were reported in [10], using a size-segregated particle dry deposition model. Fluxes greatly exceeded those expected under pristine conditions, with deposition to tropical forest found to have increased by factors of 12,2 (NO_3^-), 6,2 (PO_4^{3-}) and 2,6 (K^+) (Figure 9). Source apportionment using principal component analysis (PCA) and multiple linear regression analysis (MLRA) revealed that in central São Paulo State, biomass burning, products of secondary reactions and soil dust re-suspension contributed 43%, 31% and 21% of $PM_{2.5}$ mass, respectively. Re-suspension and biomass burning contributed 22% and 19%, respectively, to PM_{10} mass, and re-suspension accounted for approximately half the mass of coarse particles. At least 40% of NO_3^--N, 20% of phosphorus and 55% of potassium deposited originated from agriculture-related emissions.

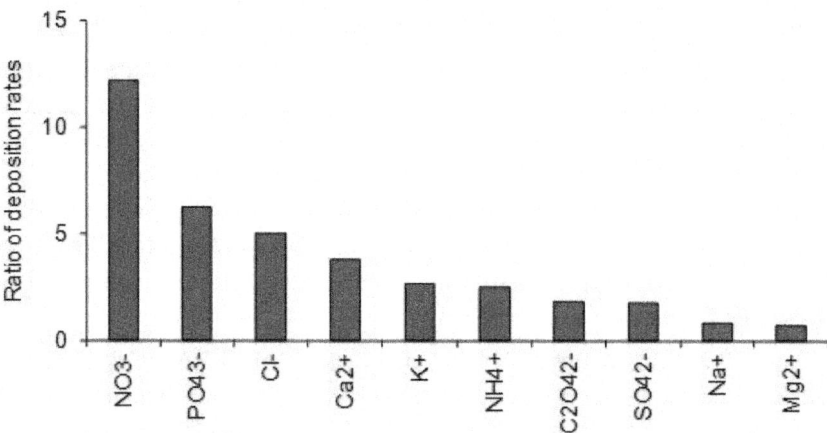

Figure 9. Graph showing the present-day increase in aerosol dry deposition rates to a tropical forest surface, compared to deposition rates estimated for pristine conditions.

Emissions of reactive nitrogen compounds are of concern due to their influence on both atmospheric acidity (production of HNO_3 from reactions involving NO_2) and the formation of photochemical oxidants such as ozone and peroxyacetyl nitrate (PAN). Reactions of acidic species with ammonia generate ammonium sulphates and nitrates, mainly in the long-lived accumulation mode size fraction. Deposition of reactive nitrogen can cause eutrophication of water bodies, as well as the release of trace metals in soils. Machado *et al.* [47] found that emissions of reactive nitrogen during sugar cane burning, in the forms of NH_3, NO_x and particulate nitrate and ammonium, were equivalent to 35% of the annual fertilizer-N application. The concentrations of nitrogen oxides showed a positive association with the number of fires, reflecting the importance of biomass burning as a major emission source, and mean concentrations of NO_x in the dry season were twice those in the wet season. During the dry season, biomass burning was the main source of NH_3, with other sources (wastes, soil, biogenic) predominant during the wet season. The estimated emission fluxes of NO_2-N, NH_3-N, NO_3^--N and NH_4^+-N from sugar cane burning in a planted area of about $2,2 \times 10^6$ ha were 11,0, 1,1, 0,2 and 1,2 Gg.N.yr^{-1}, respectively.

The sources, atmospheric transport and reactions of the main inorganic reactive nitrogen (N_r) species (NO_2, NH_3, HNO_3 and aerosol nitrate and ammonium) were investigated in a study conducted over a period of one year at six sites distributed across an area of about 130,000 km^2 in São Paulo State [11]. Oxidized forms of nitrogen were estimated to account for about 90% of dry deposited N_r, due to the emissions of nitrogen oxides from biomass burning and road transport. NO_2-N was important closer to urban areas; however, HNO_3-N was the largest individual component of dry deposited N_r. A simple mathematical model was developed to enable determination of total N_r dry deposition from knowledge of NO_2 concentrations. The model, whose error ranged from <1% to 29%, provided a new tool for the mapping of reactive nitrogen deposition.

The sugar cane burning emissions radically alter the chemistry of precipitation water. Coelho *et al.*[122] found that concentrations of soluble ions (K^+, Na^+, NH_4^+, Ca^{2+}, Mg^{2+}, Cl^-, NO_3^-, SO_4^{2-}, F^-, PO_4^{3-}, CH_3COO^-, $HCOO^-$, $C_2O_4^{2-}$ and HCO_3^-) increased by between two and six-fold during the harvest period. Principal component analysis revealed three main sources of the material in rainwater: biomass burning and soil dust re-suspension (52% of the total variance), secondary aerosols (26%) and vehicular emissions (10%). The biomass burning component diminished in the summer (non-burning period), when there was a relative increase in the importance of road transport/industrial emissions. The volume-weighted mean concentrations of ammonium (23,4 mol.L^{-1}) and nitrate (17,5 mol.L^{-1}) in rainwater samples collected during the harvest period were similar to those found in rainwater from São Paulo city, which emphasized the importance of including rural agro-industrial emissions in regional-scale atmospheric chemistry and transport models. There was evidence of a biomass-burning source throughout the year, which suggests that vegetation fires may continue to emit aerosols and their precursor gases, even after sugar cane burning is phased out.

Metropolitan São Paulo (RMSP)

In terms of trace species, the composition of the lower troposphere in the conurbation of the RMSP differs considerably from that of the interior of the State and the coastal zone. The critical air quality issue here is the scale of the emissions from road vehicles. In 2001, the vehicle fleet consisted of 17,2% hydrated ethanol-fuelled, 76,3% gasohol-fuelled and 6,5% diesel-fuelled vehicles, with ethanol contributing 34% of the total fuel consumption [31]. The figures for 2011 were 46,7% gasohol (cars and light commercial), 3,9% hydrated ethanol, 31,9% flex-fuel, 5,4% diesel and 12% motorcycles [9].

It is important to consider the relative amounts of the different fuels used, since emissions vary according to fuel, which has consequences for aerosol composition. For example, there is a larger fraction of oxygenated compounds in the secondary aerosols produced from reactions involving the aldehydes and alcohols emitted during ethanol combustion, which can affect the hygroscopicity of the particles, as well as their toxicological properties [48, 49].

The proportions of gasohol (gasoline with 22% anhydrous ethanol) and hydrated ethanol used have varied considerably in recent decades. Ethanol was first adopted as a road vehicle fuel in Brazil in 1979, due to the Brazilian National Alcohol Program (PROALCOOL), which was introduced as a response to the 1970s oil crisis. This not only reduced Brazil's dependency on oil imports, but also helped to eliminate the use of lead-containing anti-knock additives [49]. Sales of hydrated ethanol-fuelled vehicles peaked in the 1980s [50]. More recently, since around 2005, the new car market has been dominated by flex-fuel vehicles equipped with engine systems able to adjust to the gasoline/ethanol mixture present in the fuel tank [9].

In 2011, the sources of PM_{10} in metropolitan São Paulo were: heavy goods vehicles (38,6%), re-suspended dusts (25%), secondary aerosols (25%), industrial processes (10%) and light duty vehicles (1,4%). Annual mean concentrations of PM_{10} measured at the 18 automatic monitoring stations in São Paulo ranged between 31 and 50 g.m^{-3} [9]. A detailed analysis of these measurements, as well as of $PM_{2.5}$, TSP and black smoke measurements made at a smaller number of locations, are provided in the CETESB report [9] and in earlier annual reports published by CETESB.

The pollutant source profile remains fairly constant throughout the year. Use of absolute principal factor analysis showed that the contributions of different sources to $PM_{2.5}$ mass during winter and summer were: vehicle emissions (28 and 24% for the two seasons, respectively), re-suspended soil dusts (25 and 39%), oil combustion (18 and 21%), sulphates (23 and 17%) and industrial emissions (5 and 6%). Soil dusts accounted for 75-78% of the mass of coarse particles [51]. Andrade *et al.* [14] reported the results of elemental analyses, using particle-induced X-ray emission (PIXE) analysis of fine and coarse aerosols collected in 1989. Principal component analysis revealed the following sources of fine particles: oil and diesel combustion (explaining 41% of the mass), re-suspended soil dusts (18%), industrial emissions (13%), and a source associated with emissions of Cu and Mg (18%). Sources of coarse particles were: re-suspended soil dusts (59%), industrial emissions (19%), oil burning (8%) and marine aerosols (14%). Alonso *et al.* [52] used chemical mass balance (CMB) receptor modeling to show that the composition of fine particles was consistent with the presence of primary material from vehicles and secondary organic carbon and sulphate. Road dust re-suspension and vehicle emissions were the main sources of coarse particles and TSP. The same trends in source profiles were observed at geographically distinct locations in São Paulo. Sanchez-Ccoyllo and Andrade [53] used receptor modeling to identify five main sources of aerosols: vehicles, waste incineration, vegetation, suspended soil dust and fuel oil burning.

Organic and elemental carbon, emitted mainly from diesel vehicles, together with ammonium sulphate, make up most of the mass of fine particles [54, 55]. In [56] it is reported that 80% of the mass of fine ($PM_{2.5}$) particles consisted of organic material, with SO_4^{2-}, NO_3^- and NH_4^+ present in the fine fraction, and NO_3^-, SO_4^{2-}, Ca^{2+}, and Cl^- predominant in coarse particles ($PM_{2.5-10}$). Albuquerque et al. [57] found that fine particles were rich in BC, S and Pb, while elements associated with crustal aerosols and/or industrial emissions (Al, Si, Ca, Ti, and Fe), together with ammonium sulphate and BC, composed the coarse mode particles. Other species, including K, Al, Fe and soil minerals, are included as a smaller component of fine particle mass [46]. Both vehicular and industrial emissions are sources of trace metals (Zn, Pb, Cr, Mn, Cd, etc.) [58, 59], and there appear to be continuing emissions of Pb from the road vehicle fleet, despite apparently low levels of Pb in fuels [60].

Aerosol composition similar to that of São Paulo is found in other major conurbations. In Campinas, the second largest city in the State, 100 km inland from São Paulo, fine particles were found to consist of 48% elemental carbon and 22% organic carbon, together with soluble ions and trace elements [61].

The PM concentrations are influenced not only by the magnitudes of emission sources, but also by ventilation and relative humidity. Miranda and Andrade [54] reported that higher PM_{10} concentrations (105 g.m^{-3}) measured during the winter of 1999, compared to winter 2000 (60 g.m^{-3}), were due to both better ventilation of the city during the latter period, as well as an increase in particle sizes at higher humidity. Similar findings were reported in [53], with lower pollutant levels associated with increased ventilation, precipitation, and relative humidity.

Primary emissions from vehicles result in large diurnal cycles in the concentrations of PM_{10}, BC, CO, NO_x and SO_2 [51], however the diurnal trends in particle mass concentrations differ between highly polluted and less polluted periods, with concentrations higher during the daytime for the former, and during the nighttime for the latter [57]. A possible influence of humidity on both the mass and size distribution of the Sao Paulo aerosol was suggested by the observation that while the size distribution of ammonium sulphate was unimodal during the daytime (with a maximum at 0,38 m), at night, when humidity is higher, the size distribution was bimodal (with maxima at 0,38 and 0,59 m) [55]. Furthermore, particle growth, observed using a Scanning Mobility Particle Sizer (SMPS), has been found to increase under polluted conditions [57].

Although local sources are by far the most important contributors to particulate air pollution in São Paulo city, back-trajectory analysis has shown that the atmosphere of the city can also be affected by the advection of air masses from distant regions where agricultural biomass burning is practiced, especially northeast Brazil [62]. This could explain the finding that the relative contribution of ammonium sulphate is higher under less polluted conditions [57].

An important consequence of the prevalence of fine mode particles in the atmosphere of the city is that the indoor environment provides little or no protection against exposure to these pollutants, since they easily infiltrate buildings. This was observed [63] using simultaneous indoor and outdoor measurements of a range of ionic species associated with both primary emissions (potassium, magnesium, sodium and calcium) and secondary aerosol formation (chloride, acetate, nitrate, formate, pyruvate, nitrite, sulphate, oxalate and ammonium). The measurements were made in offices, restaurants and a hotel. In the fine mode, only oxalate and ammonium showed significantly lower concentrations indoors. In the coarse mode, lower concentrations were normally found indoors (with the exception of acetate, chloride and potassium), reflecting the less efficient infiltration of larger aerosols.

Polycyclic aromatic hydrocarbons are an important component of the urban aerosols. Chrysene, benzo(e)pyrene and benzo(b)fluoranthene were found to be the predominant PAHs in PM_{10}, originating from industry, vehicles and long-range transport [64]. Levels of PM_{10} similar to those in São Paulo were measured in a city (Araraquara) situated in the rural biomass burning zone, although here PAH concentrations were lower. In both cases, dry deposition appeared to be the main mechanism of removal of PAH-containing aerosols from the atmosphere [65].

Bourotte et al. [66] measured the concentrations of 13 PAHs in fine ($PM_{2.5}$) and coarse ($PM_{2.5-10}$) aerosols. In both fractions, the predominant compounds were indeno(1,2,3-cd)pyrene, benzo(ghi)perylene and benzo(b)fluoranthene and PAH ratios suggested that automobile exhaust was the main source of the compounds. Factor analysis revealed four source components for the $PM_{2.5}$ fraction: diesel emissions, stationary combustion, vehicle emissions, and combustion of natural gas and biomass. For the coarse fraction, two components were identified, corresponding to vehicles and a mixture of gas, oil, and waste combustion.

Coastal Regions

Although measurements of atmospheric aerosol are scarce in most of the coastal regions, an exception is the industrialized town of Cubatão, located near sea level at the base of the Serra do Mar scarp, where there is a large industrial complex comprising over 20 heavy industries (petrochemical, chemical, iron and steel, fertilizer, cement, coking and others). The monitoring stations in this area register regular episodes of particulate pollution, with the emissions from the industrial installations being entrained into a sea breeze circulation, when PM_{10} concentrations can increase by as much as an order of magnitude [67]. Pollutants absorbed into cloud water and precipitation are subsequently deposited to the vegetation of the Serra do Mar Atlantic rainforest, causing extensive ecological damage [68].

Due to extreme levels of pollution, air quality in the Cubatão region has been monitored by CETESB since the 1980s, and there are currently three sites where PM_{10} is continuously measured (Figure 5), and one where TSP is measured [9]. The case of Cubatão is unique, since in contrast to the RMSP, by far the largest source of particulates is industrial emissions, rather than road transport. Guideline levels of TSP and PM_{10} have been frequently exceeded in the industrial zone (Vila Parisi) of Cubatão, and there has been no improvement in PM_{10} levels in recent years. During 2011, the annual mean PM_{10}concentrations in the three zones of Cubatão were 99 g.m^{-3} (Vila Parisi), 61 g.m^{-3} (Vila Mogi) and 38 g.m^{-3} (Centro) [9]. At the industrial Vila Parisi site, the annual geometric mean TSP concentration was 236 g.m^{-3}, greatly exceeding the primary and secondary air quality standards for this pollutant species (80 and 60 g.m^{-3}, respectively).

Although industrial emissions are responsible for the largest proportion of the aerosol loading of the atmosphere near the Cubatão industrial complex, the organic fraction has an important road transport-related component, because concentrations of polycyclic aromatic hydrocarbons (PAHs) are governed by emissions from heavy duty diesel vehicles [69]. In the same work, it was reported that a shift to greater use of biodiesel might decrease emissions of the PAHs.

In regions distant from the industrial installations, the aerosol composition reflects mainly natural sources (biogenic, terrigenous and marine). Bourotte *et al.* [70] found that aerosol (PM_{10}) composition in a State Park in the Cunha region was characterized by an abundance of K^+, Ca^{2+}, Na^+, Cl^- and Pb, while Vasconcellos *et al.* [71] reported the presence of aliphatic hydrocarbons emitted from biogenic sources in the coastal region.

LIDAR OBSERVATIONS

MSP-LIDAR

In 2001, an elastic backscattering lidar system (MSP-Lidar) was installed in a suburban area of São Paulo city, on the Campus of the University of São Paulo (23°33' S, 46°44' W; Figure 6) and is being operated by the *Centro de Lasers e Aplicações* (CLA) of the *Instituto de Pesquisas Energéticas e Nucleares* (IPEN). The lidar is collocated with an AERONET sunphotometer, which provides the vertical profile of the aerosol backscatter coefficient at 532 nm up to an altitude of 4–6 km above sea level [72]. The MSP-Lidar comprises a Nd:YAG laser with a wavelength of 532 nm, and is operated with a repetition rate (PRF) of 20 Hz and an energy pulse of up to 120 mJ. The backscattering signal is captured by a Newtonian telescope with 1,5 m focal length. Attached to the telescope is a photomultiplier optimized for the visible spectrum with a 1 nm FWHM interference filter. Observations are being made whenever atmospheric conditions (absence of low or middle clouds; no rain) permit the operation of the lidar, resulting in a vast amount of data having been accumulated, which have so far been exploited in 5 MSc and 4 PhD theses, the most relevant being [73-76].

In January 2004, the IPEN MSP-Lidar system was installed for 6 weeks at IPMet in Bauru (Figure 6), located in the central part of São Paulo State, to provide the first measurements of aerosol layers in the interior of the State [77]. At the beginning of the campaign, the lidar was operated in its original configuration and the data were digitized using a digital oscilloscope with 1 GHz bandwidth and 11-bit resolution; at the end of January 2004, this device was replaced by a transient recorder, capable of simultaneous analog and photon counting measurements at higher resolution (12-bit). The system was operated on 31 different days, during periods of about 4 hours in the morning, 4 hours in the afternoon and 6-8 hours during the night, depending on the occurrence of cloud and/or precipitation. The daytime measurements had a 15-30 m spatial resolution and maximum altitude of 10 km, yielding information on the diurnal variation of the Planetary Boundary Layer (PBL), while the measurements at night had a 30-60 m resolution, reaching up to 30-35 km maximum altitude. The diurnal variation of the PBL during the austral summer could be documented, as well as some background concentrations of aerosols, because very little biomass burning takes place during the rainy period.

Figure 10 shows a typical example of the diurnal variation of the height of the PBL on a cloudless day in Bauru. It should be noted that, due to the latitude of -22,3, the lidar cannot be operated during the midday period in summer, but as a result of turbulent mixing, the PBL could easily reach a maximum height of ≥3,5 km above ground level (AGL) during the early afternoon. The top of the PBL starts decreasing well before sunset, until it stabilizes at around 1,5 km AGL during the night. Times are indicated in Local Time (LT = UT-3h).

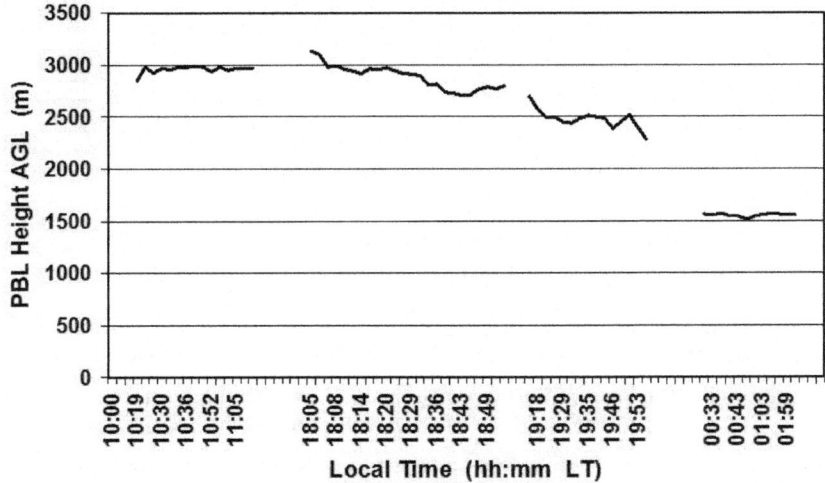

Figure 10. Height of the PBL over Bauru on 01/02 March 2004 for four different periods (10:19-11:08; 18:05-18:55; 19:13-19:54; 00:27-02:02 LT), with a vertical resolution of 30 m (after [78]).

The increased vertical range of the lidar during nocturnal operation permitted the detection of thin clouds and layers of aerosols, as shown in Figure 11. A cloud layer is clearly visible at around 4,5 km, while aerosols were detected at 3 and 5 km AGL, respectively. The top of the PBL is at about 1850 m AGL, with the faint layering being indicated in shades of green and light-blue colours.

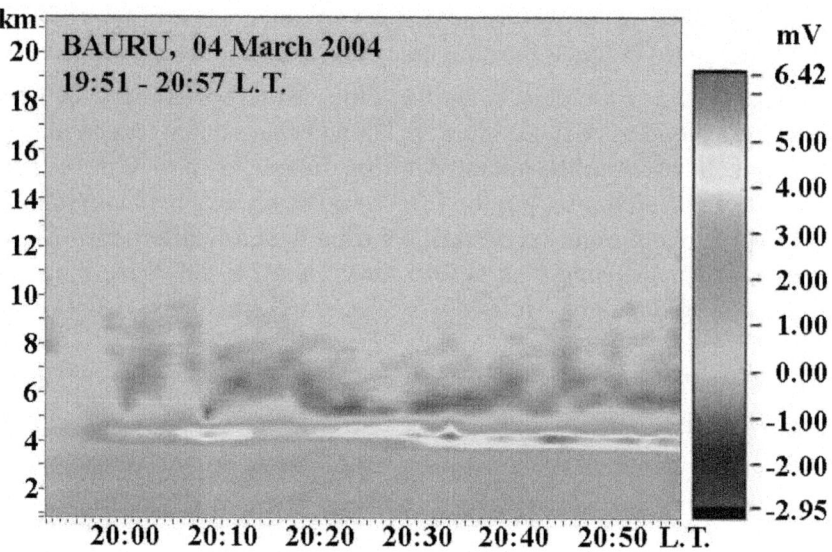

Figure 11. Nocturnal lidar observation above Bauru on 04 March 2004. Vertical range from 855 m to 21,5 km AGL, with a resolution of 30 m (after [78]).

In early 2008, the MSP-Lidar system was upgraded to a Raman lidar, and in its present 3-channel configuration it can measure elastic backscatter at 355 nm, together with nitrogen and water vapour Raman backscatters at 387 nm and 408 nm, respectively. Therefore, the PBL data now available include aerosol backscattering and extinction coefficients, as well as the Lidar Ratio (LR) and water vapour mixing ratio. Figures 12 and 13 present typical results of Raman lidar measurements recorded during night-time of 09/10 January 2008, during the austral summer season. This period of the year is characterized by a very well defined boundary layer throughout the day and relatively high humidity. The major part of aerosols and water vapour is contained within the boundary layer, while the scattering above the PBL is mainly due to molecules. Figure 12 shows the aerosol extinction and backscattering coefficient profiles at 355 nm, where one can see a residual aerosol layer between 900 m and 2000 m AGL, indicating a very pronounced presence of aerosols, overlaid by another discrete layer above it between 2500 m and 3500 m AGL. The height profile of the Lidar Ratio is shown inFigure 13a. The Lidar Ratio is about 80 sr and stable throughout the PBL up to about 3000 m AGL. The vertical profile of the lidar-derived water vapour mixing ratio can be seen in Figure 13b. The calibration of the lidar was performed using radiosonde data from the nearby São Paulo Campo de Marte airport. Although the sonde had a relatively low height resolution, integrating the water vapour content with height made such calibration possible.

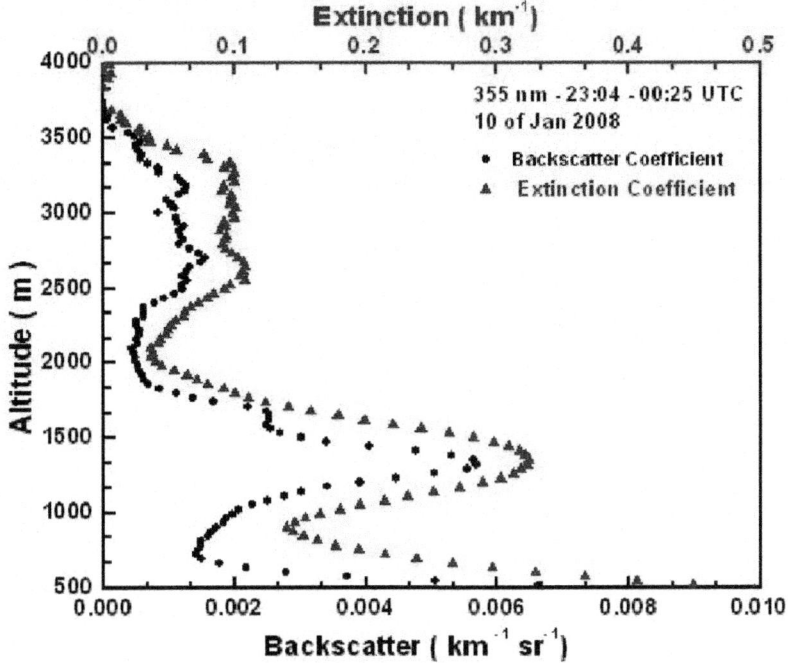

Figure 12. Aerosol backscatter and extinction profiles at 355 nm observed on 10 January 2008 at 00:25 UT (21:25 LT). The PBL top height is considered to be at 2000 m.

The MSP-Lidar system has contributed to several studies concerning the properties of aerosols and their influence on the air quality index of the city of São Paulo. Lidar measurements conducted daily provided observations of the PBL variation, which could be compared to corresponding air quality index values from local air quality monitoring and management agencies, as well as identifying potential air dispersion conditions [79]. It has also been deployed to monitor the long-range transport of aerosol plumes from different regions of Brazil to the RMSP and to evaluate the contribution of aerosol pollutants from remote sources. Landulfo and Lopes [80] have analyzed an event during the period 02 - 09 August 2007 when the AOD (Aerosol Optical Depth) and AE (Ångström Exponent) values retrieved from the AERONET sunphotometer indicated that high aerosol loads at five different locations in the Brazilian territory corresponded to biomass-burning particles. This was validated by the mean values of the Total Attenuated Backscatter Coefficient at 532 nm, the mean depolarization ratio and also the Lidar Ratio (about 70 sr) for all sites over-flown by the CALIOP sensor onboard the CALIPSO satellite.

In another case study during the dry winter season of 2008, fire plumes attributed to sugar cane fires were frequently observed by IPMet's radars in the absence of rain echoes and documented in terms of radar reflectivity, time and location [12]. On several occasions, IPEN's Elastic Backscatter Lidar in São Paulo observed layers of aerosols of varying strength and heights above the city. The most significant days were selected for calculating backward, as well as forward trajectories, deploying the Flextra 3.3 Trajectory Model [81], which was initiated with ECMWF historical data with a 0,25 x 0,25 grid spacing [12].

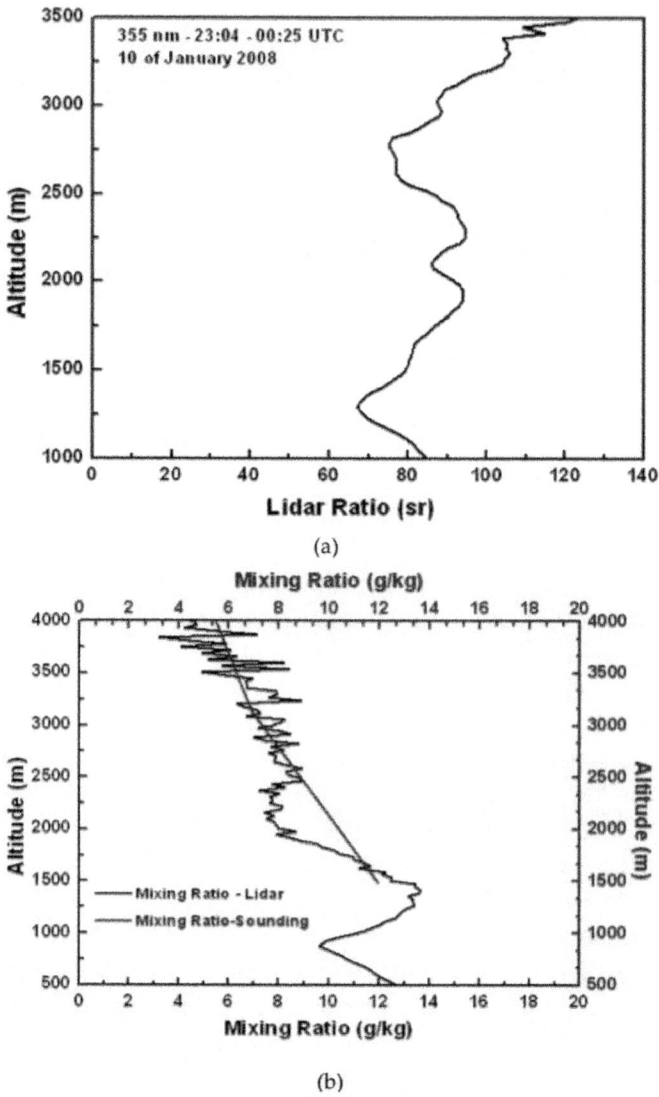

Figure 13. a) shows the 355 nm Lidar Ratio profile on 10 January 2008 at 00:25 UT (21:25 LT). (b) shows the water vapour mixing ratio extracted on the same day from the 408 nm channel (00:25 UT) and from a radiosonde ascent (00:00 UT).

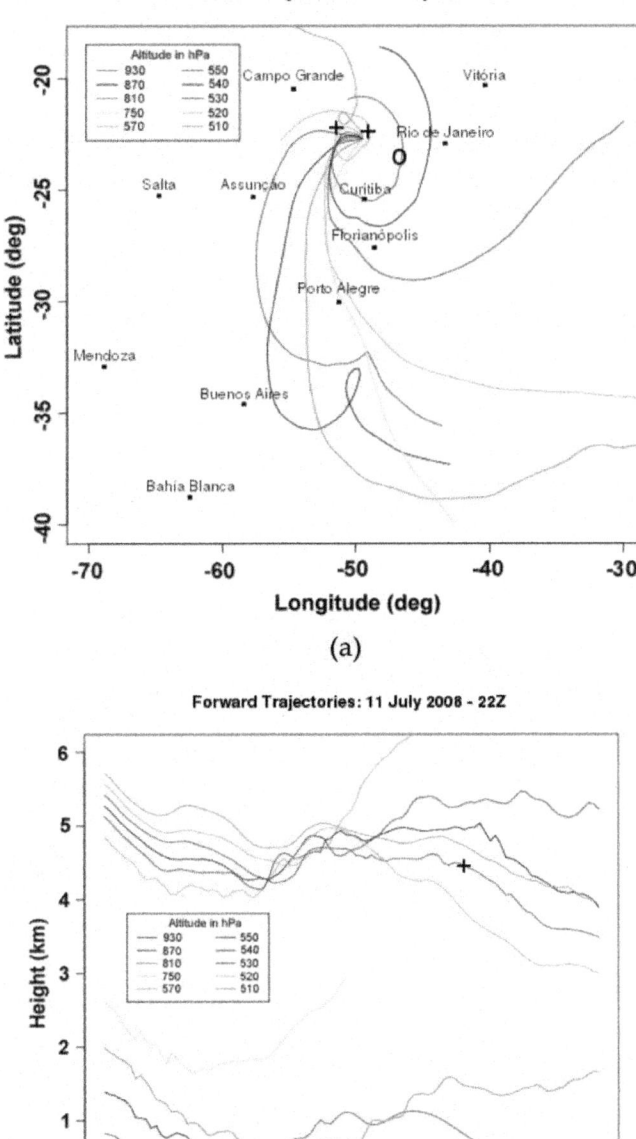

Figure 14. Forward trajectories initiated at different heights where a large fire was observed by IPMet's radars on 11 July 2008, 22:00 UT (19:00 LT). (a) The + indicates

the position of the PPR and BRU radars; o indicates the position of the lidar in São Paulo (IPEN). (b) Forward trajectories plotted against height and time. The + indicates the position of IPEN, marking height and time of arrival matching exactly with the lidar observation (Figure 15).

The results showed an excellent match between the radar-detected sources of the plumes and lidar observations in São Paulo.Figure 14 presents a typical case study, when emissions from biomass fires were identified by the radars on 11 July 2008 in the central parts of the State, and were subsequently monitored by IPEN's lidar over Metropolitan São Paulo on 14 July 2008, deploying forward and backward trajectories. The forward trajectories, initiated at different heights ranging from 930 hPa (close to ground level) up to 450 hPa (ca 6,7 km amsl) at 30 hPa intervals (only the most significant 10 heights are shown in Figure 14), indicated a transport duration of approximately 70 hours under the prevailing meteorological conditions (Figure 14b). The arrival of the plume over the RMSP on 14 July 2008, as observed by the lidar at IPEN, is shown in Figure 15.

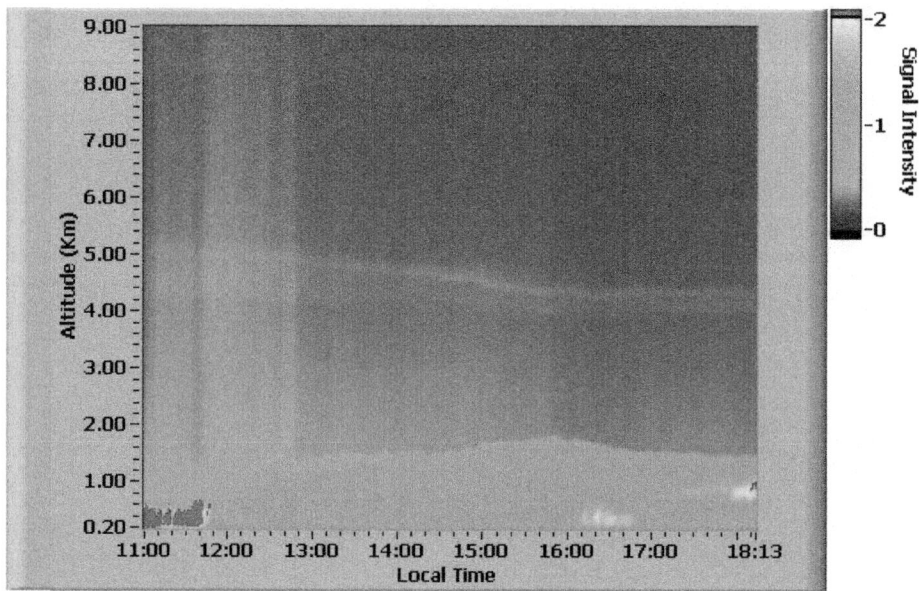

Figure 15. Lidar observations at IPEN in São Paulo, showing the range-corrected signal in arbitrary units, on 14 July 2008 between 11:00 and 18:13 LT. The plume identified in Figure 14 can be seen between 4-5 km AGL.

The MSP-Lidar system in São Paulo has also been contributing to CALIPSO satellite validation procedures [75, 82]. During 2007, correlative measurements were carried out with special attention to the dry season (May-October), when most of the days have poor dispersion conditions and long distance transport is more frequent. From a total of 28 days of measurements, on only 10 days were no clouds present below 4 km. Figure 16a presents a typical example, showing the range-corrected signal retrieved by the lidar system at São Paulo on 10 October 2007 between 03:34 and 05:35 UT, which contains the CALIPSO overpass window, beginning at 04:30 UT (Figure 16b). On this day, the closest distance of the satellite ground-track from the lidar site was about 48 km. The presence of aerosol layers above the PBL at 4-5 km, 6 km and 9 km is noticeable. The same features are also observed in the CALIOP 532 nm Total Attenuated Backscattering plot, as shown in Figure 16b. Both systems detected a cirrus structure between 12 and 13 km AGL, but the strong cirrus cloud signal observed in the CALIOP "plot-curtain" is much weaker in the lidar image. The red box in Figure 16b represents the CALIPSO ground-track region over Metropolitan São Paulo with coordinates of -22,5625 latitude and -46,0247 longitude at about 04:35 UT.

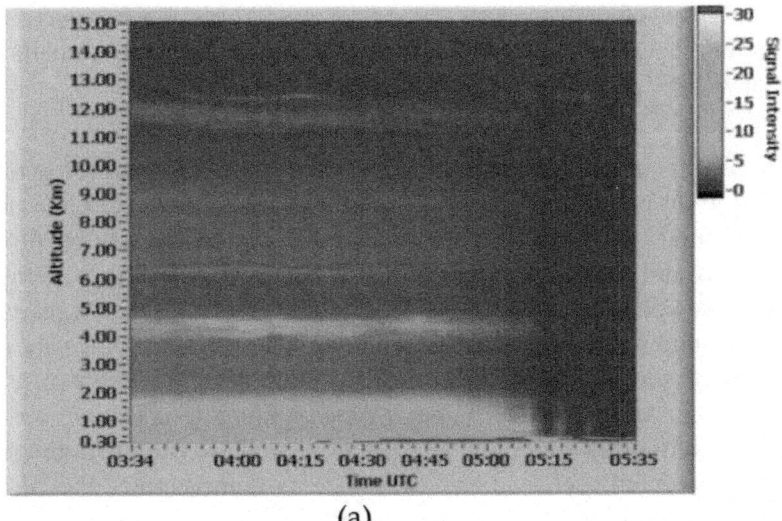

(a)

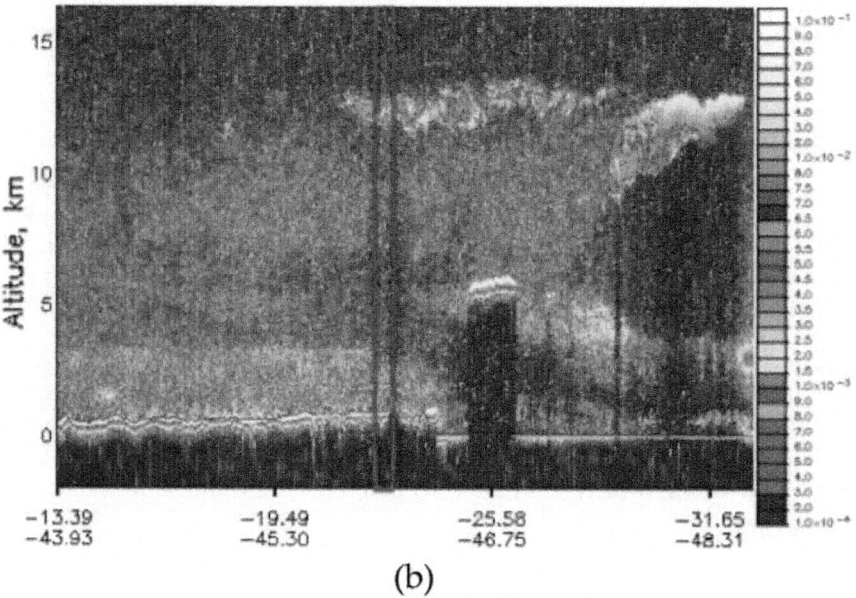

(b)

Figure 16. a) Range-corrected lidar signal (plot-curtain) measured by the MSP-Lidar on 10 October 2007, 03:34 - 05:35 UT. (b) Total Attenuated Backscattering signal measured by the CALIOP at 532 nm during the period 04:30 - 04:41 UT on the same day, when it was closest to the MSP-Lidar site (red box).

Figure 17 compares the attenuated backscatter coefficient profile retrieved by CALIOP on board the CALIPSO satellite and the corrected one obtained from the ground-based MSP-Lidar system in São Paulo. The satellite profile has a 5 km horizontal resolution. The attenuated backscatter profile from the MSP-Lidar site was derived under cloud-free conditions from the range-corrected and background noise-subtracted lidar return signal. Both profiles are in good agreement, presenting similar layer patterns in the profiles observed at 5-6 km and about 7 km AGL. Since it can be assumed with reasonable confidence that, at higher altitudes, the horizontal atmospheric structure is more homogeneous, the good agreement between the two systems demonstrates the possibility that they were probing the same air masses for this specific measurement. At lower altitudes, observation of some differences between the two profiles is more likely due to local effects. In this case, the localized effects are more pronounced, and the fact that the systems are not covering the exact same region becomes evident.

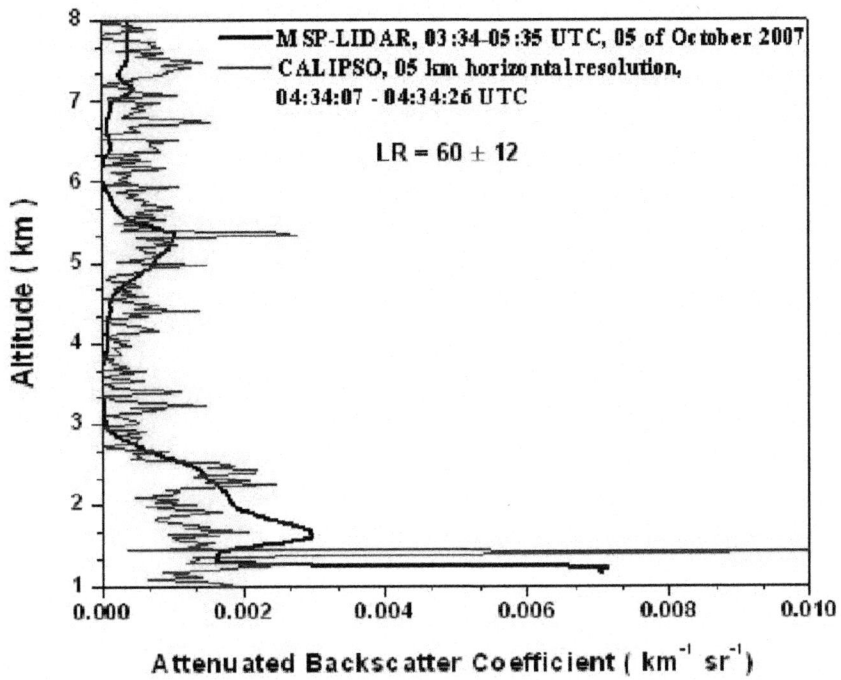

Figure 17. Total Attenuated Backscatter Coefficient profiles at 532 nm for the horizontal coverage of CALIPSO level 1 data compared to the Attenuated Backscatter Coefficient retrieved by the ground-based MSP-Lidar system in Metropolitan São Paulo on 10 October 2007.

Mobile Raman Lidar

The mobile bi-axial Raman lidar system uses a commercial pulsed Nd:YAG laser, operating at a wavelength of 532 nm in the elastic channel and 607 nm in the Nitrogen Raman channel, with a pulse energy of 130 mJ at 20 Hz PRF. The pulse width is 25 ns, yielding a spatial resolution of 7,5 m. A detailed description of the system is found in [83]. The system allows the determination of the optical properties of the atmosphere, including aerosol backscatter and extinction coefficients, as well as an indication of the type of aerosol present, based on the Lidar Ratio. This lidar has so far been deployed during specific campaigns at three different sites within the central region of São Paulo State, *viz.*, Rio Claro [84], Bauru and Ourinhos [85-87], as well as in Cubatão, an industrial hub at the coast, near Santos [88], as shown in Figure 6.

A one-month pilot study was undertaken during August 2010 in Ourinhos (Figure 6), which is situated in one of the State's major sugar cane producing regions, where biomass burning is a regular occurrence. The objective was to characterize the effects of these emissions on the atmosphere, considering the local circulation and the consequences for the region [85]. In the absence of rain, the plumes were tracked by IPMet's two S-band Doppler radars within their quantitative ranges of 240 km (BRU = Bauru, PPR = Presidente Prudente; Figure 6), using the TITAN (*Thunderstorm Identification, Tracking, Analysis, and Nowcasting*) Radar Software [89]. A large range of meteorological, physical and chemical instrumentation, including the mobile Raman lidar, was used to observe elevated layers and the type of aerosols. A medium-sized sodar, as well as 6 automatic weather stations, were also deployed in the region. Various gases and aerosol size fractions were sampled, providing an atmospheric chemistry database and thus documenting the impact of the harvesting practice on the region. The aerosol load of the atmosphere was quantified by hourly mean AOD values and hourly mean backscatter profiles. Several case studies have already been analyzed, but the one of 25-26 August 2010 will be shown in this Chapter to illustrate how the various remote sensing instruments are being deployed to generate a complete picture of events.

During the second half of August 2010, the weather was dominated by a high pressure system, resulting in a rise in temperatures, with low humidity favoring the accumulation of pollutants in the atmosphere of the region [25]. IPMet's radars have a 2° beam width and a quantitative range of 240 km, generating a volume-scan every 7,5 minutes, with a resolution of 250 m radially and 1° in azimuth. Reflectivities and radial velocities are recorded at 16 elevations. However, in order to detect and track the biomass burning plumes, a special scanning cycle was configured to provide a better vertical resolution up to the anticipated detectable top of the plumes: 10,0, 8,0, 6,5, 5,0, 4,0, 3,2, 2,4, 1,6, 0,8 and 0,3, with each "sweep" (Plan Position Indicator - PPI) having 360 rays with 957 range bins each. Two different software systems were deployed, *viz.*, IRIS (*Interactive Radar Information System*) Analysis was used first to generate CAPPIs (Constant Altitude PPIs) at 1,5 and 2,0 km amsl, in order to identify all smoke plumes within the 240 km range of the radars. Once a plume was identified as likely to pass over the monitoring site, it was tracked using TITAN Software to determine its intensity (based on radar reflectivity in dBZ), horizontal and vertical dimensions, and the velocity of approach. The thresholds used for tracking were 10 dBZ with a minimum volume of 2 km^3. It should be noted that TITAN uses Universal Time (Local Time LT = UT-3h).

A typical case study of a sugar cane fire in the Ourinhos region is now presented, demonstrating the integration of all types of data into one coherent event. The first echo of a smoke plume was detected by the Bauru radar on 26

August 2010 at 00:08 LT, about 35 km north-northeast of Ourinhos and about 85 km southwest of the radar (Figure 18), rapidly gaining in area and intensity (≤40 dBZ near its origin). By 00:22 LT, the TITAN Software could already identify its centroid of 10 dBZ reflectivity and tracked it until 02:45 LT, when the plume had already spread over Ourinhos, where the Raman lidar and sodar were located. As the plume moved southwards with the northerly winds, the aerosols spread out (dispersed) and the reflectivity dropped gradually, but it could still be detected by the radar until 03:46 LT, >20 km south of Ourinhos, using a reflectivity threshold of –6 dBZ [85].

Furthermore, it can be deduced from Figure 18a that while the plume was at a low height during the initial phase of transport, it moved very slowly (3-4 km.h^{-1}), since the wind speed in the first few hundred meters was very low (5 m.s^{-1}), as observed by the sodar. There was also a shift of the wind direction from easterly to northerly winds above 300 m AGL. These northerly winds were above the nighttime surface inversion, confirmed by the "Skew T x Log P" profiles of the Meso-Eta model in the 900-800 hPa layer (650–1650 m AGL) as shown in [85]. The vertical velocity (w), measured by the sodar, indicated that downward mixing of the pollutants (aerosols), trapped above the inversion, only commenced at around 09:00-09:30 LT, since from 00:00-09:00 LT the atmosphere was extremely stable below 300 m AGL (w = 0 m.s^{-1}).

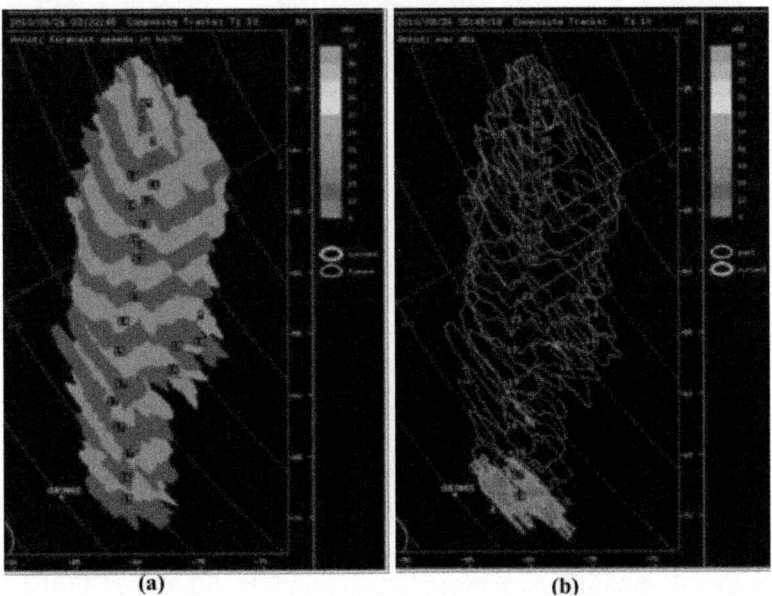

(a)   (b)

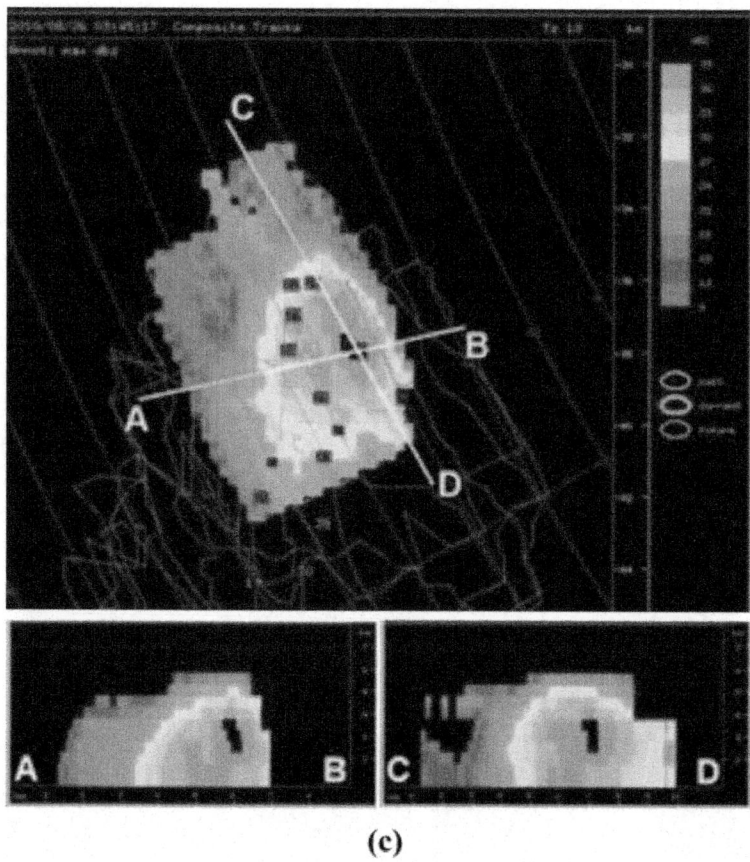

(c)

Figure 18. Examples of the tracks generated by TITAN on 26 August 2010. The envelopes (10 dBZ reflectivity) show the position of the *queimada* (smoke plume) in intervals of 7,5 min (blue = actual time; green = future; yellow = past). (a) First TITAN centroid of the *queimada* (actual fire, blue) at 03:22 UT (00:22 LT; annotation: propagation velocity in km.h^{-1}); (b) The *queimada* reached the Ourinhos region at 05:45 UT (02:45 LT, blue; annotation: maximum reflectivity in dBZ).(c) Vertical cross-sections at 03:45 UT (00:45 LT), showing the horizontal and vertical extent along the base lines A-B and C-D.

The lidar observed the arrival of the plume at 02:40 LT between 350 and 600 m AGL (Figure 19a). The top of the PBL extended to about 2,6 km AGL, above which a very dry and relatively warm and clean air mass was advected from the west, creating an elevated inversion which blocked further upward mixing. The lowest layer ≤250 m AGL appeared clean, being trapped within the surface inversion, inhibiting downward mixing, also confirmed by the sodar measurements, indicating a very stable layer. Lidar data from the Raman

Channel (non-elastic signal at 607 nm) were integrated into hourly means until 09:00 LT to obtain the AOD. The results confirmed a high aerosol load of the atmosphere, with hourly mean values of AOD varying between 0,265 and 0,288 until 07:00 LT, after which they increased to 0,433 by 09:00 LT. Hourly means of the Lidar Ratio confirmed the arrival of the plume between 02:00 and 03:00 LT (example shown in Figure 19b), while an almost 20% increase of LR to 72 sr after 07:00 LT was probably due to downward mixing of the aerosols accumulated above the inversion, also confirmed by an increase of AOD values from the Raman signal [85]. LR values of around 70 sr suggest aerosols originating from biomass burning [90, 91].

Visual images from overpasses of the MODIS-AQUA satellite on 25 and 26 August 2010 (at 17:35 and 16:40 UT, respectively; 14:35 and 13:40 LT) showed intense smoke plumes to the west and south of the Ourinhos region, with AOD values of up to about 1,0. In the Ourinhos region, the AOD increased during the period 25-26 August, from about 0,2 to about 0,6 (Figure 20a), which is in agreement with the early afternoon lidar measurements (Figure 20b), which provided an AOD value of 0,380 during the period from 13:00 to 14:00 LT.

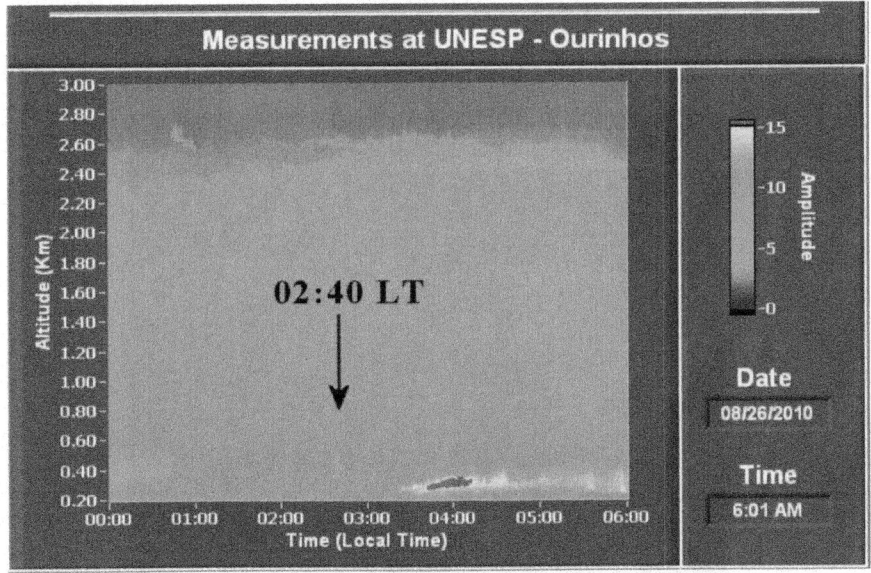

(a)

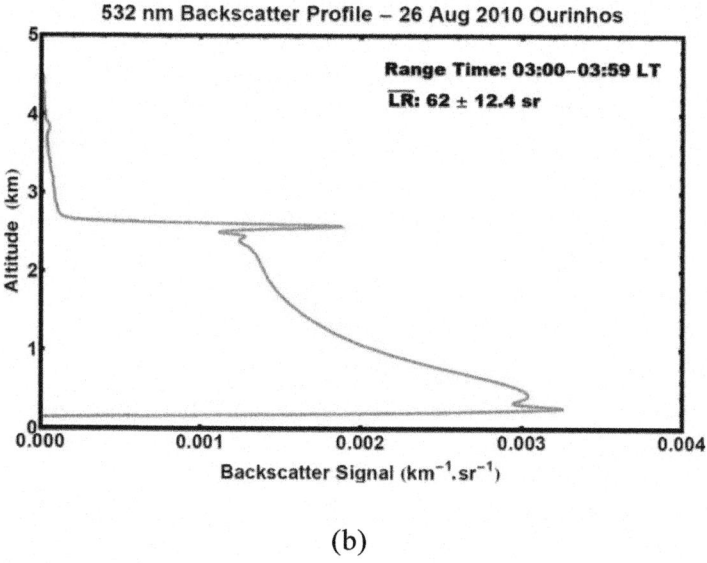

(b)

Figure 19. (a) Lidar signal (arbitrary units) visualized for 00:00-06:00 LT, up to 3 km AGL.(b) Backscatter Profile at 532 nm for the hourly mean period 03:00-03:59 LT on 26 August 2010.

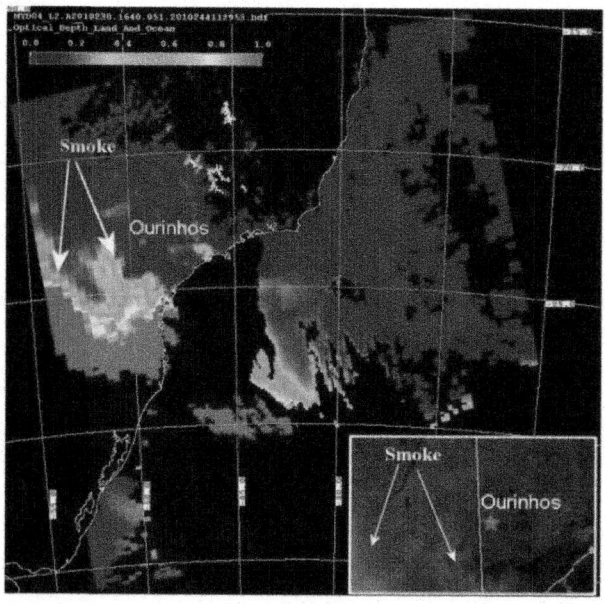

(a)

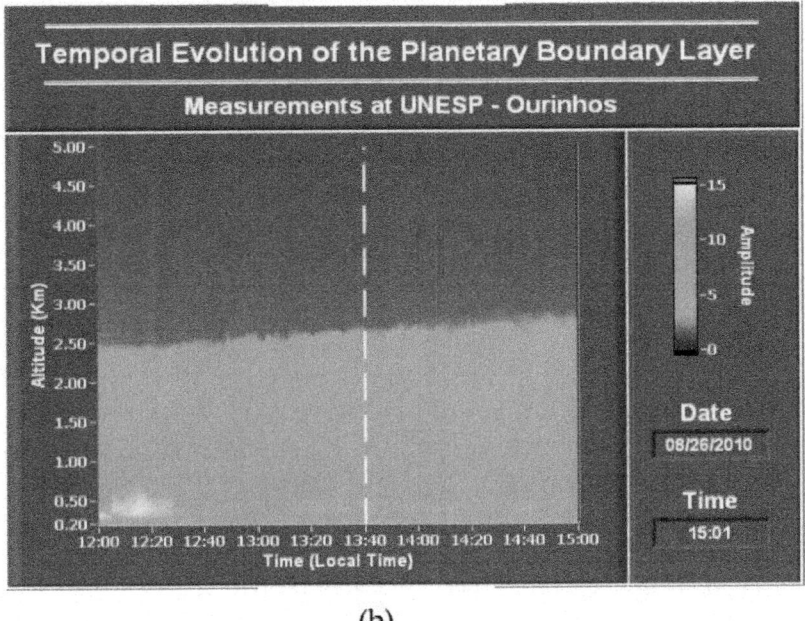

(b)

Figure 20. a) AOD image from MODIS-AQUA on 26 August 2010, 16:40-16:45 UT (13:40-13:45 LT). The inset shows a simultaneous visual image of the Ourinhos region. (b) Lidar measurements on 26 August 2010, 12:01-15:01 LT. The time of the MODIS-AQUA overpass is indicated by the dashed white line.

Aerosols collected during daytime and nighttime periods at the lidar site [85-87, 92], using low-volume filter samplers, were chemically characterized by means of ion chromatography. A higher concentration of K^+ during the period from 22:00 on 25 August to 16:00 on 26 August 2010 indicated the presence of biomass-burning material (Figure 21), since K^+ is a plant macronutrient released during the combustion process. Levoglucosan, a very specific chemical marker of biomass combustion, was well above average concentration during day sampling on 26 August and even higher during the following night, indicating a strong presence of biomass smoke on both days.

In the study region, ions such as magnesium (Mg^{2+}) and calcium (Ca^{2+}) are associated with the re-suspension of soil dust, which often accompanies biomass fires due to the intense updrafts created. On 26 August, concentrations of these species were higher during the daytime, due to the increased emissions from barren fields and unsealed roads associated with higher wind speeds (Figure 21).

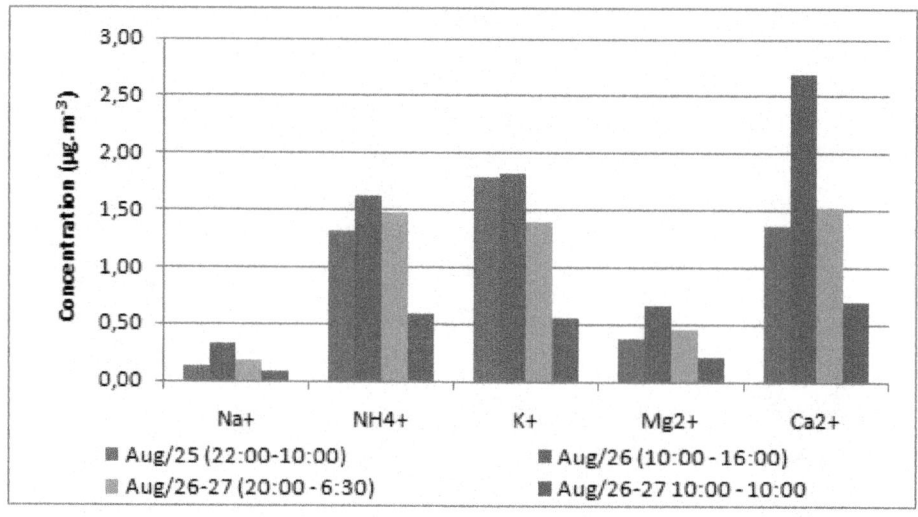

Figure 21. Soluble major cation concentrations for the period 25-27 August 2010 (sampling periods are indicated in local time; after [85]).

Further evidence of the impact on the Ourinhos region of emissions from sugar cane fires was obtained by comparing the concentrations of organic compounds in aerosol particles collected on 26 August with those collected one day earlier. Ambient levels of polycyclic aromatic hydrocarbons (PAH), as well as PAH derivatives, such as oxy-PAH, were significantly higher on 26 August 2010 than on the previous day, confirming that emissions from sugar cane fires affected the urban atmosphere of Ourinhos.

Scanning Lidar in Cubatão

An elastic backscatter lidar system, with similar characteristics to the mobile lidar, was installed in 2011 at CEPEMA-USP (*Centro de Pesquisas em Meio Ambiente,* a Center for Environmental Research and Training, under the responsibility of the Universidade de São Paulo) in the Cubatão industrial area, with the ultimate goal of remotely monitoring industrial emissions. It also uses a commercial pulsed Nd:YAG laser, operating at three wavelengths (355, 532 and 1064 nm) with pulse energies of 100, 200 and 400 mJ, respectively, at 20 Hz PRF. A detailed description of the system and its location is found in [33]. The system allows the determination of the optical properties of the atmosphere, including aerosol backscatter and extinction coefficients, as well as an indication of the type of aerosol present, based on the Lidar Ratio. The lidar is co-located with a sodar / RASS system and an air quality monitoring station.

During May 2011, the system was deployed in a vertical pointing mode during an intensive field campaign. A 24-hour period was selected that demonstrated the complexity of the local situation, which is dominated by topographical effects and prevailing meteorological conditions [33]. Vertical profiles of the Backscatter Coefficient (BSC) and the Colour Ratio were calculated for 30-minute periods from 17:30 – 19:59 and 21:42 – 23:36 LT. The BSC was highest for all frequencies between 19:30 and 19:59 LT (Figure 22a), indicating a strong inflow of aerosols, while after 21:42 LT the BSC showed much lower values (Figure 22b), representing a relatively clean air mass. At the same time, the Colour Ratio between all frequencies increased significantly, indicating the presence of small particles, especially between 0,8 and 1,3 km AGL [33]. Ground-level observations of PM_{10} and $PM_{2.5}$ for the 24-hour period indicate that PM_{10} concentrations were almost twice as high as those of $PM_{2.5}$ until about 18:00 LT (Figure 23). During the same period, the sodar observed extremely low wind speeds from varying directions. However, this resulted in very stable PBL conditions, and a temperature inversion began to develop from 18:30 onwards, reaching its greatest depth and intensity at 21:30. Thereafter, it gradually dropped in height and began to erode, as the air flow from the interior intensified, until it totally dissipated by 01:00 LT [33], due to the katabatic warming of the descending northerly airflow, which then also reduced the aerosol concentrations at ground level (Figure 23).Figure 24a shows the development of the surface inversion at 20:00 LT, overlaid by warm air flowing from the interior, with simultaneous downward motion below 240 m AGL (Figure 24b), highlighting the complex interaction of meteorology and topography in this region. This situation clearly demonstrates the need for solid environmental impact studies *before* locating industrial developments, in order to avoid any negative health impacts in the local population due to the accumulation of pollutants.

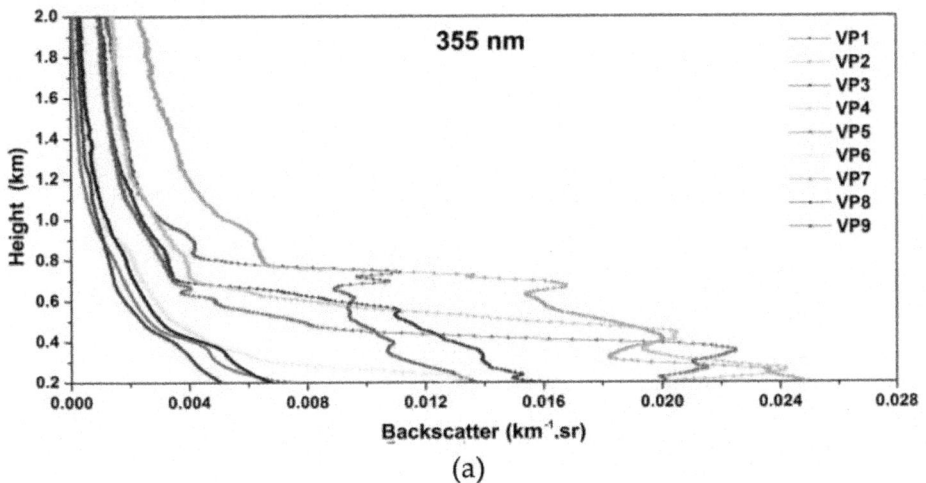

(a)

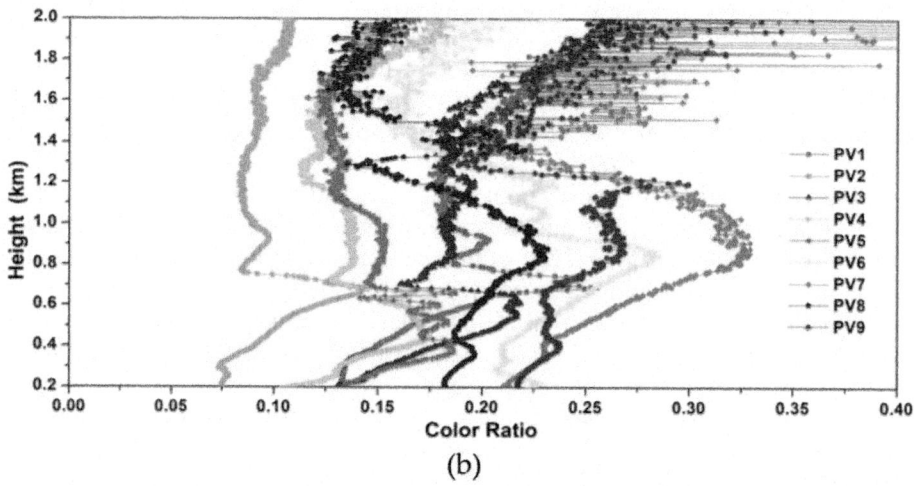

(b)

Figure 22. a) Vertical profile of Backscatter Coefficient (BSC) at 355 nm; (b) Colour Ratio 532/355 nm. (After [33]).

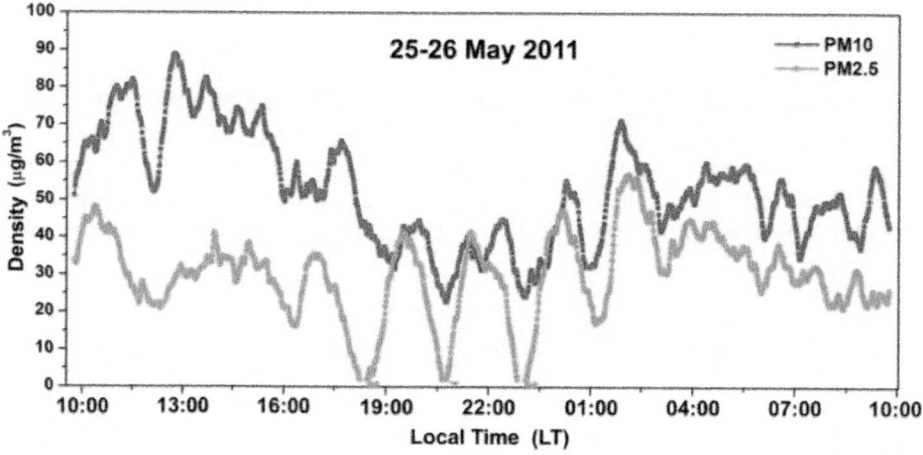

Figure 23. Concentrations of PM_{10} and $PM_{2.5}$ from 10:00 LT on 25 May to 10:00 LT on 26 May 2011. (After [33]).

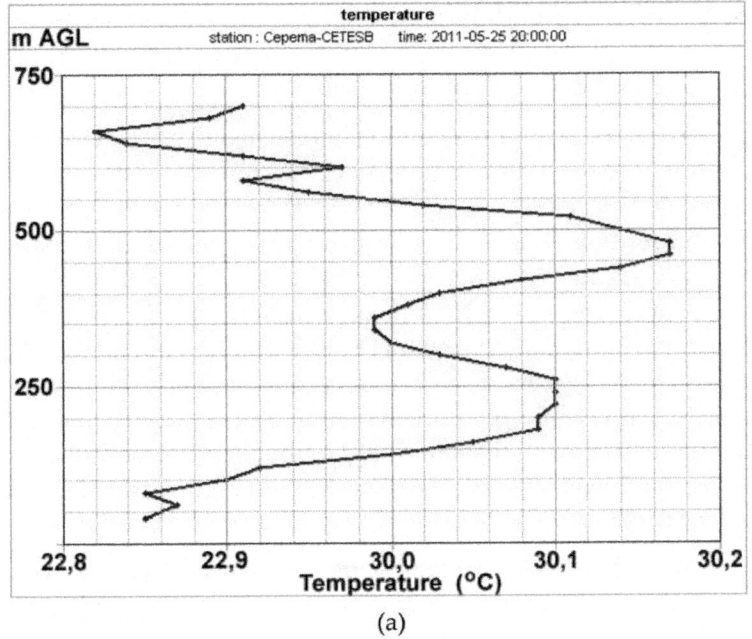

(a)

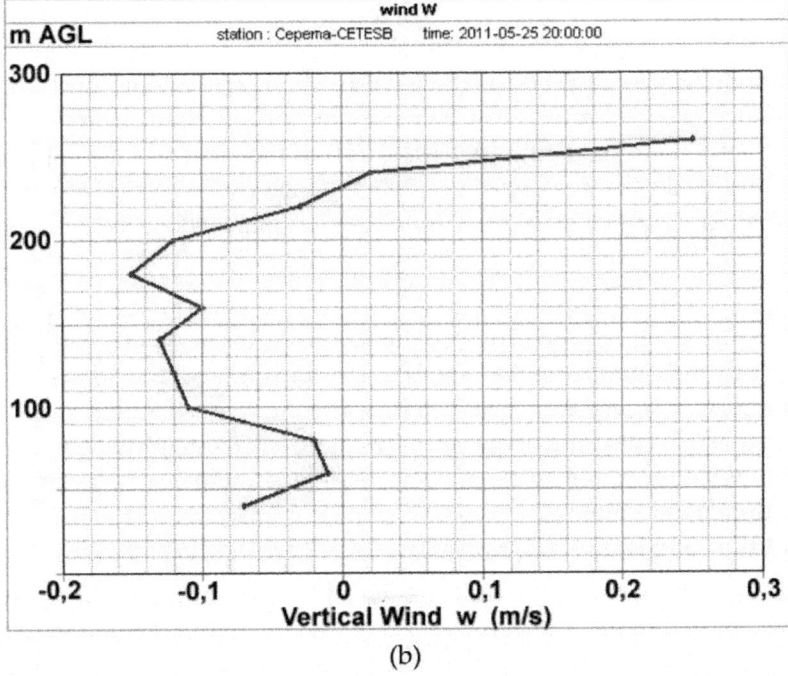

(b)

Figure 24. Sodar/RASS measurements on 25 May 2011, mean profiles for 19:30-20:00 LT. (a) Vertical profile of temperature; (b) Vertical profile of vertical wind velocity.

IMPACTS OF AEROSOLS

Impact of Aerosols on Human Health

The impact of anthropogenic aerosols on human health has been acknowledged in both metropolitan and rural regions [93], and references therein]. In general terms, as pointed out in [93], in recent years the population of São Paulo State has suffered from either acute (short-term, high concentration) or chronic (long-term, lower concentration) exposure to particulate air pollutants, depending on location. In rural regions, there is acute exposure to high concentrations of biomass burning particulates present in plumes, as well as chronic exposure to these aerosols on a regional basis throughout the dry season. In metropolitan São Paulo, there is chronic exposure to particulates derived from road transport and industrial emissions, together with periodic acute exposure to extremely high levels of pollutants under conditions of thermal inversions and stationary air masses [94-96].

There have been many studies of the correlation between aerosol concentrations and human health impacts in the metropolitan regions, especially in São Paulo city [97-102]. Typical effects include asthma and pneumonia, as well as other cardiovascular and respiratory symptoms. Increased levels of PM_{10} were associated with increases of 6,7% and 2,2% in hospital admissions of children due to respiratory illness [96, 103]. Increments of 10 $g.m^{-3}$ in PM_{10} concentrations resulted in increases in hospital admissions of between 0,9% and 6,7% in Sao Paulo [97, 103-105]. In the elderly, a 5,4% increase in the number of deaths was linked to a 10 $g.m^{-3}$ increase in PM_{10} [105]. Industrial emissions in Cubatão have been found to seriously affect the lung function of children, with respiratory airflow rates correlated with PM_{10} concentrations obtained for the preceding month [106].

Bourotte *et al.* [56] investigated the relationships between peak expiratory flow (PEF) measurements and soluble ions in fine and coarse aerosols, and found a negative correlation between PEF and the coarse fraction ions Cl^-, Na^+, Mg^{2+} and NH_4^+, as well as between PEF and fine fraction Mg^{2+}. The findings suggested that increased levels of coarse particles could be of especial concern for asthmatic individuals.

In these heavily polluted regions, atmospheric particles contain components known to be carcinogenic and mutagenic, including ketones, aldehydes, quinolines, carboxylic acids, polycyclic aromatic compounds (PAHs), and nitro-PAHs. These substances have been associated with exhaust emissions from road vehicles in southeast Brazil [69, 107-109]. Benzo[a]pyrene equivalent values suggest that the cancer risk is greater for the São

Paulo city aerosol than elsewhere in the State, although concentrations may not exceed World Health Organization guidelines [65].

Biomass burning emissions in rural regions also have a recognized influence on human health, as well as environmental impacts including modification of nutrient cycling [10, 11, 47], and effects on climate including alterations of the radiative properties of the lower atmosphere, cloud formation and precipitation [110, 111]. For these reasons, as well as due to the need to meet certification requirements of importing countries, there has been increased pressure for mechanization of harvesting, since the mechanized process does not necessarily require prior burning of the crop. Nonetheless, until recently burning has continued to be employed in mechanized areas (using simpler machinery) because it can improve economic efficiency by around 30-40% [50, 112].

A clear relationship between particulate air pollution and the occurrence of respiratory illness in sugar cane burning regions of the State has been reported [45, 113-116]. Particulate material from sugar cane burning was demonstrated to have the greatest detrimental effect on the respiratory systems of the most sensitive population groups. Cançado et al. [45] measured black carbon and trace elements in fine and coarse aerosol fractions, and related the concentrations to daily records of hospital admissions for respiratory illness of children (<13 years old) and the elderly (>64 years), in the town of Piracicaba. Increases of 10,2 $g.m^{-3}$ ($PM_{2.5}$) and 42,9 $g.m^{-3}$ (PM_{10}) were associated with increases in hospital admissions of 21,4% (children) and 31,03% (elderly people).

Carcinogenic and mutagenic compounds are emitted during biomass burning [109]. Concentrations of PAHs in a rural sugar cane burning region during the harvest period were in the range 0,5-8,6 $ng.m^{-3}$[38]. The mutagenic activity of PM_{10} was much higher during the harvest season, when the PM_{10} concentration was 67 $g.m^{-3}$, and the mutagenic potency was 13,45 revertants m^{-3}. During the summer (non-burning period), the PM_{10} concentration was 20,9 $g.m^{-3}$, and the mutagenic potency was 1,30 revertants m^{-3} [117].

Impact of Aerosols on Rainfall

Aerosols derived from all of the sources described above are able to alter the radiative properties of the troposphere, and can modify the processes that lead to the development of cloud condensation nuclei, cloud droplets, and ultimately precipitation [118-120]. The magnitudes of these effects depend on the size distribution, number concentration and chemical composition of the particles, and can therefore vary widely within the same region.

Dufek and Ambrizzi [121] used daily precipitation data collected at 59 locations in São Paulo State to investigate rainfall trends for the period 1950-1999. Although some of the findings were contradictory, an overall trend towards a wetter climate was identified, with rainfall concentrated into a smaller number of more intense events. It was suggested that these changes could be related to the presence of biomass burning aerosols, as well as changes in land use. Evidence that the aerosols probably act as cloud condensation nuclei was provided in [122], where a relationship was identified between water-soluble organic carbon (WSOC) in the particles and dissolved organic carbon (DOC) in rainwater.

An important point is that sugar cane production in São Paulo State has increased over this period. Between the 1990/91 and 2000/01 seasons, the harvest increased from 132 Mt to 194 Mt [123]. It can therefore be supposed that there was also a large increase in emissions of aerosols from the burning of the crop, since manual harvesting of the cane (which requires prior burning) was the norm over the period. Mechanization of the crop (which does not involve burning) has only been introduced recently (from around 2005). The main conclusion to be drawn from this is that the trend towards a smaller number of more intense precipitation events, as reported in [121], could now be reversed in the interior of São Paulo State, as sugar cane burning is progressively phased out.

The climatological characterization of storm properties, such as area, volume, maximum echo top and reflectivity during two summer seasons, viz., 1998-1999 and 1999-2000, based on observations from the Bauru S-band Doppler radar, has for the first time shown the spatial distribution of these parameters in central São Paulo State. Gomes and Escobedo [124] showed that some preferential areas of precipitation, taking into account a precipitation envelope area defined by the 25 dBZ threshold, were located along the Tietê River valley. The mean maximum reflectivity field (>40 dBZ), representing the cores of convective precipitation systems, has highlighted some preferential regions for convection to develop over urban and industrialized areas, such as metropolitan Campinas (Figure 25a). A climatology of flash density (Figure 25b; [21]) also identified Campinas as one of three regions with a higher concentration of lightning discharges, attributed to the occurrence of heat islands due to anthropogenic activities. Thus, the spatial distribution of the reflectivity field exceeding 40 dBZ in the Campinas region reinforces results showing a strong correlation between the frequency of cloud-to-ground lightning strokes and precipitation intensity.

The influence of anthropogenic aerosols on precipitation patterns in the region is the subject of research currently in progress as part of a thematic climate change research programme sponsored by FAPESP (the São Paulo State Research Foundation). The quantitative evaluation of changes in the rainfall pattern, such as increases or decreases of area rainfall totals, number and volume of convective cells, duration of rain events, distribution of echo top heights, etc., is in progress for a 10-year period of integrated radar observations (Bauru and Presidente Prudente radars), using the TITAN Software.

Impact of Aerosols on the Frequency of Lightning

Westcott [125] documented for the first time an impact of large cities on the cloud-to-ground (CG) lightning frequency in the Midwest of the United States. This was followed up by various researchers around the world, including in Brazil [20] and ultimately summarized in [22], using 10 years of observations from the Brazilian Lightning Detection Networks (1999-2008). This research confirmed the impact of anthropogenic activities on lightning, but it also highlighted the complexity of the correlation between urban heat islands, concentrations of PM_{10} and SO_2 in terms of weekly cycles and meteorological conditions, such as CAPE (Convective Available Potential Energy) and other microphysical parameters. One of the most important findings was that the CG frequency increases with increasing concentrations of PM_{10} up to a certain threshold of PM_{10} concentration (saturation), after which it decreases with further increases of PM_{10} concentrations. As the CG frequency increases due to urban impacts, the percentage of positive strokes is reduced.

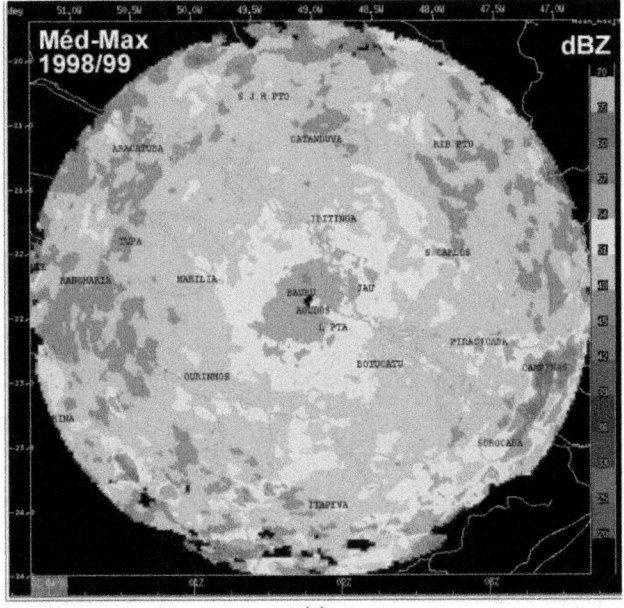

(a)

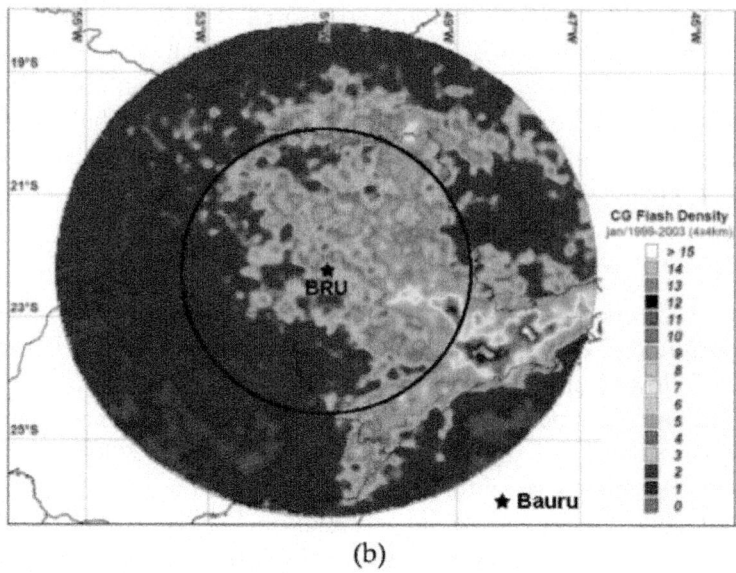

(b)

Figure 25. a) Spatial distribution of the average maximum reflectivity (dBZ), during the period October 1998 to March 1999. TITAN storm threshold was defined as reflectivity >40 dBZ within the 240 km range of the Bauru Doppler radar (after [124]). (b) Flash density within the 450 km range of the Bauru radar (after [21]). The circle indicates the 240 km quantitative range as in the TITAN image above.

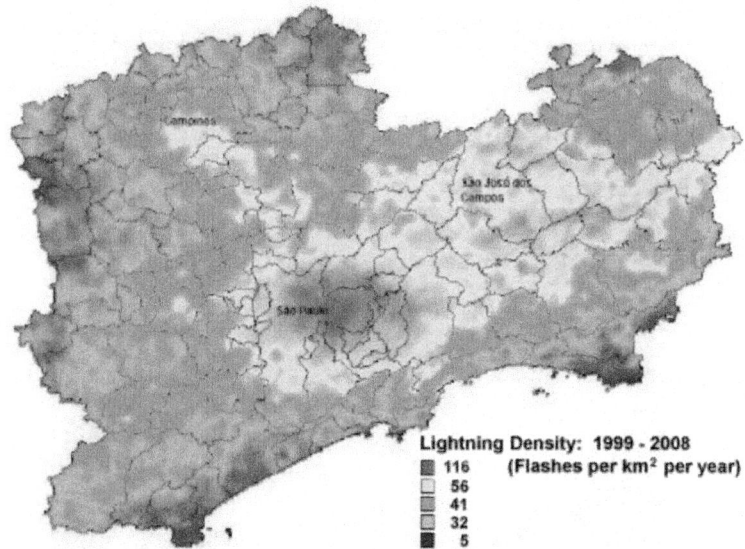

Figure 26. Lightning density (flashes per km² per year) during the period 1999 - 2008 for the eastern part of the State of São Paulo, which includes the major urban complexes, as well as the highly industrialized Paraiba Valley (after [22]).

FINAL CONSIDERATIONS

This Chapter provides a review of all relevant historical data concerning the nature, concentrations and impacts of atmospheric aerosols in southeast Brazil. Highlights are the characterization of chemical, physical and optical properties of aerosols, as well as their geographic distribution within the State of São Paulo.

A significant reduction of mean annual PM_{10} concentrations could be noticed from 1998 onwards, confirming the success of the implementation of stringent air quality control measures, administered by CETESB. However, within the industrial suburb of Cubatão, confined in a valley, concentrations are still about twice the PQAr. After 2002, the annual mean PM_{10} concentrations in the RMSP and the interior of the state show relatively little year to year variation, but remain mostly below the Annual Standard (PQAr = 50 $\mu g.m^{-3}$).

At present, sugar cane burning, together with the re-suspension of soil dust that is inevitable during the harvesting process, is a major influence on aerosol concentrations, size distribution and dry deposition in rural regions of São Paulo State. However, in this region (and elsewhere in Brazil), the practice of pre-harvest burning is being eliminated. Recent legislation (State Law no. 11.241/02) envisages the complete cessation of the practice in mechanizable areas by 2021 and in non-mechanizable areas by 2031. Furthermore, an agreement between sugar cane producers and the State government has been reached, which involves elimination of burning in mechanizable areas by 2014, and in non-mechanizable areas by 2017 [9, 126]. This will have major environmental implications, including improvements in air quality and changes in the rates of deposition of nutrient species from the atmosphere to vegetation, soils and freshwater bodies [10]. Nonetheless, at present burning continues in 44% of the area planted with sugar cane [9]. Tsao *et al.* [127] suggest, using a life cycle analysis, that pollutant emissions in sugar cane regions are still increasing, due an expansion of the planted area, and that the burning step still contributes the largest fraction of the total emission.

Improvements in air quality in the metropolitan regions are likely to proceed at a slower pace than in the interior of the State, largely due to the dominant influence of emissions from the road transport sector. Nonetheless, emissions of aerosols and other pollutants are ultimately expected to be attenuated following progressive modernization of the vehicle fleet, and implementation of better controls on emissions from both vehicular and industrial sources.

Examples of case studies presented have demonstrated the capability of weather radars to detect, track and quantify emissions from biomass fires in the absence of rain echoes, deploying a special elevation scanning procedure to generate Volume-Scans every 7,5 min. Furthermore, satellites orbiting with lidar systems on board (e.g., MODIS-AQUA, CALIPSO, CloudSat) also have the capability to detect and quantify optical properties of aerosols.

With the gradual introduction of lidars in Brazil during recent years, it has also become possible to quantify in situ the vertical distribution and optical properties of suspended aerosols. However, in the State of São Paulo there are currently only three lidar systems available, viz., one fixed lidar each in São Paulo city and in Cubatão, supplemented by the mobile lidar for periodic deployment in the interior of the State. Additional fixed lidar installations are therefore suggested for Campinas, Rio Claro, Bauru and São José do Rio Preto (situated in an important sugar cane production region in the north of the State) as a minimum configuration for a network, together with a second mobile system in São José dos Campos to cover the industrial activities in the Paraiba Valley.

REFERNCES

1.   TRACE-A1992http://www-gte.larc.nasa.gov/trace/tra_hmpg.htm
2.   J. A. Lindesay, M. O. Andreae, J. G. Goldammer, G. Harris, H. J. Annegarn, M. Garstang, R. J. Scholes, B. W. van Wilgen, 1996International Geosphere-Biosphere Programme/International Global Atmospheric Chemistry SAFARI-92 field experiment: Background and overview, J. Geophys. Res., 101, D192352123
3.   LBA2012http://lba.inpa.gov.br/lba/#
4.   S. Kocinas, P. Artaxo, 1992O monitoramento contínuo de elementos traços em aerossóis atmosféricos da bacia Amazônica. Proceedings, VII CBMET, São Paulo, 901903http://www.cbmet.com/cbm-files/19-02759b1503f7d21b92c9a5774714f2ba.pdf
5.   A. G. Allen, A. A. Cardoso, Rocha. G. O. Da, 2004Influence of sugarcane burning on aerosol soluble ion composition in Southeastern Brazil. Atmos. Environ., 3850255038
6.   Rocha. G. O. Da, A. G. Allen, A. A. Cardoso, 2005Influence of agricultural biomass burning on aerosol size distribution and dry deposition. Environ. Sci. Technol., 3952935301
7.   Lara L L, Artaxo P, Martinelli L A, Victoria R L, Ferraz E S B2005Characteristics of aerosols from sugar-cane burning emissions in Southeastern Brazil. Atmos. Environ., 3946274636
8.   IBGE2011Levantamento Sistemático da Produção Agrícola. Instituto Brasileiro de Geografia e Estatística (IBGE), Rio de Janeiro, December 2011: 0010-34430103443X.

9.   CETESB2012Relatório 2011: Qualidade do Ar no Estado de São Paulo. CETESB, São Paulo, 124p.http://www.cetesb.sp.gov.br/ar/qualidade-do-ar/31 -publicacoes-e-relatorios

10.   A. G. Allen, A. A. Cardoso, A. Wiatr, C. M. D. Machado, W. C. Paterlini, J. Baker, 2010Influence of intensive agriculture on dry deposition of aerosol nutrients. J. Braz. Chem. Soc., 218797

11.   Allen A G, Machado C M D, Cardoso A A2011Measurements and modeling of reactive nitrogen deposition in southeast Brazil. Environ. Pollut., 15911901197

12.   G. Held, E. Landulfo, F. J. S. Lopes, J. Arteta, V. Marecal, J. M. Bassan, 2011Emissions from sugar cane fires in the central & western State of São Paulo and aerosol layers over metropolitan São Paulo observed by IPEN´s lidar: Is there a connection? Opt. Pura Apl., 448391

13.   Orsini C M Q, Artaxo P1983Algumas caracteristicas da materia particulada inalavel em Cubatao. In: Seminário sobre uma Sintese do Conhecimento sobre a Baixada Santista, CETESB, São Paulo, SP, 1983, 516p.

14.   M. F. Andrade, C. Orsini, W. Maenhaut, 1994Relation between aerosol sources and meteorological parameters for inhalable atmospheric particles in Sao Paulo City, Brazil. Atmos. Environ., 2823072315

15.   Wikipedia2011http://en.wikipedia.org/wiki/Largest_cities_in_the_worldAccessed in December 2011).

16.   IBGE2010http://www.censo2010.ibge.gov.br/resultados_do_censo2010.phpAccessed in December 2011).

17.   P. H. N. Saldiva, C. A. Pope, J. Schwartz, D. W. Dockery, A. J. Lichtenfels, J. M. Salge, I. Barone, G. M. Bohm, 1995Air pollution and mortality in elderly people: A time-series study in São Paulo, Brazil. Arch. Environ. Health, 50159163

18.   Nestpdg=RINDAT (2012). http://www.rindat.com.br/

19.   ELAT/INPE2012http://www.inpe.br/webelat/homepage/

20.   K. P. Naccarato, O. Pinto Jr, I. R. C. A. Pinto, 2003Evidence of thermal and aerosol effects in the cloud-to-ground lightning density and polarity over large urban areas in Southeastern Brazil- Overview and Comparison to the Campaign Period. Geophysical Res. Letters, 301316741677doi:GLO17496.

21.   K. P. Naccarato, O. Pinto Jr, G. Held, 2004Climatology of Lightning in Brazil- Overview and Comparison to the Campaign Period. Proceedings, HIBISCUS / TroCCiBras / TROCINOX Workshop, Bauru, SP, 16-19

November 2004, 10http://www.ipmet.unesp.br/ipmet_html/troccibras/publicacoes.html

22. Gomes W R2010Estudo das caraterísticas da atividade dos raios na região metropolitana de São Paulo. PhD thesis, Geofísica Espacial, INPE, São José dos Campos, Brazil, 157p.

23. INMET2012http://www.inmet.gov.br/html/clima/mapas/?mapa=prec

24. Kousky V E1988Pentad outgoing longwave radiation climatology for the South American sector. Rev. Bras. Meteor., 3217231

25. Held G, Nery J T, Gomes A M, Lopes F J S, Ramires T, Lima B R O2012Study of biomass emissions in the central State of São Paulo: meteorological conditions during August 2010 cause an accumulation of pollutants in the Ourinhos region. Proceedings, XI Congresso Argentino de Meteorologia, Mendoza, Argentina, 28 May- 01 June 2012, 8p.

26. G. Held, J. M. Bassan, R. S. Frascarelli Jr, 2011Continuous Monitoring of the Lower Boundary Layer in the central State of São Paulo, Brazil, with a SODAR. Geophysical Research Abstracts, 13, EGU General Assembly 2011, Vienna, Austria, 0308April 2011.http://presentations.copernicus.org/EGU2011-13634_presentation.pdf

27. Feliz G S2012Aspectos sobre a análise meteorológica do jato de baixo nível na cidade de Bauru. Monografia para BSc (Geografia), USC, Bauru, 2012, 70p.

28. Karam H A2002Estudo do Jato de Baixo Nível de Iperó e das Implicações no Transporte de Poluentes no Estado de São Paulo. Tese de doutorado, IAG/USP, São Paulo, 2002, 213p.

29. G. Held, I. R. Danford, Y. Hong, G. R. Tosen, A. R. Preece, 1990The life cycle of the low-level wind maximum in the Eastern Transvaal Highveld: A cross-sectional study. A Report to the CSIR Executive and the Management of Eskom Engineering Investigations, CSIR Report EMA-C 90146, Pretoria, September 1990, 50p.

30. CETESB2012http://sistemasinter.cetesb.sp.gov.br/Ar/mapa_qualidade/mapa_qualidade_interior.asp?id=350431

31. CETESB2002Relatório de qualidade do ar do Estado de São Paulo em 2001. CETESB, São Paulo, 132p.http://www.cetesb.sp.gov.br/ar/qualidade-do-ar/31 -publicacoes-e-relatorios

32. WHO2005WHO Air Quality Guidelines: Global update 2005. Report on Working Group Meeting, Bonn/Germany, 1820October 2005, 30p.

33. J. Steffens, Costa. R. F. Da, E. Landulfo, R. Guardani, P. F. Moreira, G. Jr Held, 2011Remote sensing detection of atmospheric pollutants

using lidar, sodar and correlation with air quality data in an industrial area. Proceedings, SPIE Remote Sensing Conference, Prague, Czech Republic, 19-22 September 2011, 8182Z; doi:

34. Rocha. G. O. Da, A. G. Allen, A. A. Cardoso, 2004Influence of sugar cane burning on aerosol soluble ion composition in southeastern Brazil. Atmos. Environ., 3850255038

35. Campos M L A M, Urban R C, Da Silva L C, Souza M L, Allen A G2012Use of levoglucosan, potassium, and water-soluble organic carbon to characterize the origins of biomass burning aerosols. Submitted to Atmos.Environ., March 2012.

36. Rocha. G. O. Da, A. Franco, A. G. Allen, A. A. Cardoso, 2003Sources of atmospheric acidity in an agricultural-industrial region of São Paulo State, Brazil. J. Geophys. Res., 108(D7), 1-11.

37. R. H. M. Godoi, A. F. L. Godoi, A. Worobiec, S. J. Andrade, J. de Hoog, M. R. Santiago-Silva, R. Van Grieken, 2004Characterisation of sugar cane combustion particles in the Araraquara region, Southeast Brazil. Microchim. Acta, 1455356

38. A. F. L. Godoi, K. Ravindra, R. H. M. Godoi, S. J. Andrade, M. Santiago-Silva, L. Van Vaeck, R. Van Grieken, 2004Fast chromatographic determination of polycyclic aromatic hydrocarbons in aerosol samples from sugar cane burning. J. Chromatog., A 10274953

39. Kirchhoff V W J H, Marinho E V A, Dias P L S, Pereira E B, Calheiros R, André R, Volpe C1991Enhancements of CO and O3 from burnings in sugar cane fields. J. Atmos. Chem., 1287102

40. C. Oppenheimer, V. I. Tsanev, A. G. Allen, A. J. S. Mc Gonigle, A. A. Cardoso, A. Wiatr, W. Paterlini, C. M. Dias, 20042emissions from agricultural burning in São Paulo, Brazil. Environ. Sci. Technol., 3845574561

41. CETESB2002Companhia de Tecnologia de Saneamento Ambiental. Avaliação dos compostos orgânicos provenientes da queima de palha de cana-de-açúcar na região de Araraquara e comparação com medições efetuadas em São Paulo e Cubatão (relatório final- 2002). CETESB, São Paulo, 97p.http://www.cetesb.sp.gov.br/Ar/publicacoes.asp

42. G. C. M. Zamperlini, M. Santiago-Silva, W. Vilegas, 1997Identification of polycyclic aromatic hydrocarbons in sugar-cane soot by gas chromatography mass spectrometry. Chromatographia, 46655663

43. G. C. M. Zamperlini, M. Santiago-Silva, W. Vilegas, 2000Solid-phase extraction of sugar-cane soot extract for analysis by gas chromatography with flame ionisation and mass spectrometric detection. J. Chromatog.,

A 889281286

44. S. J. De Andrade, J. Cristale, F. S. Silva, G. J. Zocolo, M. R. R. Marchi, 2010Contribution of sugar-cane harvesting season to atmospheric contamination by polycyclic aromatic hydrocarbons (PAHs) in Araraquara city, Southeast Brazil. Atmos. Environ., 4429132919

45. J. E. D. Cançado, P. H. N. Saldiva, L. A. A. Pereira, L. B. L. S. Lara, P. Artaxo, L. A. Martinelli, Zanobetti. A. Arbex, A. L. F. Braga, 2006The impact of sugar cane-burning emissions on the respiratory system of children and the elderly. Environ. Hlth. Persp., 114725729

46. R. M. Miranda, M. D. Andrade, A. Worobiec, R. Van Grieken, 2002Characterisation of aerosol particles in the São Paulo Metropolitan Area. Atmos. Environ., 36345352

47. Machado C M D, Cardoso A A, Allen A G2008Atmospheric emission of reactive nitrogen during biofuel ethanol production. Environ. Sci. Technol., 42381385

48. Gaffney J S, Marley N A2009The impacts of combustion emissions on air quality and climate- From coal to biofuels and beyond. Atmos. Environ., 432336

49. E. M. Martins, G. Arbilla, 2003Computer modeling study of ethanol and aldehyde reactivities in Rio de Janeiro urban air. Atmos. Environ., 3717151722

50. O. Braunbeck, A. Bauen, F. Rosillo-Calle, L. Cortez, 1999Prospects for green cane harvesting and cane residue use in Brazil. Biomass Bioenergy, 17495506

51. D. A. Castanho, P. Artaxo, 2001Wintertime and summertime São Paulo aerosol source apportionment study. Atmos. Environ., 3548894902

52. Alonso C D, Martins M H R B, Romano J, Godinho R1997São Paulo aerosol characterization study. J. Air Waste Mgt. Assoc., 4712971300

53. Sanchez-Ccoyllo O R, Andrade M D2002The influence of meteorological conditions on the behavior of pollutants concentrations in São Paulo, Brazil. Environ. Poll., 116257263

54. Miranda R M, Andrade M F2005Physicochemical characteristics of atmospheric aerosol during winter in the São Paulo Metropolitan area in Brazil. Atmos. Environ., 3961886193

55. Ynoue R Y, Andrade M D2004Size-resolved mass balance of aerosol particles over the São Paulo metropolitan area of Brazil. Aerosol Sci. Technol., 385262

56. Bourotte C, Curl-Amarante A P, Forti M C, Pereira L A A, Braga A L,

Lotufo P A2007Association between ionic composition of fine and coarse aerosol soluble fraction and peak expiratory flow of asthmatic patients in São Paulo city (Brazil). Atmos.Environ., 4120362048

57. Albuquerque A T T, Andrade M F, Ynoue R Y2012Characterization of atmospheric aerosols in the city of São Paulo, Brazil: Comparisons between polluted and unpolluted periods. Environ. Monit. Assess., 184969984

58. Gioda A, Sales J A, Cavalcanti P M S, Maia M F, Maia L F P G, Aquino Neto F R2004Evaluation of air quality in Volta Redonda, the main metallurgical industrial city in Brazil. J. Braz. Chem. Soc., 14856864

59. V. E. Toledo, Júnior. P. B. Almeida, S. L. Quiterio, G. Arbilla, A. Moreira, V. Escaleira, J. C. Moreira, 2008Evaluation of levels, sources and distribution of toxic elements in PM10 in a suburban industrial region of Rio de Janeiro, Brazil. Environ. Monit. Assess., 1394959

60. Gioia S M C L, Babinski M, Weiss D J, Kerr A A F S2010Insights into the dynamics and sources of atmospheric lead and particulate matter in São Paulo, Brazil, from high temporal resolution sampling. Atmos. Res., 98. 478485

61. R. Miranda, E. Tornaz, 2008Characterization of urban aerosol in Campinas, São Paulo, Brazil. Atmos. Res., 87147157

62. O. R. Sanchez-Ccoyllo, Dias. P. L. Silva, M. D. Andrade, S. R. Freitas, 2006Determination of O3, CO, and PM10 transport in the metropolitan area of São Paulo, Brazil through synoptic-scale analysis of back trajectories. Meteorol. Atmos. Phys., 928393

63. Allen A G, Miguel A H1995Indoor organic and inorganic pollutants- In-situ formation and dry deposition in southeastern Brazil. Atmos. Environ., 2935193526

64. P. C. Vasconcellos, D. Z. Souza, D. Magalhaes, Rocha. G. O. Da, 2011Seasonal variation of n-alkanes and polycyclic aromatic hydrocarbon concentrations in PM10 samples collected at urban sites of São Paulo State, Brazil. Water Air Soil Poll., 222325336

65. P. C. Vasconcellos, D. Z. Souza, S. G. Avila, M. P. Araujo, E. Naoto, K. H. Nascimento, F. S. Cavalcante, Santos. M. Dos, P. Smichowski, E. Behrentz, 2011Comparative study of the atmospheric chemical composition of three South American cities. Atmos. Environ., 4557705777

66. C. Bourotte, M. C. Forti, S. Taniguchi, M. C. Bicego, P. A. Lotufo, 2005A wintertime study of PAHs in fine and coarse aerosols in São Paulo city, Brazil. Atmos. Environ., 3937993811

67. A. G. Allen, A. J. S. Mc Gonigle, A. A. Cardoso, C. M. D. Machado, B. Davison, W. Paterlini, Rocha. G. O. Da, J. B. De Andrade, 2009Influence of sources and meteorology on surface concentrations of gases and aerosols in a coastal industrial complex. J. Braz. Chem. Soc., 20214221

68. W. Vautz, S. Pahl, H. Pilger, M. Schilling, D. Klockow, 2003Deposition of trace substances via cloud droplets in the Atlantic rain forest of the Serra do Mar, São Paulo State, SE Brazil. Atmos. Environ., 3732773287

69. A. G. Allen, Rocha. G. O. Da, A. A. Cardoso, W. C. Paterlini, C. M. D. Machado, 2008Atmospheric particulate polycyclic aromatic hydrocarbons from road transport in southeast Brazil. Transportation Res., 13483490

70. C. Bourotte, M. C. Forti, A. J. Melfi, Y. Lucas, 2006Morphology and solutes content of atmospheric particles in an urban and a natural area of São Paulo State, Brazil. Water Air Soil Poll., 170301316

71. P. C. Vasconcellos, D. Z. Souza, O. Sanchez-Ccoyllo, J. O. V. Bustillos, H. Lee, F. C. Santos, K. H. Nascimento, M. P. Araujo, K. Saarnio, K. Teinila, R. Hillamo, 2010Determination of anthropogenic and biogenic compounds in atmospheric aerosol collected in urban, biomass burning and forest areas in São Paulo, Brazil. Sci. Tot. Environ., 40858365844

72. E. Landulfo, A. Papayannis, P. Artaxo, A. D. A. Castanho, A. Z. Freitas, R. F. De Souza, N. D. V. Junior, M. Jorge, O. R. Sánchez-Ccoyllo, D. S. Moreira, 2003Synergetic measurements of aerosols over São Paulo, Brazil using lidar, sunphotometer and satellite data during the dry season. Atmos. Chem. Phys, 315231539doi:10.5194/acp-3-1523-2003.

73. Costa. R. F. Da, 2010Study of the optical properties of aerosols in the State of São Paulo with the Raman Lidar technique. Master Dissertation, Centro de Lasers e Aplicações, Instituto de Pesquisas Energéticas e Nucleares, Universidade de São Paulo, São Paulo, Brazil, 90p.

74. Torres A S2008Development of a methodology for an independent water vapor Raman Lidar calibration to study water vapor atmospheric profiles, PhD Thesis, Centro de Lasers e Aplicações, Instituto de Pesquisas Energéticas e Nucleares, Universidade de São Paulo, São Paulo, Brazil, 144p.

75. Lopes F J S2011Validation of elastic backscatter lidar data from the CALIPSO satellite using the AERONET sun photometer network. PhD thesis, Centro de Lasers e Aplicações, Instituto de Pesquisas Energéticas e Nucleares, Universidade de São Paulo, São Paulo, Brazil, 169p.

76. Larroza E G2011Caracterização das Nuvens Cirrus na Região Metropolitana de São Paulo (RMSP) com a Técnica de Lidar de

Retroespalhamento Elástico. PhD thesis, Centro de Lasers e Aplicações, Instituto de Pesquisas Energéticas e Nucleares, Universidade de São Paulo, São Paulo, Brazil, 118p.

77. E. Landulfo, G. Held, A. Z. De Freitas, A. Papayannis, A. F. De Souza, 2007Results from first lidar measurements in the central State of São Paulo during the TroCCiBras 2004 Campaign with IPEN´S aerosol lidar. Abstracts, 4th Workshop on Lidar Measurements in Latin America, Ilhabela, S.1722June 2007. http://www.ipen.br/sitio/LWS_Brasil/p4w_abstracts.htm

78. G. Held, J. Pommereau-P, U. Schumann, 2008TroCCiBras and its partner projects HIBISCUS and TROCCINOX: The 2004 Field Campaign in the State of São Paulo. Opt.Pura. Apl., 41 (2), 207-216. http://www.sedoptica.es/Menu_Volumenes/pdfs/299.pdf

79. E. Landulfo, F. J. S. Lopes, G. L. Mariano, A. S. Torres, W. C. Jesus, W. M. Nakaema, M. Jorge, R. Mariani, 2010Study of the properties of aerosols and the air quality index using a backscatter lidar system and AERONET sunphotometer in the city of São Paulo, Brazil. J. Air Waste Manage. Assoc., 60386392doi:

80. Landulfo E, Lopes F J S2009Initial approach in biomass burning aerosol transport tracking with Calipso and Modis satellites, sunphotometer and a backscatter lidar system in Brazil. Proceedings of SPIE- The International Society for Optical Engineering, 7479doi:

81. A. Stohl, 1999The FLEXTRA Trajectory Model, Version 3.0: User Guide. Lehrstuhl für Bioklimatologie und Immissionsforschung, University of Munich, Freising, Germany, 1999, 41p. Available from: http://www.forst.uni-muenchen.de/LST/METEOR/stohl/flextra.htm

82. F. J. S. Lopes, E. Landulfo, E. Giannakaki, 2008One-Year of CALIPSO Measurements Over the City of São Paulo, In: Reviewed and Revised Papers presented at the 24th International Laser Radar Conference 2008, Boulder, Colorado, 2

83. E. Landulfo, M. P. Jorge, G. Held, R. Guardani, J. Steffens, S. A. F. Pinto, I. R. Andre, A. G. Garcia, F. J. S. Lopes, G. L. Mariano, Costa. R. F. Da, P. F. Rodrigues, 2010Lidar observation campaign of sugar cane fires and industrial emissions in the State of São Paulo, Brazil. SPIE Digital Library, Proc. SPIE, 7832p; doi:

84. Mariano G L, Lopes F J S, Steffens J, Jorge M P P M, Landulfo E, Held G, Pinto S A F2011Aerosols monitoring in Rio Claro, Brazil, using LIDAR and air pollution analyzers. Opt. Pura Apl., 445564

85. Held G, Lopes F J S, Bassan J M, Nery J T, Cardoso A A, Gomes A M,

Ramires T, Lima B R O, Allen A G, Da Silva L C, Souza M L, De Souza K F, Carvalho L R F, Urban R C, Landulfo E, Decco A M, Campos M L A A, Nassur M E Q, Nogueira R F P2011Raman lidar monitors emissions from sugar cane fires in the State of São Paulo: A pilot project integrating radar, sodar, aerosol and gas observations. Revista Boliviana de Física, 202426Full paper submitted to Opt. Pura Apl. in February 2012].

86. G. Held, E. G. Larroza, F. J. S. Lopes, Costa. R. F. Da, E. Landulfo, 2011Raman lidar and sodar measurements in the State of São Paulo, Brazil. Proceedings, 2011 NDACC Symposium, Reunion Island, 0710November 2011, 4p.

87. F. J. S. Lopes, G. Held, W. M. Nakaema, P. F. Rodrigues, J. M. Bassan, E. Landulfo, 2011Initial analysis from a lidar observation campaign of sugar cane fires in the central and western portion of the São Paulo State, Brazil In: Lidar Technologies, Techniques, and Measurements for Atmospheric Remote Sensing VII, 2011. Czech Republic: Proceedings of SPIE- The International Society for Optical Engineering, 8182doi:

88. J. Steffens, R. Guardani, E. Landulfo, F. J. S. Lopes, P. F. Moreira, A. Moreira, 2011Capability of atmospheric air monitoring in the urban area of Cubatão using lidar technique. Opt. Pura Apl., 446570

89. M. Dixon, G. Wiener, 1993TITAN: Thunderstorm Identification, Tracking, Analysis & Nowcasting- A radar-based methodology, J. Atmos. Oceanic Technol., 10785797

90. C. Cattrall, J. Reagan, K. Thome, O. Dubovik, 2005Variability of aerosol and spectral lidar and backscatter and extinction ratios of key aerosol types derived from selected Aerosol Robotic Network locations, Journal of Geophysical Research, 110, D10S11. doi:10.1029/2004JD005124.

91. A. H. Omar, D. M. Winker, C. Kittaka, M. A. Vaughan, Z. Liu, Y. Hu, C. R. Trepte, R. R. Rogers, R. A. Ferrare, K. P. Lee, R. E. Kuehn, C. A. Hostetler, 2009The CALIPSO Automated Aerosol Classification and Lidar Ratio Selection Algorithm, Journal of Atmospheric and Oceanic Technology, 2619942014doi:10.1175/2009JTECHA1231.1.

92. Silva. L. C. Da, A. G. Allen, A. A. Cardoso, G. Held, 2011Aerosol physical and chemical characteristics in the region of Ourinhos (São Paulo State). Proceedings, Second Conference of the Brazilian Association for Aerosol Research, Rio de Janeiro, 15August 2011.

93. B. F. A. De Oliveira, E. Ignotti, S. S. Hacon, 2011A systematic review of the physical and chemical characteristics of pollutants from biomass burning and combustion of fossil fuels and health effects in Brazil. Cadernos de Saúde Pública, 2716781698

94. H. A. D. Castro, M. F. D. Cunha, G. A. S. Mendonça, W. L. Junger, J. Cunha-Cruz, Leon. A. Ponce de, 2009Effect of air pollution on lung function in schoolchildren in Rio de Janeiro, Brazil. Rev. Saúde Pública, 432634

95. CETESB2008Companhia de Tecnologia de Saneamento Ambiental. Relatório de qualidade do ar no estado de São Paulo 2007. CETESB, São Paulo.http://www.cetesb.sp.gov.br/Ar/publicacoes.asp

96. N.Gouveia,C.U.Freitas,L.C.Martins,I.O.Marcilio,2006Hospitalizações por causas respiratórias e cardiovasculares associadas à contaminação atmosférica no Município de São Paulo, Brasil. Cad. Saúde Pública, 2226692677

97. A. L. Braga, P. H. Saldiva, L. A. Pereira, J. J. Menezes, G. M. Conceição, C. A. Lin, A. Zanobetti, J. Schwartz, D. W. Dockery, 2001Health effects of air pollution exposure on children and adolescents in São Paulo, Brazil. Pediatr. Pulmonol., 3110613

98. Conceição G M S, Miraglia S G E K, Kishi H S, Saldiva P H N, Singer J M2001Air pollution and child mortality: a time-series study in São Paulo, Brazil. Environ. Hlth. Persp., 109347350

99. N. Gouveia, T. Fletcher, 2000Respiratory disease in children and outdoor air pollution in São Paulo, Brazil: a time-series analysis. Occup. Environ. Med., 57477483

100. Martins L C, Latorre M R D O, Saldiva P H N, Braga A L F2001Relação entre poluição atmosférica e atendimentos por infecção das vias aéreas superiores no município de São Paulo: avaliação do rodízio de veículos. Rev. Brás. Epidemiol., 4220229

101. Martins L C, Latorre M R D O, Saldiva P H, Braga A L2002Air pollution and emergency room visits due to chronic lower respiratory disease in the elderly: an ecological times-series study in São Paulo, Brazil. J. Occup. Environ. Med., 44622627

102. Miraglia S G E K, Saldiva P H N, Böhm G M2005An evaluation of air pollution health impacts and costs in São Paulo, Brazil. Environ. Mgt,. 35667676

103. N. Gouveia, G. A. S. Mendonça, Leon. A. Ponce de, J. E. M. Correia, W. L. Junger, C. U. Freitas, R. P. Daumas, L. C. Martins, L. Guissepe, G. M. S. Conceição, A. Manerich, J. Cunha-Cruz, 2003Poluição do ar e efeitos na saúde nas populações de duas grandes metrópoles brasileiras. Epidemiol. Serv. Saúde, 122940

104. C. Freitas, S. A. Bremner, N. Gouveia, L. A. Pereira, P. H. N. Saldiva, 2004Internações e óbitos e sua relação com a poluição atmosférica em

São Paulo, 1993 a 1997. Rev. Saúde Pública, 38751757

105. Martins M C H, Fatigati F L, Véspoli T C, Martins L C, Pereira L A A, Martins M A, Saldiva P H N, Braga A L F2004Influence of socioeconomic conditions on air pollution adverse health effects in elderly people: an analysis of six regions in São Paulo, Brazil. J. Epidemiol. Community Health, 584146

106. D. M. Spektor, V. A. Hofmeister, P. Artaxo, J. A. P. Brague, F. Echelar, D. P. Nogueira, C. Hayes, G. D. Thurston, M. Lippmann, 1991Effects of heavy industrial pollution on respiratory function in the children of Cubatão, Brazil- A preliminary report. Environ. Hlth. Persp., 945154

107. B. S. De Martinis, N. Y. Kado, L. R. F. Carvalho, R. A. Okamoto, L. A. Gundel, 1999Genotoxicity of fractionated organic material in airborne particles from São Paulo, Brazil. Mutation Res., 4468394

108. B. S. De Martinis, R. A. Okamoto, N. Y. Kado, L. A. Gundel, L. R. F. Carvalho, 2002Polycyclic aromatic hydrocarbons in a bioassay-fractionated extract of PM10 collected in São Paulo, Brazil. Atmos. Environ., 36307314

109. P. C. Vasconcellos, O. Sanchez-Ccoyllo, C. Balducci, R. Mabilia, A. Cecinato, 2008Occurrence and concentration levels of nitro-PAH in the air of three Brazilian cities experiencing different emission impacts. Water Air Soil Poll., 1908794

110. E. P. Vendrasco, Dias. P. L. Silva, E. D. Freitas, 2009A case study of the direct radiative effect of biomass burning aerosols on precipitation in the Eastern Amazon. Atmos. Res., 94409421

111. J. A. Martins, Dias. M. A. F. Silva, 2009The impact of smoke from forest fires on the spectral dispersion of cloud droplet size distributions in the Amazonian region. Env. Res. Lett., 4, art. 015002

112. J. R. Moreira, J. Goldemberg, 1999The Alcohol Program. Energy Policy, 27229245

113. Arbex M A, Böhm G M, Saldiva P H N, Conceição G M S, Pope III A C, Braga A L F2000Assessment of the effects of sugar cane plantation burning on daily counts of inhalation therapy. J. Air Waste Mgt. Assoc., 5017451749

114. Arbex M A, Martins L C, Oliveira R C, Pereira L A, Arbex F F, Cançado J E, Saldiva P H, Braga A L2007Air pollution from biomass burning and asthma hospital admissions in a sugarcane plantation area in Brazil. J. Epidemiol. Comm. Hlth., 61395400

115. F. S. Lopes, H. Ribeiro, 2006Mapeamento de internações hospitalares por problemas respiratórios e possíveis associações à exposição humana aos produtos da queima de palha de cana-de-açúcar no estado de São Paulo. Rev. Brás. Epidemiol., 9215225

116. F. Mazzoli-Rocha, C. B. Magalhães, O. Malm, P. H. Saldiva, W. A. Zin, D. S. Faffe, 2008Comparative respiratory toxicity of particles produced by traffic and sugarcane burning. Environ. Res., 1083541

117. S. J. De Andrade, S. D. Varella, G. T. Pereira, G. J. Zocolo, M. R. R. De Marchi, E. A. Varanda, 2011Mutagenic activity of airborne particulate matter (PM10) in a sugarcane farming area (Araraquara city, southeast Brazil). Environ. Res., 111545550

118. T. L. Bell, D. Rosenfeld, K. M. Kim, J. M. Yoo, M. I. Lee, M. Hahnenberger, 2008Midweek increase in US summer rain and storm heights suggests air pollution invigorates rainstorms. J. Geophys. Res., 113, D2, Art. D02209

119. IPCC2007Fourth Assessment Report "Climate Change 2007", International Panel on Climate Change: www.ipcc.ch.

120. D. Rosenfeld, 2006Aerosol-cloud interactions control of earth radiation and latent heat release budgets. Space Sci. Rev., 125149157

121. A. S. Dufek, T. Ambrizzi, 2008Precipitation variability in São Paulo State, Brazil. Theoret. Appl. Climatol., 93167178

122. Coelho C H, Allen A G, Fornaro A, Orlando E A, Grigoletto T L B, Campos M L A M2011Wet deposition of major ions in a rural area impacted by biomass burning emissions. Atmos. Environ., 4552605265

123. UNICA2012http://english.unica.com.br/dadosCotacao/estatistica/ Accessed 29/4/2012).

124. Gomes A M, Escobedo J F2010Climatologia de Tempestades na Área Central do Estado de São Paulo Usando Radar Meteorológico. Revista Energia na Agricultura, 25, 11120ISSN- 1808-875.

125. Westcott N E1995Summertime cloud-to-ground lightning activity around major Midwestern urban areas. J. Appl. Meteor., 3416331642

126. UNICA2011Accessed 17/4/2012) http://english.unica.com.br/website/userFiles/protocolo-agroambiental.pdf

127. C. C. Tsao, J. E. Campbell, M. Mena-Carrasco, S. N. Spak, G. R. Carmichael, Y. Chen, 2012Increased estimates of air-pollution emissions from Brazilian sugar-cane ethanol. Nature Climate Change, 25357

CITATION

CHAPTER 1

Yan-Hong Shi, Zhi-Yong Xie, Rui Wang, Shan-Jun Huang, Yi-Ming Li and Zheng-Tao Wang, "Quantitative and Chemical Fingerprint Analysis for the Quality Evaluation of Isatis indigotica based on Ultra-Performance Liquid Chromatography with Photodiode Array Detector Combined with Chemometric Methods," Int. J. Mol. Sci. 2012, 13(7), 9035-9050; doi:10.3390/ijms13079035.

CHAPTER 2

Yan-Bin Wu, Li-Jun Zheng, Jun Yi, Jian-Guo Wu, Ti-Qiang Chen and Jin-Zhong Wu, "Quantitative and Chemical Fingerprint Analysis for the Quality Evaluation of Receptaculum Nelumbinis by RP-HPLC Coupled with Hierarchical Clustering Analysis," Int. J. Mol. Sci. 2013, 14(1), 1999-2010; doi:10.3390/ijms14011999.

CHAPTER 3

Teng-Hua Wang, Jing Zhang, Xiao-Hui Qiu, Jun-Qi Bai, You-Heng Gao and Wen Xu, "Application of Ultra-High-Performance Liquid Chromatography Coupled with LTQ-Orbitrap Mass Spectrometry for the Qualitative and Quantitative Analysis of Polygonum multiflorum Thumb. and Its Processed Products," Molecules 2016, 21(1), 0040; doi:10.3390/molecules21010040.

CHAPTER 4

Shao-Dan Chen, Chuan-Jian Lu and Rui-Zhi Zhao, "Qualitative and Quantitative Analysis of Rhizoma Smilacis glabrae by Ultra High Performance Liquid Chromatography Coupled with LTQ OrbitrapXLHybrid Mass Spectrometry," Molecules 2014, 19(7), 10427-10439; doi:10.3390/molecules190710427.

CHAPTER 5

Zarrin Es'haghi (2011). Photodiode Array Detection in Clinical Applications; Quantitative Analyte Assay Advantages, Limitations and Disadvantages, Photodiodes - Communications, Bio-Sensings, Measurements and High-Energy Physics, Associate Professor Jin-Wei Shi (Ed.), ISBN: 978-953-307-277-7, InTech, DOI: 10.5772/18244.

CHAPTER 6

Mohammed Mahabubur Rahman and Rahman Md. Motiur (2012). Quantitative Chemical Defense Traits, Litter Decomposition and Forest Ecosystem Functioning, Forest Ecosystems - More than Just Trees, Dr Juan A. Blanco (Ed.), ISBN: 978-953-51-0202-1, InTech, DOI: 10.5772/39003.

CHAPTER 7

Lay-Harn Gam (2012). Tandem Mass Spectrometry for Simultaneous Qualitative and Quantitative Analysis of Protein, Tandem Mass Spectrometry - Applications and Principles, Dr Jeevan Prasain (Ed.), ISBN: 978-953-51-0141-3, InTech, DOI: 10.5772/31504.

CHAPTER 8

Michael J. Haugh and Marilyn Schneider (2011). Quantitative Measurements of X-Ray Intensity, Photodiodes - Communications, Bio-Sensings, Measurements and High-Energy Physics, Associate Professor Jin-Wei Shi (Ed.), ISBN: 978-953-307-277-7, InTech, DOI: 10.5772/18261.

CHAPTER 9

Gerhard Held, Andrew G. Allen, Fabio J.S. Lopes, Ana Maria Gomes, Arnaldo A. Cardoso, Eduardo Landulfo (2012). Review of Aerosol Observations by Lidar and Chemical Analysis in the State of São Paulo, Brazil, Atmospheric Aerosols - Regional Characteristics - Chemistry and Physics, Dr. Hayder Abdul-Razzak (Ed.), ISBN: 978-953-51-0728-6, InTech, DOI: 10.5772/50737.

INDEX